W0079389

**Progress in Inorganic Biochemistry
and Biophysics
Vol. 2**

**Edited by
Harry Gray
Ivano Bertini**

Birkhäuser
Boston · Basel · Stuttgart

Advanced Magnetic Resonance Techniques in Systems of High Molecular Complexity

Neri Niccolai
Gianni Valensin
editors

1986

Birkhäuser
Boston · Basel · Stuttgart

Editors:

Neri Niccolai
Department of Chemistry
University of Siena
Siena, Italy

Gianni Valensin
Department of Chemistry
University of Siena
Siena, Italy

CIP-Kurztitelaufnahme der Deutschen Bibliothek

**Advanced magnetic resonance techniques in systems
of high molecular complexity** / Neri Niccolai ;
Gianni Valensin, ed. – Boston ; Basel ; Stuttgart :
Birkhäuser, 1986.
  (Progress in inorganic biochemistry and
  biophysics ; Vol. 2)

NE: Niccolai, Neri [Hrsg.]; GT

ISBN 978-1-4615-8523-7   ISBN 978-1-4615-8521-3 (eBook)
DOI 10.1007/978-1-4615-8521-3

All rights reserved.
No part of this publication may be reproduced, stored in a retrieval
system, or transmitted, in any form or by any means, electronic,
mechanical, photocopying, recording or otherwise, without prior
permission of the copyright owner.

© 1986 Birkhäuser Boston, Inc.

Softcover reprint of the hardcover 1st edition 1986

CONTENTS

Foreword

The second volume of the series on inorganic biochemistry and biophysics is singularly devoted to magnetic resonance on systems of high
molecular complexity. Recently, there have been important advances in
magnetic resonance studies of polymers; these advances touch on all
aspects of magnetic resonance, both theoretical and applied. Particular
emphasis is placed here on multipulse experiments.

We believe such an report will be of considerable interest to
the readers of our series owing to the importance of magnetic resonance
techniques in the investigation of biopolymers.

Ivano Bertini
Harry Gray

Series Editors

Preface

    This book is a record of the Proceedings of the International Symposium on "Advanced Magnetic Resonance Techniques in Systems of High Molecular Complexity", which was held in Siena between 15 and 18 May 1985. The idea of the meeting is due to Proff. N.M. Atherton, G. Giacometti and E. Tiezzi with the aim of honouring the scientific personality of Prof. S.I. Weissman. The meeting has been organized with the assistance of a National Committee formed by R. Basosi, I. Bertini, P. Bucci, C. Corvaia, A. Gamba, G. Martini, G.F. Pedulli, P.A. Temussi, and C.A. Veracini. The invited lecturers responded enthusiastically and a comprehensive picture of the theoretical and practical aspects of magnetic resonance could be therefore provided.

    The book contains all the plenary lectures delivered during the meeting and also a wide selection among the huge amount of contributions collected by the organizers.

    The editors acknowledge all the contributors to this book, that contains some of the most exciting advances in several fields of magnetic resonance, such as new pulse methods, spatial localization, conformational and structural investigations, liquid crystalline mesophases, charge-transfer complexes, triplet state molecules, computer simulations, and several others. We hope the reader will enjoy in learning the progress in the area.

Neri Niccolai
Gianni Valensin

PBB, Vol. 2
Advanced Magnetic Resonance Techniques
in Systems of High Molecular Complexity
ⓒ 1986 Birkhäuser Boston, Inc.

SPATIAL LOCALIZATION IN NMR

A.J. Shaka and Ray Freeman

Oxford University, England

Magnetic resonance took on an entirely new character with the realizat-
ion that it could be used for medical imaging (1) in a similar way to
X-ray tomography. It was not long before it became apparent that NMR
methods can have an even more fundamental impact on medicine through the
observation of high resolution spectra in vivo. Often these are
phosphorus-31 spectra of molecules involved in metabolism, because of
the inherent simplicity of such spectra, but proton and carbon-13 high
resolution spectroscopy is also feasible, and the latter offers the pro-
mise of work with isotopically enriched drugs. Medicine will never be
quite the same; nor will NMR spectroscopy, for it is just now expe-
riencing a wide publicity never dreamed of in the early days.

All this is based on one simple concept that has been with us for many
years — that a field gradient may be used to code the NMR signals
according to their spatial coordinates in that particular dimension.
Imposition of three successive orthogonal gradients then provides a
three dimensional map of an object or an animal. A minor extension of
this concept allows a small 'active volume' to be defined so that a high
resolution spectrum may be extracted from a restricted region of some
organ, observed without invasion of the animal under investigation.

The majority of these experiments have been carried out by applying
strong gradients of the static field B_0. Once the technique moved up to

human patients, questions were naturally asked about possible undesirable physiological effects of the experiment – the intense polarizing
field B_0, the weak radiofrequency field B_1, and the rapidly switched B_0
gradients. There appears to be little cause for concern, although, of
the three factors involved, the switched B_0 gradients might be the most
likely to have a deleterious effect on the patient.

As an alternative, Hoult (2) has proposed using the B_1 field intensity
to define spatial coordinates. This has the advantage that the gradients
of B_1 are comparable with the nominal value B^o_1, whereas B_0 gradients
are always very much weaker than the applied static field B_0. Furthermore, for experiments aimed at the measurement of chemical shifts, B_1
gradients are less likely to cause complications than B_0 gradients.
These experiments are often carried out with a 'surface coil' placed in
contact with the animal but designed to probe the NRM signal below the
surface, and the B_1 field from such a coil is necessarily inhomogeneous
in space. It is therefore tempting to make a virtue out of a necessity
and use this for the purpose of localizing the NMR response.

The present work explores the feasibility of using pulse sequences to
enhance the sensitivity of the NMR response to small changes in B_1 intensity. The aim is to discover a pulse scheme which only excites an NMR
response when B_1 is near to its nominal value B^o_1, other signals being
suppressed. Similar ideas have been explored by Bendall et al. (3,4) and
Tycko and Pines (5,6) as an alternative to experiments that use B_0 gradients (7,8). Our approach has been to find composite pulse sequences
which have a narrow 'B_1 profile' that is to say, a high sensitivity to
variations in B_1 intensity. The concept of a composite pulse was first
introduced by Levitt (9) in an attempt to compensate some of the intrinsic imperfections of a simple radiofrequency pulse, principally
pulse length error and radiofrequency offset. Normally a composite pulse

is devised to compensate one or other of these imperfections; in certain circumstances two sources of error may be compensated simultaneously (10,11). Spatial localization makes different demands; the dependence on B_1 intensity should be <u>accentuated</u> rather than compensated, and the B_1 profile should be made independent of resonance offset ΔB over a reasonable range of offsets, otherwise the active volume will vary with the chemical shift, a most undesirable effect. It is also convenient to arrange for the signal phase to be some simple (linear) function of resonance offset, so that a pure absorption-mode spectrum can be easily obtained. These new requirements mean that new families of composite pulses need to be discovered.

Our first experiments concentrated on the use of 180° 'refocussing' pulses. The idea was to excite transverse nuclear magnetization M_{XY} and then act upon it with a sequence of one or more composite 180° pulses. A phase cycling scheme formerly used in two-dimensional NMR called EXORCYCLE (12) is then used to cancel all signals not properly inverted by the 180° pulse. Thus if the composite 180° pulse has a suitably narrow B_1 profile, spatial localization can be achieved. The definition of the active sample volume can be improved by cascading two (or more) of the composite 180° pulses, since the effect is cumulative, narrowing the B_1 profile.

Such 'retrograde compensation' of the effect of B_1 intensity can be achieved by the sequence

$$180°(+X) \ 180°(-Y) \ 180°(-X)$$

which behaves like a refocussing pulse only when the actual B_1 intensity is reasonably close to its nominal value $B°_1$, otherwise it has little net effect on the nuclear spins (13). This B_1 profile can be monitored simply by measuring the spin inversion efficiency as a function of pulse flip angle, achieved in practice by varying the pulse width at constant B_1 rather than by varying B_1 at constant pulse width. The results are shown in Figure 1, and compared with the B_1 profile of a simple 180° (X) pulse.

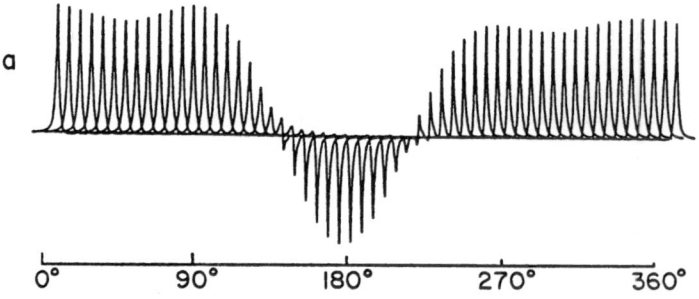

Figure 1 - Operation of the composite pulse sequence 180°(+X)180°(-Y)180°(-X) designed to accentuate sensitivity to variation in the B_1 intensity. Near the nominal pulse width, where each element is close to the 180° condition, the sequence inverts the proton signal. For most other pulse widths the sequence leaves proton magnetization near the +Z axis. For the purposes of spatial localization the pulse widths would be fixed, but spatial inhomogeneity in the B_1 field would change the flip angles.

Pulse sequences like this, which are not symmetrical with respect to time reversal, may have a resonance offset dependence which is asymmetric with respect to the sign of the offset ΔB. Symmetry can be restored, and the offset behaviour improved, if two such sequences are used in cascade with the second sequence in time-reversed order:

$$60°(X)\left[180°(+X)\ 180°(-Y)\ 180°(-X)\right]_a\left[180°(-X)\ 180°(-Y)\ 180°(+X)\right]_b$$

The first pulse simply creates transverse nuclear magnetization. In this case, the EXORCYCLE phase-cycling scheme must be operated independently for each composite pulse, the incremental radiofrequency phases being cycled according to the 16-step sequence shown in Table I.

The performance of this sequence was tested (14) by setting up a simple one-turn 'surface coil' for proton resonance at 200 MHz and a phantom consisting of three small glass bulbs containing (a) benzene (b) water (c) cyclohexane. The coil had a diameter of 30 mm and the bulbs were placed at 4 mm intervals along the coil axis as shown in Figure 2.

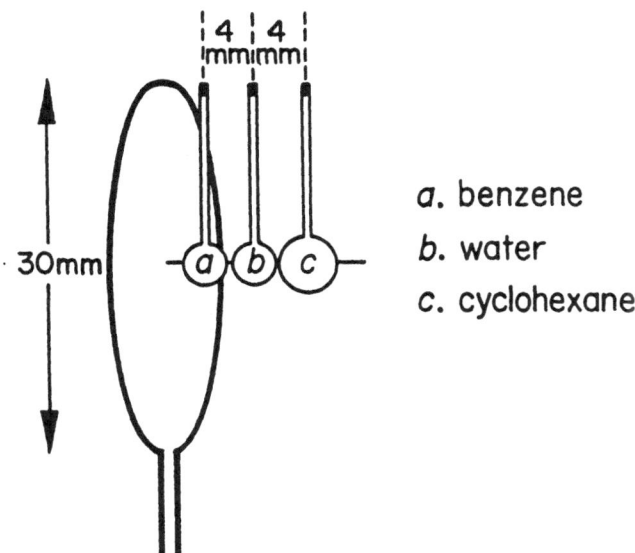

Figure 2 - Sketch of the surface coil with a phantom sample made up of three small glass bulbs containing ben- zene,water,cyclohexane

Table I

EXORCYCLE phase–cycling scheme used with two independent composite 180° pulses.

a	b	Receiver phase
0°	0°	+
90°	0°	−
180°	0°	+
270°	0°	−
0°	90°	−
90°	90°	+
180°	90°	−
270°	90°	+
0°	180°	+
90°	180°	−
180°	180°	+
270°	180°	−
0°	270°	−
90°	270°	+
180°	270°	−
270°	270°	+

The most distant bulb has only weak coupling to the coil and was therefore made somewhat larger by way of compensation.

The method works by cancelling signals from regions of space where the individual pulses are not near 180°. Consequently, by varying the pulse widths t_p the active volume can be moved to different parts of the B_1 field, in the present case further out along the coil axis. At short pulse widths, only the nearest sample (benzene) is excited, but as the pulse widths are increased the benzene signal decreases in intensity and the water signal begins to grow. At very large pulse widths both benzene and water disappear while the distant cyclohexane sample gives the only signal (Figure 3). Spatial localization has thus been demonstrated, albeit only in one dimension.

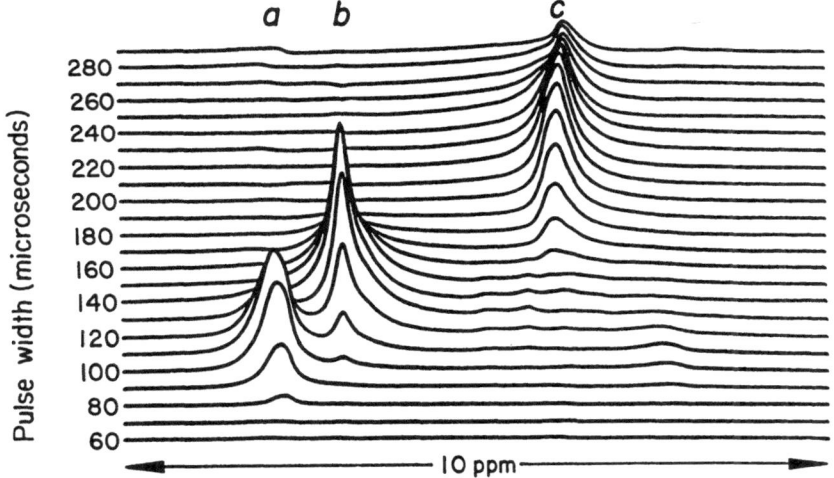

Figure 3 – Spatial localization experiments carried out on the bulb sample sketched in Fig.2. The B_1 field was strongly inhomogeneous and as the pulse widths were incremented (all in proportion) the three samples, (a) benzene (b) water (c) cyclohexane, reached the nominal value (180°) in sequence, showing that the NMR signals could be separated according to their distance along the axis of the surface coil.

If different component pulses in the composite pulse sequence were to be applied to physically separate radiofrequency coils, localization should be feasible in a second spatial dimension. Clearly these B_1 localization experiments may be combined with applied B_0 gradients to restrict the active sample volume in the third dimension (Z).

These ideas have not yet been tested in practice because there are some more pressing experimental complications. The most serious of these seems to be the effect of resonance offset. Since the B_1 intensity is finite and indeed quite low at appreciable distances from the surface coil, offsets cause a significant tilting of the effective field in the rotating frame. Not all composite pulse sequences tolerate this kind of imperfection. What is required is a B_1 profile that not only remains narrow but is also essentially unchanged as a function of offset over some reasonable offset range. If this requirement is not met, for example if the peak of the B_1 profile shifts as a function of ΔB, then the active sample volume will be different for different chemical shifts, an undesirable state of affairs. Furthermore, it is advantageous to be able to adjust the phase for pure-absorption mode signals across the entire spectrum, which is easily accomplished if the frequency dependence of the phase angle is reasonably linear.

There is another practical problem not yet mentioned. When the sequence has been set up to observe a sample region remote from the surface coil, there will always be another region closer to the coil where the B_1 field is three times the nominal value B^o_1. Unfortunately this region will also excite an NMR response because the initial pulse will be 270° instead of 90°, and the refocussing pulses will behave like 180° pulses once more. These 'harmonic' responses will also occur for higher odd multiples of B^o_1, generating more spurious responses. This was the reason for using an initial pulse of 60° (rather than 90°) so that the

third harmonic was attenuated. In practice there are some more sophisticated schemes for harmonic suppression to be described below.

Finally, care must be taken to ensure that the duration of the pulse sequence is not too long. For the kind of biological molecules implicit in
this kind of study, spin-spin relaxation times can be quite short, impo-
sing a strict upper limit on the duration of the pulse sequence.

The problem of resonance offset has to be solved by suitable design of
the composite pulses. The sequence illustrated above is reasonably good
in this respect, but is by no means optimum. A rather better 180° composite pulse is provided by the sequence

$$R_2 = 25°(-X) \; 90°(+X) \; 135°(-X) \; 250°(+X)$$

which was derived by modifying the 'WALTZ' element used in broadband
decoupling experiments (15,16). The performance can be judged by plotting the efficiency for spin inversion

$$E = 0,5 \left[1 - (M_Z/M_0) \right]$$

in the form of a contour map as a function of radiofrequency offset $\Delta B/$
$B°_1$ and the intensity of the radiofrequency field $B_1/B°_1$. Figure 4 com-
pares the performance of R_2 with that of a simple 180° pulse. Clearly R_2
operates over a much wider range of offsets ΔB, without much change in
the B_1 profiles, whereas the simple 180° pulse only works well close to
exact resonance (17). Sequences like R_2 may be further improved by computer optimization programs. One such composite 180° sequence is

$$R_3 = 51°(+X) \; 55°(-X) \; 142°(+X) \; 26°(-X) \; 172°(+X) \; 169°(-x) \; 64°(+x)$$

and there are of course many others.

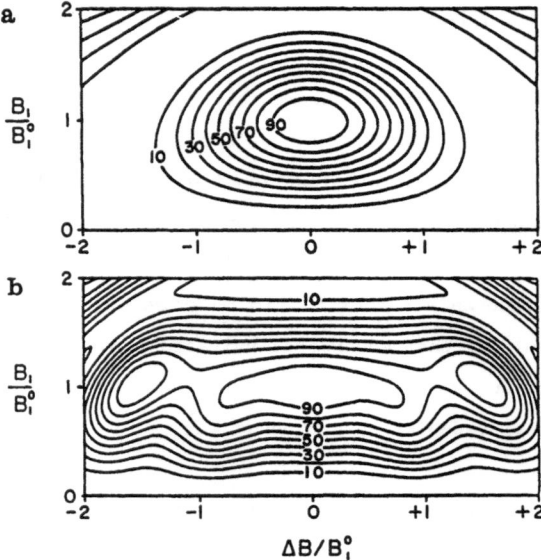

Figure 4 – Comparison of the performance of (a) the simple
180° pulse and (b) the composite pulse 25°(-X)90°
(+X)135°(-X)250°(+X). Contours represent the efficiency $(0 \geq E \geq 100\%)$ for spin inversion. The aim
is to achieve a high tolerance of radiofrequency
offset $\Delta B/B_1$° without significant changes in the
B_1 profile.

PREPULSES

Considerable improvements can be achieved by introducing a new concept,
the 'prepulse'. The ordering of the sequence is changed around such that
the last pulse of all is the 'read' pulse which generates observable
transverse nuclear magnetization. This is preceded by one or more prepulses designed to change the Z-component of magnetization according to
some function of the spatial coordinates. For example, we might choose
to leave the equilibrium magnetization M_0 unchanged at a point where B_1
has its nominal value $B°_1$, but attenuate it elsewhere. Prepulses act
only to change M_Z; any transverse magnetization after the pulse is

suppressed. This is usually achieved by adding the free induction decays from two separate experiments carried out with opposite phases of B_1, such that the X and Y components of magnetization cancel.Alternatively, the transverse magnetization after the prepulse may be dispersed by the application of a B_0 gradient pulse, leaving no net resultant.

Defined in this manner, prepulses have some useful properties. If several prepulses are applied in a sequence, they act independently and the effects are cumulative, since in each case only M_Z is affected. They may be separated by appreciable time intervals, enough to allow the imposition of B_0 gradient pulses, a change in the frequency of the transmitter, or the switching of B_1 to another coil. Spin-spin relaxation is ineffective in these intervals since only M_Z is retained. The only constraint is that the intervals should not become comparable with the spin-lattice relaxation time otherwise the B_1 profile will be appreciably distorted; in the limit the effect of the prepulse will be nullified.

This provides an advantage over the earlier technique where 180° refocussing pulses were used to operate on transverse magnetization in conjunction with the EXORCYCLE scheme, since any intervals incorporated in these experiments are affected by spin-spin relaxation, often quite rapid in biological samples. It turns out that the efficiency of the two rival techniques is the same if the prepulse scheme of Table II is employed. This is because the efficiency for spin inversion is measured by $E = \frac{1}{2}\left[1 - (M_Z/M_0)\right]$ while the efficiency of a prepulse is measured by (M_Z/M_0). Thus by introducing two extra stages of signal acquisition without a prepulse (and changing the receiver phase) the two schemes achieve the same efficiency, that is to say, the same B_1 profile for a given composite 180° pulse. This is the 'NOBLE' technique introduced by Tycko and Pines (5). There is a further practical advantage. Since in

the prepulse scheme only the last pulse generates transverse magnetization, the phase of the detected signal depends only on the properties of this read pulse. Consequently the signal phase is an essentially
linear function of resonance offset and is easily adjusted for pure absorption-mode spectra.

Table II

**Comparison of the 'Prepulse' and 'EXORCYCLE' schemes for
spatial localization**

(a) Prepulse	(b) EXORCYCLE
180°(+X) 90°(X) Acq (−)	90°(X) 180°(+X) Acq (+)
180°(−X) 90°(X) Acq (−)	90°(X) 180°(−Y) Acq (−)
— 90°(X) Acq (+)	90°(X) 180°(−X) Acq (+)
— 90°(X) Acq (+)	90°(X) 180°(+Y) Acq (−)

Prepulses also provide a convenient method for suppressing the spurious
'harmonic' responses which arise because some regions of the sample close to the surface coil experience fields of $3B^\circ_1$, $5B^\circ_1$, etc. A 30° prepulse suppresses the third harmonic response by setting M_z to the null
condition for $B_1 = 3B^\circ_1$, while attenuating the main response by only
13%. This is effective over a wide range of resonance offsets. In a similar fashion a 18° prepulse eliminates the fifth harmonic, and a 12.9°
prepulse the seventh harmonic. Removal of these spurious harmonic
responses is illustrated by the experimental results (18) shown in
Figure 5.

All these spatial localization experiments hinge on being able to achie-

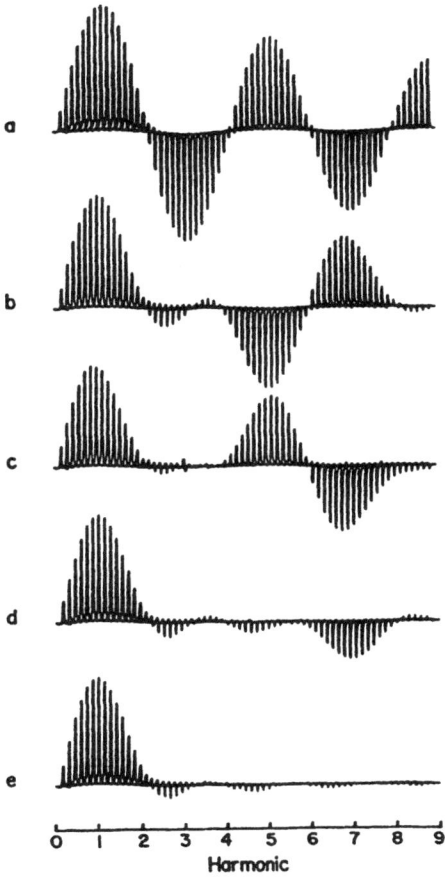

Figure 5 – Suppression of harmonic responses by means of
prepulses. (a) The 90° read pulse generates large
NMR signals at all the odd harmonics where $B_1 =
(2m+1)B_1°$. (b) A 30° prepulse suppresses the
third and ninth harmonics. (c) A cascade of two
30° prepulses eliminates the third harmonic more
effectively. (d) Prepulses of 30° and 18° suppress the 3rd, 5th and 9th harmonics. (e) An additional 12.9° prepulse destroys the 7th harmonic.

14

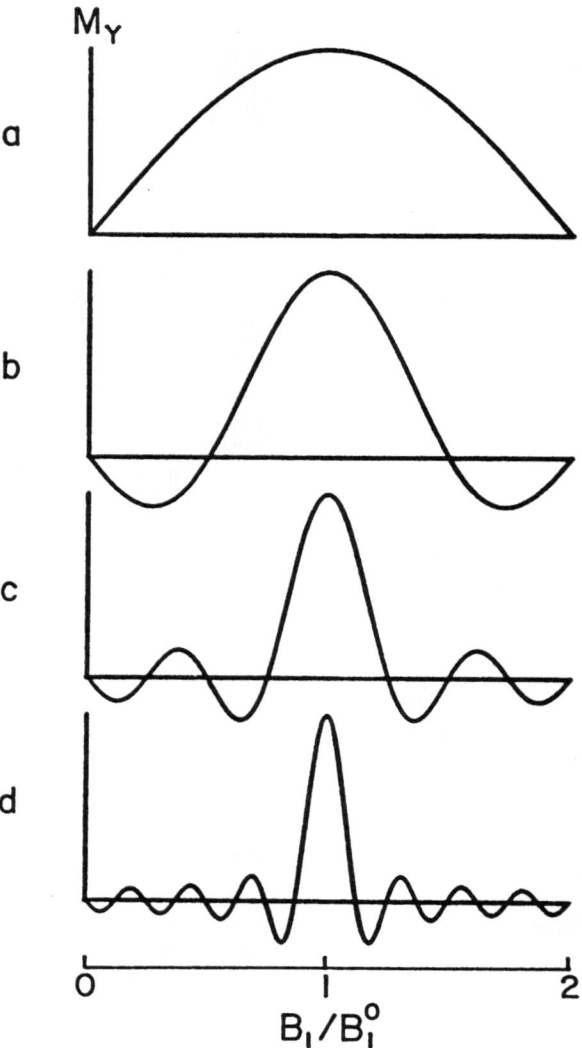

Figure 6 - Narrowing of the B_1 profile with cascaded prepulses (simulation). (a) A 90^d read pulse only. (b) $180°$ prepulse.(c) $180°$ and $360°$ prepulses. (d) $180°$, $360°$ and $720°$ prepulses

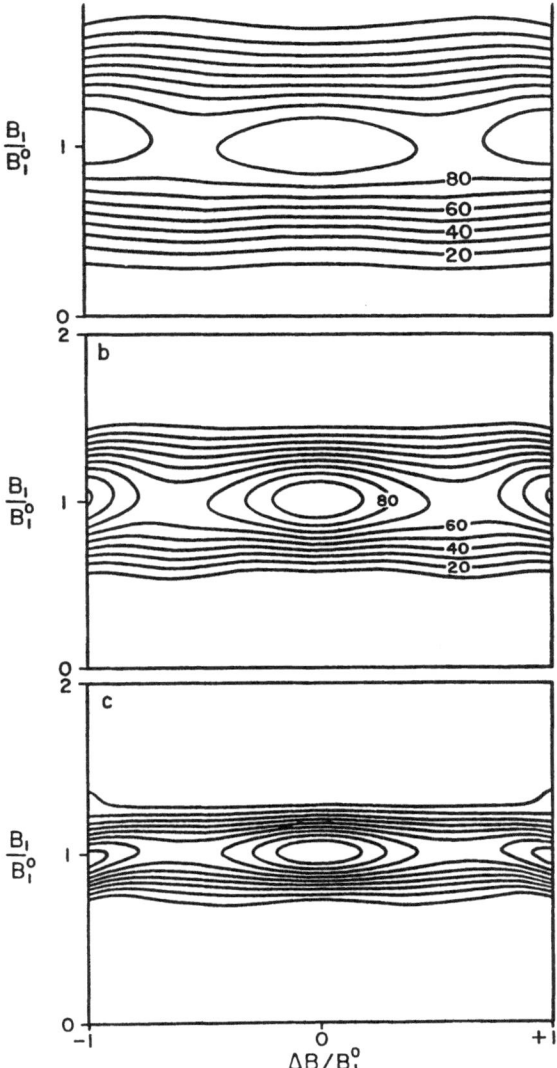

Figure 7 – Contours representing spin inversion efficiency (%) for prepulse schemes. Top: one prepulse. Centre: two prepulses. Bottom: three prepulses

ve a sufficiently narrow B_1 profile, that is to say, a signal response curve that peaks sharply at the nominal value B^o_1 and falls off rapidly for other values of B_1. Choice of a suitable composite pulse goes some way towards achieving a narrow profile, but a much greater narrowing can be obtained by cascading pulses. This holds for prepulses and refocussing pulses. Figure 6 illustrates the progressive improvement in the B_1 profile of a 90° read pulse as more and more prepulses are added. For the final sequence, which incorporates 720°, 360° and 180° prepulses, the sensitivity to B_1 is very high indeed. (It is interesting to note the analogy with the frequency-domain excitation spectra obtained with the DANTE selective excitation sequence (19,20)). Provided that each composite prepulse has a suitable tolerance of resonance offset effects, the cascade of prepulses retains this property. There is always the possibility of using one prepulse to compensate for an undesirable offset dependence of another prepulse, but this option has not been explored in practice. Figure 7 shows contours of inversion efficiency as a function of offset $\Delta B/B^o_1$ and radiofrequency field intensity B_1/B^o_1 as the number of prepulses in the cascade is increased from one to three. Note particularly that the B_1 profiles (vertical sections through the diagram) become narrower but remain essentially unaffected by resonance offset over an offset range of about $2B^o_1$. Experimental verification (21) of this uniformity of the B_1 profiles is provided by the proton signals illustrated in Figure 8, obtained at exact resonance and for offsets ΔB equal to 0.4 B^o_1 and 0.8 B^o_1. Note that the signal phase is well-behaved throughouth this series of experiments. Signals from sample regions remote from the chosen 'active volume' would be suppressed quite effectively by such a sequence. Had B_0 gradients been employed in order to define a similar active volume, there would normally be broad NMR signals of considerable intensity from the rest of the sample, and these are not easily separable.

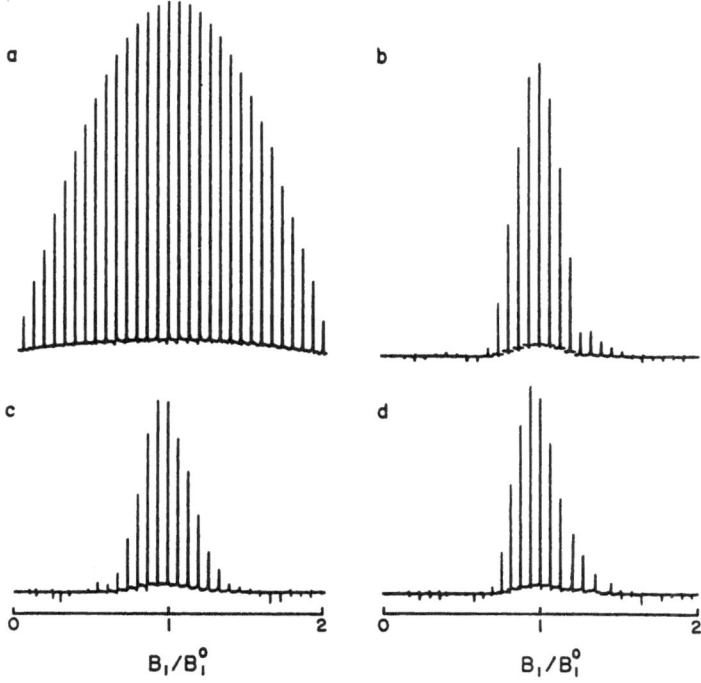

Figure 8 - Experimental proton signals as a function of pulse width (equivalent to variation of B_1/B^o_1). (a) The 90° read pulse. (b) Composite prepulse scheme at exact resonance. (c) Same, at an offset $\Delta B = 0.4\ B^o_1$. (d) Same, at an offset $\Delta B = 0.8\ B^o_1$

This technique is of course in its infancy and no true _in vivo_ studies have yet been attempted. However the feasibility of the method seems to be well established, and the problems of resonance offset and spurious harmonic responses can be solved. Considerable work will be needed on methods for extending the localization into two and three dimensions. The geometrical design of surface coils has only just begun to be explored and there are many possibilities for more complicated arrays of radiofrequency transmitter coils. One serious limitation of the simple loop surface coil is that the penetration of the radiofrequency field falls off quite rapidly, and beyond a depth of about one coil radius the field is too weak to be of much use. Of course these methods are not limited to surface coils; it is quite feasible to design coils which surround the animal completely and which have a built-in B_1 gradient for spatial localization. With laboratory animals another approach would be to use a very small coil surgically inserted near to the organ in question; it would still be important to exert some control on the active volume excited by the coil. Whichever method is adopted, localization within the B_1 field distribution promises to be a useful approach, either as an alternative to the more common B_0 gradient techniques or in conjunction with them.

REFERENCES

1) P.C.Lauterbur, Nature (London), **242**, 190 (1973).

2) D.I.Hoult, J.Magn.Reson. **33**, 183 (1979); ibid. **35**, 69 (1979); ibid. **38**, 369 (1980).

3) M.R.Bendall and R.E.Gordon, J.Magn.Reson. **53**, 365 (1983).

4) M.R.Bendall, Chem.Phys.Letters **99**, 310 (1983).

5) R.Tycko and A.Pines, J.Magn.Reson. **60**, 156 (1984).

6) R.Tycko and A.Pines, Chem.Phys.Letters **111**, 462 (1984).

7) W.P.Aue, S.Muller, T.A.Cross and J.Seelig, J.Magn.Reson. **56**, 350 (1984).

8) T.A.Cross,S.Muller and W.P.Aue, J.Magn.Reson. **62**, 87 (1985).

9) M.H.Levitt and R.Freeman, J.Magn.Reson. **33**, 473 (1979).

10) M.H.Levitt and R.R.Ernst, J.Magn.Reson. **55**, 247 (1983).

11) A.J.Shaka and R.Freeman, J.Magn.Reson. **55**, 487 (1983).

12) G.Bodenhausen, R.Freeman and D.L.Turner, J.Magn.Reson. **27**, 511 (1977).

13) A.J.Shaka and R.Freeman, J.Magn.Reson. **59**, 169 (1984).

14) A.J.Shaka, J.Keeler, M.B.Smith and R.Freeman, J.Magn.Reson. **61**, 175 (1985).

15) A.J.Shaka,J.Keeler, T.Frenkiel and R.Freeman, J.Magn.Reson. **52**, 335 (1983).

16) A.J.Shaka,J.Keeler and R.Freeman, J.Magn.Reson. **53**, 313 (1983).

17) A.J.Shaka and R.Freeman, J.Magn.Reson. (in press).

18) A.J.Shaka and R.Freeman, J.Magn.Reson. **62**, 326 (1985).

19) G.Bodenhausen, R.Freeman and G.A.Morris, J.Magn.Reson. **23**, 171 (1976).

20) G.A.Morris and R.Freeman, J.Magn.Reson. **29**, 433 (1978).

21) A.J.Shaka and R.Freeman, J.Magn.Reson. (in press).

PBB, Vol. 2
Advanced Magnetic Resonance Techniques
in Systems of High Molecular Complexity
© 1986 Birkhäuser Boston, Inc.

HOMONUCLEAR HARTMANN–HAHN MAGNETIZATION TRANSFER:

NEW ONE- AND TWO-DIMENSIONAL NMR METHODS FOR STRUCTURE

DETERMINATION AND SPECTRAL ASSIGNMENT

A.Bax and D.G.Davies

Laboratory of Chemical Physics

National Institute of Arthritis, Diabetes,

and Digestive and Kidney Diseases

National Institutes of Health

Bethesda, Md. 20205, USA.

INTRODUCTION

The idea of two-dimensional (2D) Fourier transform NMR, first introduced
by Jeener (1) has proven to be very powerful for analyzing the NMR spectra of compounds of high complexity. The general theory of 2D NMR was
described by Aue et al. (2) in 1976, and has provided the theoretical
basis for the development of a large number of new one- and two-dimensional pulse schemes. Nonetheless, the pulse scheme originally proposed
by Jeener, and described in great detail by Aue et al. is currently
still one of the most widely used 2D methods. Some minor modifications
have been made to this experiment to facilitate data handling and pro-
cessing (3,4) and to enhance the resolution by recording double quantum
filtered absorption mode spectra (5,6). Jeener's experiment is generally
referred to as the COSY (Correlated Spectroscopy) method, and has been
applied to a variety of problems, varying from small organic molecules
to biological macromolecules. The COSY spectrum displays the presence of
scalar coupling between coupled spins A and X by the presence of a so-
called cross peak centered about the frequency coordinates (F_1, F_2) =
(ν_A, ν_X) and (ν_X, ν_A), where ν_A and ν_X are the chemical shift frequencies

of spins A and X, respectively. Unfortunately, a more detailed look at such a cross multiplet shows that the individual components of this 2D multiplet are 180° out of phase relative to one another (2,3,7), i.e. no net magnetization is transferred between spins in the COSY experiment. For spin systems with a short transverse relaxation time, T_2, or digital resolution that is on the order of J_{AX} or coarser, partial cancellation of the 2D cross multiplet components invariably occurs, resulting in decreased cross peak intensity, and hence, lower sensitivity. For diagonal multiplets, centered around (ν_A, ν_A) and (ν_X, ν_X), such a mutual cancellation does not occur and the redundant diagonal resonances will often have much stronger intensity than the informative cross peaks, sometimes obscuring nearby cross peaks.

Recently, Braunschweiler and Ernst (8) introduced a new method that provides net magnetization transfer between coupled spins, and that relies on a rather different mechanism than the COSY experiment. With this method, many of the limitations of the COSY experiment can be overcome. They describe their experiment in terms of isotropic mixing, a technique first introduced in solid state NMR by Weitekamp, Garbow and Pines (9) and also used by Caravatti et al. (10). We describe new methods that rely on similar principles but that are experimentally easier to execute since they appear less sensitive to instrument imperfections. Our experiments are most conveniently analysed by considering them as homonuclear Hartmann-Hahn (HOHAHA) coherence transfer methods.

It was first demonstrated by Hartmann and Hahn (11) that net magnetization transfer between heteronuclear coupled spins occurs when one simultaneously switches on two coherent rf fields for spins I and spins S in such a way that

$$H_{1I}\gamma_I = H_{1S}\gamma_S \qquad (1)$$

often referred to as the Hartmann-Hahn match condition. In this expression H_{1I} and H_{1S} denote the effective rf field strengths, experienced by spins I and S, and γ_I and γ_S are the respective magnetogyric ratios. Under this condition complete magnetization exchange between two coupled spins occurs in a time $1/J_{IS}$. This Hartmann-Hahn magnetization transfer mechanism is the basis of heteronuclear cross polarization in solids (12), and has also been applied in 1D and 2D heteronuclear NMR experiments in liquids (13,14). Unfortunately, the heteronuclear Hartmann Hahn type magnetization transfer in liquids is very sensitive to the match condition (Eq.(1)); so despite its elegance the method has found little use in liquids. As will be shown below, a homonuclear Hartmann Hahn match is much easier to establish, even though the scalar couplings involved are often much smaller than in the heteronuclear case.

PRINCIPLES OF CROSS POLARIZATION

For the homonuclear case, γ_I and γ_S are identical and a match occurs when the effective rf field strengths, H_{1I} and H_{1S}, experienced by the two coupled spins are identical. In contrast with heteronuclear cross polarization experiments in liquids, this homonuclear Hartmann Hahn match is very insensitive to rf inhomogeneity: since H_{1I} and H_{1S} are generated by the same rf field, originating from a single coil and a single rf amplifier, the ratio H_{1I}/H_{1S} is to a good approximation independent of the location within the sample coil and of the rf power used. Hence, two scalar coupled spins will always experience the same rf field (not necessarily the same _effective_ rf field). For reasons that will become clear later, the magnetization exchange occurs a factor of two faster than in the heteronuclear case, i.e. complete transfer of magnetization from I to S can be achieved in a time, $1/(2J_{IS})$. Hartmann Hahn cross polarization has often been described in the spin temperature formalism, or by using a rather complex mathematical description. Here, we will attempt to present a simpler quasi-vector-picture of this phenomenon. Since Hartmann Hahn transfer will turn out to be closely related to strong coupling in homonuclear spin systems, this latter more familiar case will be addressed first.

Consider two homonuclear coupled spins, I and S, in a static magnetic field of strength, B_o, with no rf fields applied. The angular Larmor frequencies are Ω_I and Ω_S and the coupling equals J_{IS}. The spin operators for x, y and z magnetization of spin I are denoted by I_x, I_y and I_z, and similar operators are defined for spin S. In the Heisenberg formalism, the transverse spin operators, I_x and I_y, precess with frequency Ω_I about the z axis, and operators S_x and S_y precess with frequency Ω_S. The scalar IS interaction part of the Hamiltonian is given by

$$\mathcal{H}_J = J_{IS}(I_x S_x + I_y S_y + I_z S_z) \tag{2}$$

If the angular frequencies, Ω_I and Ω_S are vastly different, the product terms, $I_x S_x$ and $I_y S_y$, oscillate rapidly (in the Heisenberg formalism) and average to zero. Thus, for the weak coupling limit only the term $J_{IS} I_z S_z$ remains and is reponsible for the multiplet splittings commonly observed in NMR spectra. In the case where the difference between Ω_I and Ω_S is small (on the order of $2\pi J_{IS}$ or less), the terms $I_x S_x$ and $I_y S_y$ also have to be considered. As is well known from the strong coupling analysis (15), the stationary wave functions are no longer simple basic product functions but linear combinations thereof. If one suddenly changes the Hamiltonian from the weak coupling to the strong coupling limit or vice versa, this can cause non-equilibrium populations of the energy levels to be converted into off diagonal density matrix elements (and vice versa). As will be shown below, this is essentially the basis of Hartmann Hahn cross polarization.

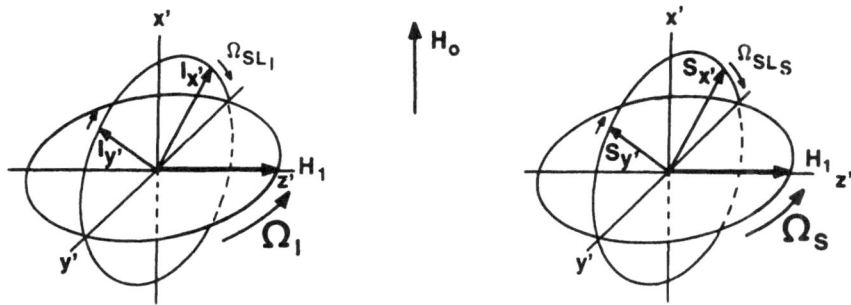

Figure 1 – Rotating frames for two coupled spins, I (left) and S
(right). The z' axes are chosen parallel to the respective
effective rf fields. The x' axes are parallel to the static
magnetic field. $I_{x'}, I_{y'}, S_{x'}$ and $S_{y'}$ are the Heisenberg spin
operators of angular momentum, as discussed in the text.

Consider first a heteronuclear two-spin-1/2 system, IS. If a $90°_x$ pulse
is applied to the I spins, followed by a spin lock rf field of strength
Ω_{SLI} along the y axis, the I spin magnetization will be aligned along
the I spin rf field, which is the new z axis for this system, and will
be labelled z' (Figure 1). The spin operators $I_{x'}$ and $I_{y'}$ in the
Heisenberg formalism, rotate with angular frequency Ω_{SLI} about the z'
axis. If simultaneously to the I spin spin-lock field a S spin rf field
of strength Ω_{SLS} is applied, the difference in Zeeman energy for the two
spins will be zero if $\Omega_{SLI} = \Omega_{SLS}$. However, the scalar coupling is
still present (<u>vide infra</u>). Because the rotating frames of spins I and S
rotate at different frequencies, the product $I_{z'}S_{z'}$ averages to zero;

however, the $I_{x'}$, $I_{y'}$, $S_{x'}$ and $S_{y'}$ operators rotate about their respective z' axis, at angular frequency Ω_{SLI} and Ω_{SLS} (Fig. 1). Since the components that are perpendicular to the static magnetic field precess at different Larmor frequencies (Ω_I and Ω_S) about the static magnetic field, only the components of $I_{x'}$ and $S_{x'}$ (and of $I_{y'}$ and $S_{y'}$) that are parallel to the static magnetic field can contribute to the average scalar product. For simplicity we assume that at the beginning of the spin lock period (t=0) both the $I_{x'}$ and the $S_{x'}$ operators are aligned parallel to the static magnetic field. Since these operators rotate at angular frequency Ω_{SLI} and Ω_{SLS} about their respective z' axis, the z components of $I_{x'}$ and $S_{x'}$ are proportional to $\cos(\Omega_{SLI}t)$ and $\cos(\Omega_{SLS}t)$, and hence their product is proportional to $\cos(\Omega_{SLI}t)$ $\cos(\Omega_{SLS}t)$. Similarly, the product of $I_{y'}S_{y'}$ is proportional to $\sin(\Omega_{SLI}t)\sin(\Omega_{SLS}t)$. For calculational purposes it is simpler to work in the Schroedinger formalism, chosen with the z' axis parallel to the spin lock field and the x' axis parallel to the static magnetic field. In this case one is dealing with two rotating frames that precess with different Larmor frequencies about the static magnetic field axis. The scalar interaction, I.S, then reduces to $I_{x}S_{x'}$ because the y and z components rotate at different Larmor frequencies. Writing $I_x=(I_+ +I_-)/2$, etc., and $\nu_I=\Omega_{SLI}/2\pi$, $\nu_S=\Omega_{SLS}/2\pi$, one obtains for the Hamiltonian under spin-locked conditions:

$$\mathcal{H}' = \nu_I I_{z'} + \nu_S S_z + J_{IS}(I_+S_- +I_-S_+ +I_+S_+ +I_-S_-)/4 \qquad (3)$$

In matrix form, this Hamiltonian is given by:

$$\mathcal{H}' = \begin{bmatrix} \nu_I+\nu_S & 0 & 0 & J/4 \\ 0 & \nu_I-\nu_S & J/4 & 0 \\ 0 & J/4 & \nu_S-\nu_I & 0 \\ J/4 & 0 & 0 & -\nu_I-\nu_S \end{bmatrix} \qquad (4)$$

and this matrix operates in the basic operator product function basis. Therefore, if spins I have unity magnetization at the beginning of the spin lock period and spins S are saturated, the deviations of the average populations of the energy levels $|\alpha\alpha\rangle, |\alpha\beta\rangle, |\beta\alpha\rangle$ and $|\beta\beta\rangle$ are $1,1,-1$ and -1, respectively, in units $h\,\Omega_I/4\pi kT$ (h is Planck's constant and k is Boltzmann's constant). It is then easily found that the terms H_{23}, and H_{32}, induce oscillatory changes in the populations of levels $|2\rangle$ and $|3\rangle$. For exactly matched rf fields, i.e. $\nu_I = \nu_S$, after a mixing period of duration $1/J$, the populations of levels $|2\rangle$ and $|3\rangle$ are interchanged, and all I spin magnetization has been transferred to the S spins. Note that matrix elements H_{14}, and H_{41}, have very little effect because the energies of levels $|1\rangle$ and $|4\rangle$ differ by $2\nu_I + 2\nu_S$, and these elements are therefore nonsecular.

The analysis of heteronuclear cross polarization given above is not new and a number of rather similar descriptions can be found in the literature. The main point we wish to make is that in order to induce magnetization transfer between the two coupled spins, one has to reduce the difference in apparent Zeeman energies (ν_I and ν_S) for the two spins to less than the size of the interaction between them, i.e. one has to modify a weakly coupled spin system into a strongly coupled one.

So far, we have discussed cross polarization for the heteronuclear case. However, the treatment for the homonuclear case is completely analogous and even slightly simpler since all discussions relate to a single rotating frame. Consider two weakly coupled ($J_{IS} \ll |\Omega_I - \Omega_S|/2\pi$) protons, I and S, and the longitudinal magnetization of spin S has ben destroyed by presaturation. Again a 90°_x pulse followed by a spin lock along the y axis is applied. Assuming that the nominal strength of the spin lock rf field, ν, is much larger than the offset frequencies, $\Omega_I/2\pi$ and $\Omega_S/2\pi$, the effective rf fields for the two spins are approximately parallel and their magnitudes, ν_I and ν_S, are nearly identical. Since the spin lock frames for the two homonuclear spins almost coincide (in contrast with the heteronuclear case), the size the scalar interaction between I and S is unaffected by the spin lock field. In the Schroedinger representation, we find for the Hamiltonian under spin locked conditions:

$$\mathcal{H}' = \nu_I I_{z'} + \nu_S S_{z'} + J_{IS}(I_{x'}S_{x'} + I_{y'}S_{y'} + I_{z'}S_{z'}) \tag{5}$$

where ν_I and ν_S are the effective spin lock field strengths for spins I and S, respectively. In matrix form (in the product basis) this yields

$$\mathcal{H}' = \begin{bmatrix} \nu_I + \nu_S + J/2 & 0 & 0 & 0 \\ 0 & \nu_I - \nu_S - J/2 & J/2 & 0 \\ 0 & J/2 & \nu_S - \nu_I - J/2 & 0 \\ 0 & 0 & 0 & -\nu_I - \nu_S + J/2 \end{bmatrix} \tag{6}$$

Because the product $I_{y'}S_{y'}$ does not average to zero (in contrast with the heteronulear case), matrix elements H'_{23} and H'_{32} are a factor of two larger than in Eq.(4), and magnetization transfer between spins occurs twice as fast, i.e. for $\nu_I = \nu_S$, complete transfer from I to S is obtained in a time $1/(2J_{IS})$. Again, the form of the Hamiltonian is identical to that of a strongly coupled homonuclear AB spin system.

THE EFFECT OF RF OFFSET

If the two spins, I and S, have different offsets from the carrier frequency, the magnitude of the effective spin lock fields will differ for the two spins. For offsets, Δ, small relative to the nominal rf field strength, ν, the effective rf field strength is to a good approximation given by

$$\nu_{eff} = \nu + \Delta^2/2\nu \tag{7}$$

Therefore, if the two offsets $|\Delta_I|$ and $|\Delta_S|$ differ, this causes a mismatch of the Hartmann Hahn condition, which results in an incomplete magnetization transfer between the two spins. Simple matrix calculation, based on Eq.(6), shows that for $|\nu_I - \nu_S| = |J|$ only half the I spin magnetization can be transferred to S and vice versa.

For a system of three coupled spins, AMX, the situation is somewhat more complicated. In this case, an 8x8 Hamiltonian and an 8x8 density matrix have to be considered. However, as is the case for the two spin system, magnetizazion transfer occurs only between levels of equal total spin angular momentum, $\underline{M}(\alpha\beta$ and $\beta\alpha$ for the 2-spin case). For the 3-spin case, one has to consider two sets of energy levels: $\underline{M}=\frac{1}{2}(\alpha\alpha\beta, \alpha\beta\alpha, \beta\alpha\alpha)$ and $\underline{M}=-\frac{1}{2}(\alpha\beta\beta, \beta\alpha\beta, \beta\beta\alpha)$. We therefore can consider separately the reduced Hamiltonians, $\mathcal{H}_{r'}$, for $\underline{M}=\pm\frac{1}{2}$. For $\underline{M}=\frac{1}{2}$ and $J_{AX}=0$, one obtains:

$$
\mathcal{H}_{r'} = \begin{bmatrix} \nu_A+\nu_M-\nu_X+(J_{AM}-J_{MX})/2 & J_{MX}/2 & 0 \\ J_{MX}/2 & \nu_A-\nu_M+\nu_X-(J_{AM}+J_{MX})/2 & J_{AM}/2 \\ 0 & J_{AM}/2 & -\nu_A+\nu_M+\nu_X-(J_{AM}-J_{MX})/2 \end{bmatrix} \quad (8)
$$

In this 3-level system, the energies of levels $\alpha\alpha\beta$ and $\alpha\beta\alpha$ differ by $J_{AM}+2\nu_M-2\nu_X$. In other words, the coupling between spins A and M induces a mismatch between spin M and X if $\nu_M=\nu_X$. This mismatch can be reduced by choosing the carrier position such that $\nu_X-\nu_M=J_{AM}/2$ (Eq.(7)), but in this case the magnetization transfer in the second subset ($\underline{M}=-\frac{1}{2}$) between the levels $\beta\alpha\beta$ and $\beta\beta\alpha$ is even further mismatched, i.e. significant transfer will only involve one of the two X-spin doublet components under spin locked conditions. Upon the switchoff of the rf field, the total X spin-locked magnetization will be symmetrically distributed over the two doublet components and this multiplet effect becomes invisible.

The qualitative analysis presented above suggests that the presence of other spins broadens the match condition between M and X (provided that A and M are mismatched) and simultaneously decreases the total amount of transfer that is obtainable between two coupled spins. Nevertheless, significant magnetization transfer between coupled spins I and S remains limited to $|\nu_I - \nu_S| < \Sigma J/2$ where ΣJ denotes the sum of the multiplet widths of spins I and S (in the absence of heteronuclear coupling).

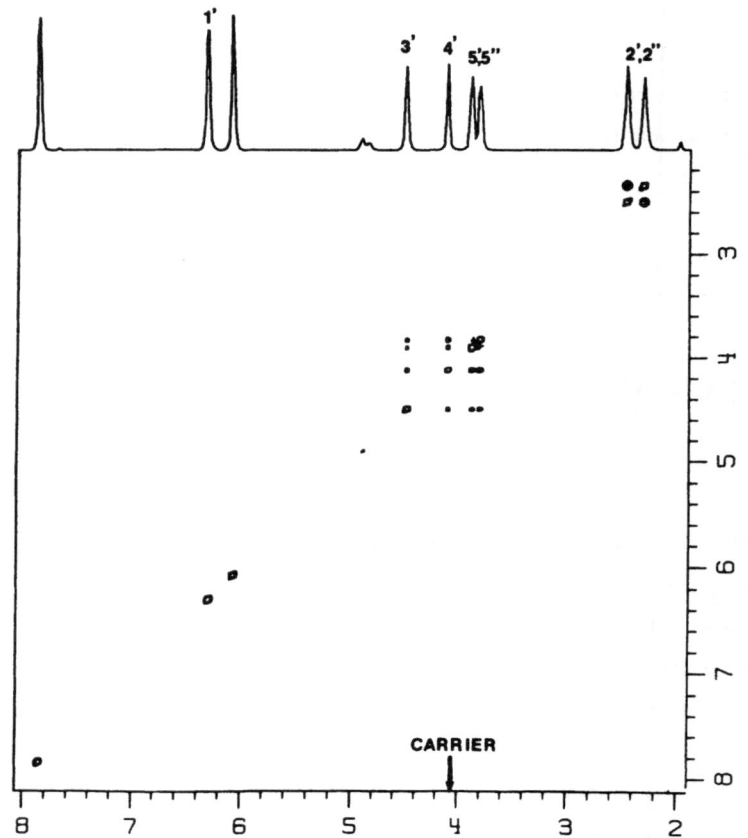

Figure 2 – Part of the two-dimensional spectrum of 2'-deoxycytidine, obtained with a mixing time of 100 ms. The position of the carrier frequency has been indicated on the F_1 axis. Magnetization transfer is restricted to protons that experience similar magnitudes of the effective spin lock field

TABLE I

Phase, φ, of the initial 90° pulse, and the way data
are added to and subtracted from memory in successive
scans of the experiments sketched in Figure 3

scan	φ	Acq[a]
1	x	+
2	y	+
3	-x	-
4	-y	-

[a]Data in odd- and even-numbered scans are stored in separate memory
locations

This limits the straightforward application of homonuclear Hartmann Hahn
spectroscopy to spins that have a relatively large coupling and similar
magnitudes for the effective rf field strength, i.e. similar magnitudes
for the rf offset from the carrier frequency. An example of such a case
in Figure 2, for the nucleotide 2' deoxycytidine. The pulse scheme
$90°_\varphi -t_1-SL_y-$ acquire(t_2) has been employed, with standard phase cycling
of φ (Table 1) to allow a hypercomplex Fourier transformation (13,16,17)
for obtaining an absorption mode 2D spectrum. In this experiment the
carrier frequency has been positioned in the center of the 3', 4' and
5', 5" region and a 100 ms spin lock pulse with a rf field strength of
approximately 6 kHz has been used. Despite the fact that the coupling
between the 3' and 4' protons is rather small (several Hertz), an intense cross peak between the 3' and 4' protons is observed. Cross peaks

between the 3' and the 5', 5" protons are also observed, and are due to relay effects: magnetization that has been transferred from the 3' to the 4' proton during the first part of the mixing period is subsequently transferred to the 5', 5" protons during the second part of the mixing period. For very short durations of the mixing period ($\ll 1/J$), these relay effects are absent. Cross peaks between the 2', 2" protons are also observed (large J and small difference in offset frequency), but no cross peak is observed between the 3' and the 2', 2" protons. This is caused by the Hartmann Hahn mismatch, induced by the difference in off-set frequencies of the 3' and the 2', 2" protons. The selectivity of this type of Hartmann Hahn experiment can be beneficial: a mismatch between the 2',2" protons enhances the cross peak intensity between the 3' and 4' protons because no 3' magnetization is drawn away by the 2', 2" protons. A major advantage of the homonuclear Hartmann Hahn experiment over the COSY experiment is that net magnetization transfer is obtained, allowing phase sensitive spectra to be recorded, and yielding improved sensitivity in the case of poorly resolved resonances. Several points should be mentioned here:

1. The individual components of a cross multiplet are generally not pure absorptive, but consist of a mixture of absorptive and dispersive components (8,13,14). The dispersive components are in antiphase re-lative to one another and these unwanted components therefore largely cancel. However, at very high digital resolution and for well resol-ved couplings this phase distortion is clearly visible. For simple spin systems (AX, AX_2, AX_3, AMX) explicit expressions can be derived from density matrix calculations. For a two-spin system these results have been presented by Braunschweiler and Ernst (8). The AX_2 case has recently been discussed by Chandrakumar and Subramanian (18).

2. Because the total amount of spin locked magnetization is a constant of motion, magnetization transfer between spins is an exchange process, and the absorptive part of the cross multiplet will be in

phase with the diagonal resonances. As has been pointed out by Bothner-By and co-workers (19), one can use the same pulse sequence as discussed above for measuring transverse NOE effects. Transverse NOE effects are positive for all values of the motional correlation time, τ_c, and therefore give rise to negative cross peaks. Hartmann Hahn and spin locked NOE cross peaks are thus easily distinguished (20,21). In practice, we have always found that for the mixing times most commonly used in the Hartmann Hahn experiment (15–100 ms), the size of NOE cross peaks is usually very small and does not present any serious problems. Cross peaks due to chemical exchange during the spin lock period (22,23) also give rise to positive cross peaks that should not be confused with Hartmann Hahn cross peaks. Note that NOE peaks have opposite phase relative to exchange peaks, independent of the motional correlation time, τ_c. This feature immediately distinguishes these two types of magnetization transfer mechanisms (24).

REMOVAL OF THE HARTMANN HAHN MISMATCH

Above, it was pointed out that the effectiveness of homonuclear Hartmann Hahn type coherence transfer rapidly decreases if the difference in effective rf field strengths for two coupled spins becomes larger than the scalar coupling. Consequently, only relatively narrow spectral bandwidths can be covered. A solution to this problems is to alternate the rf phase of the spin lock field along the $\pm y$ axis (Figure 3b) (21). We will discuse the reasoning behind this phase alternation in a qualitative manner, based on the Heisenberg vector picture.

As pointed out before, strong homonuclear coupling occurs if the Heisenberg spin operators I_x and S_x precess at similar rates about the static magnetic field (or the spin lock field) and diverge slowly relative to $1/J$. As was found several decades ago, one can induce strong

34

coupling effects by the application of a series of 180° pulses that are
spaced closely relative to the difference in resonance frequency,
$|\Delta_I - \Delta_S|$ (25,26). This type of mixing scheme is also used in the TOCSY
scheme, proposed by Branschweiler and Ernst (8). In the Heisenberg vector picture, this strong coupling effect is easily understood: the 180°
pulses prevent the I_x and S_x (and the I_y and S_y) components from dephasing, and hence the $I_x S_x$ (and the $I_y S_y$) term does not disappear, i.e.
strong coupling is induced.

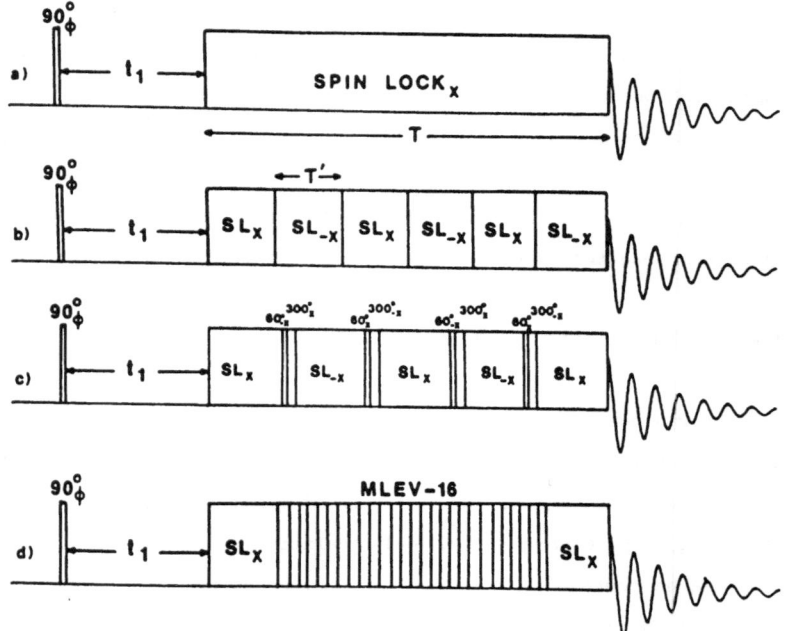

Figure 3 - Various 2D pulse schemes for establishing HOHAHA
magnetization transfer.
(a) using a continuous spin-lock field,
(b) a phase alternated spin lock field,
(c) a compensated phase alternated spin lock field,
(d) an MLEV sequence with trim pulses

In the homonuclear Hartmann Hahn experiment, where the spin lock field
is applied along the y axis (relabeled for convenience as z'), the $I_{x'}$

and $S_{x'}$ operators precess at different rates about the z' axis if the effective field strengths differ. However, the dephasing of $I_{x'}$ and $S_{x'}$ can be refocused by changing the phase of the spin lock field (Fig. 3). This causes the precession of the $I_{x'}$ and $S_{x'}$ vectors to reverse, and prevents the product $I_{x'}S_{x'}$ from averaging to zero. For effective Hartmann Hahn mixing, the $I_{x'}$ and $S_{x'}$ vectors should not diverge more than about $\pi/2$, and hence, one wants to alternate the phase of the spin lock field at rate, $1/\tau'$, given by

$$1/\tau' > 4|\nu_I - \nu_S| \qquad (9)$$

For $1/\tau' = |\nu_I - \nu_S|$ no Hartmann Hahn type coherence transfer will occur, and for $1/\tau' = 2|\nu_I - \nu_S|$, the rate will be reduced by about 30% relative to the matched Hartmann Hahn transfer rate. Note that in Eq. (8) the size of the coupling does not appear. As mentioned before, however, the transfer rate is directly proportional to the size of the scalar interaction.

In order to obtain 2D absorption mode spectra, it is important that only the magnetization components parallel to the spin lock axis remain, i.e. transverse components have to be defocused. Phase alternation of the spin lock field causes the introduction of rotary echoes (27) for magnetization that is transverse to the spin lock field. In principle, these echoes contain useful signal (10) since they provide the information about magnetization that is perpendicular to the spin lock field, i.e. this signal provides the complementary half needed for quadrature detection in the t_1 dimension. In practice, however, we find that the decay of these echoes occurs much faster than $T_{1\varrho}$ (caused by experimental imperfections) and we therefore try to eliminate them in order to allow the recording of absorption mode spectra. These rotary echoes can be effectively removed by applying a "trim pulse" of approximately 3 ms at the end of the mixing period, with the same rf phase as the spin lock pulses. The inhomogeneity of the rf field will defocus magnetization that is perpendicular to the axis along which the trim pulse is applied (with the exception of certain antiphase multiplet components) and spectra are therefore modulated in amplitude as a function of t_1, and have the required format for obtaining 2D absorption mode spectra (13,16,17).

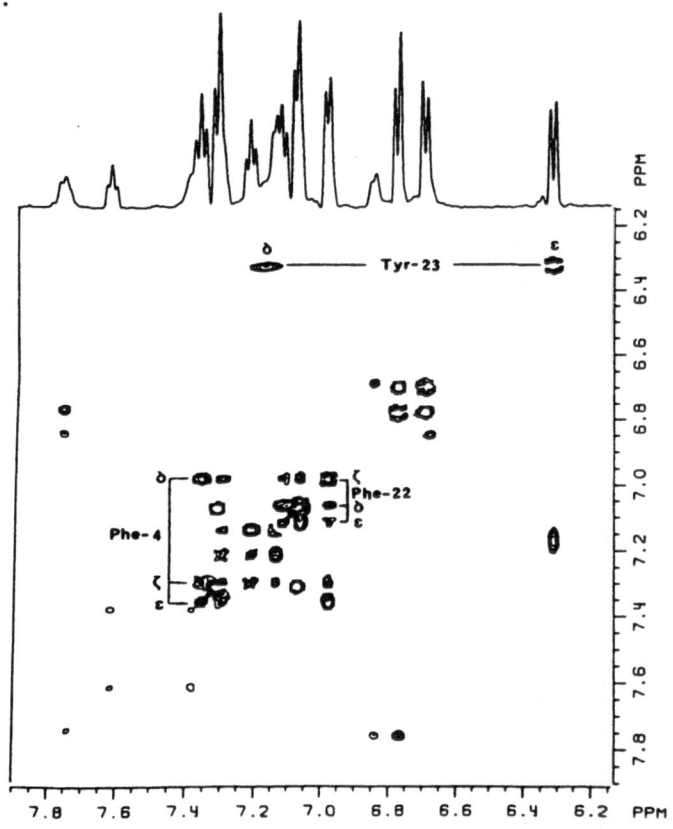

Figure 4 – 2D HOHAHA spectrum of the aromatic region of a 5 mM soluti-on of bovine trypsin inhibitor in D_2O, recorded at 500 MHz, 42°C, pH 4.6, using a 35 ms spin lock period. The spectrum confirms the reassignment of the Phe-4 protons by Rance et al.(6), and additionally suggests that resonances of Phe-22 should be reassigned as indicated

The phase alternated spin lock version of the Hartmann Hahn experiment makes it possible to cover fairly large spectral widths. On our 500 MHz spectrometer, and 8 W linear amplifier was used for generating the spin lock rf field of approximately 7.5 kHz rf field strength. Although the amplifier generated 8 W, 6 dB of this power was lost in the electronic circuitry (tuned diodes, bandpass filter, gating circuitry) and only 2 W rf power was dissipated in the high resolution probehead of our 500 MHz ^{1}H probe. With a 7.5 kHz rf field strength we can cover a bandwidth of approximately ± 1 kHz, i.e. $\pm$ 2 ppm on our instrument, if the spin lock phase is alternated every 5 ms (21). As an example, Fig. 4 shows the Hartmann Hahn spectrum obtained for the aromatic region of a 8 mM solution of bovine trypsin inhibitor, a compound extensively investigated in the past by Wüthrich, Wagner, and co-wokers (28). The spectrum, recorded at 42 °C and using a 35 ms mixing time, shows the connectivities within each of the aromatic rings of the Tyr and Phe residues. The spectrum is displayed in the absorption mode and shows the excellent resolution that is obtained with this method. The net magnetization transfer provides high sensitivity because no mutual cancellation of the absorptive components of the cross multiplet occurs. This allows the observation of a cross peak between the δ and ε protons of Tyr-23, even while the resonances of the δ protons are broadened by slow flipping of the ring. In a COSY spectrum this cross peak could not be observed at this temperature. For the Phe residues, relay peaks are also observed. For example, for Phe-22 the ζ proton (7.00 ppm) shows connectivity to both the ε protons (7.15 ppm) and relay peaks to the δ protons (7.09 ppm). The fact that both the δ protons and the ε protons have identical chemical shifts suggests, contrary to earlier assumptions (29), that this residue is flipping fast relative to the difference in chemical shift frequency.

COMPENSATED PHASE-ALTERNATED SPIN LOCKING

As mentioned above, with the phase alternated spin lock mixing sequence, a substantial bandwidth can be covered. The typical spectral width for many compounds of interest is about 8-10 ppm. On the basis of Eq.(8), one then calculates that one has to alternate the rf phase as often as every millisecond or faster. Sice the effective rf field vectors for spin locking along the x and -x axes are not 180° out of phase relative

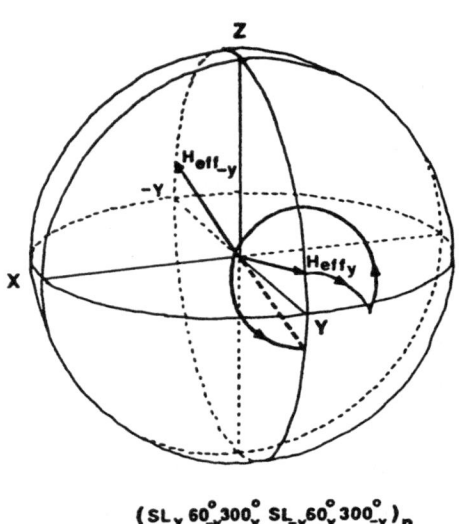

$$(SL_Y \, 60^\circ_{-Y} 300^\circ_Y \, SL_{-Y} 60^\circ_Y 300^\circ_{-Y})_n$$

Figure 5 - The rotation of magnetization, initially parallel to the y effective field, to the effective field direction for a -y rf field (parallel to the boldface broken line). The 60° 300° pulse pair rotates magnetization from the y effective field to the -y effective field direction. A counterclockwise rotation of magnetization about the magnetic field is assumed

to one another in the case of resonance offset (Fig.5), during each phase alternation some magnetization is lost. For resonance offset, Δ , the effective field vectors are at an angle of approximately $\pi - 2\Delta/\nu$, and during each phase alternation a fraction $1 - \cos(2\Delta/\nu)$ is lost. For large offsets and relatively rapid phase alternation this leads to a rapid decay of spin locked magnetization. One way to avoid this problem is to rotate the magnetization that is initially locked along the +x effective rf field to the −x effective rf field axis. A simple way to accomplish this is the insertion of a $60°_{-x}$ $300°_{x}$ pulse pair after the spin lock along the +x axis (Fig. 3c). The rotation by such a pulse pair of magnetization, initially parallel to the effective spin lock field, to the new effective field vector with inverted rf phase, is sketched in Fig. 5. Similarly, a $60°_{x}$ $300°_{-x}$ pulse pair is inserted when the phase is changed from −x to +x. In pratice, we "fine tune" both the 60° and the 300° pulse by maximizing the spin locked signal for an isolated resonance that is approximately 2 kHz off resonance, using for example, 1.5 ms τ' values and about 40 phase alternations. This approach allows us to cover about ±2 kHz (with a 7.5 kHz rf field), and is sufficient for most applications. For covering narrower spectral widths, no fine tuning of the 60° and 300° pulses appears necessary.

As an example, Fig. 6 shows the entire 2D spectrum of the bovine trypsin inhibitor, recorded in 3 h., and using a mixing time of 34 ms. The sample has been dissolved in D_2O, at pH 4.6, and only the non-exchanging amide protons are visible in this spectrum. Connectivity between the amide and α protons can be seen, and significant amounts of relay to the β protons is also observed. At lower contour levels (not shown) nearly all amide-$H(C_\beta)$ relay connectivities are observed. The inset shows the connectivity between the δ and ε protons of Tyr-23, and shows the broadening of the δ protons due to slow ring flipping. Similar broadened cross peaks are observed for Phe-45 (at about 8ppm). The low intensity negative resonances in the inset are not due to spin locked NOE but are a common feature in many types of 2D absorption mode spectra. These negative "ditches" running parallel to the F_2 axis, are caused by the fact that the first couple of points of the free induction decay (acquired during t_2) are too low in intensity, caused by the filter bandwidth of the audio filters. Using a wider filter bandwidth and minimizing the delay between the end of the mixing period and the opening of the receiver gating can be used to minimize these artefacts.

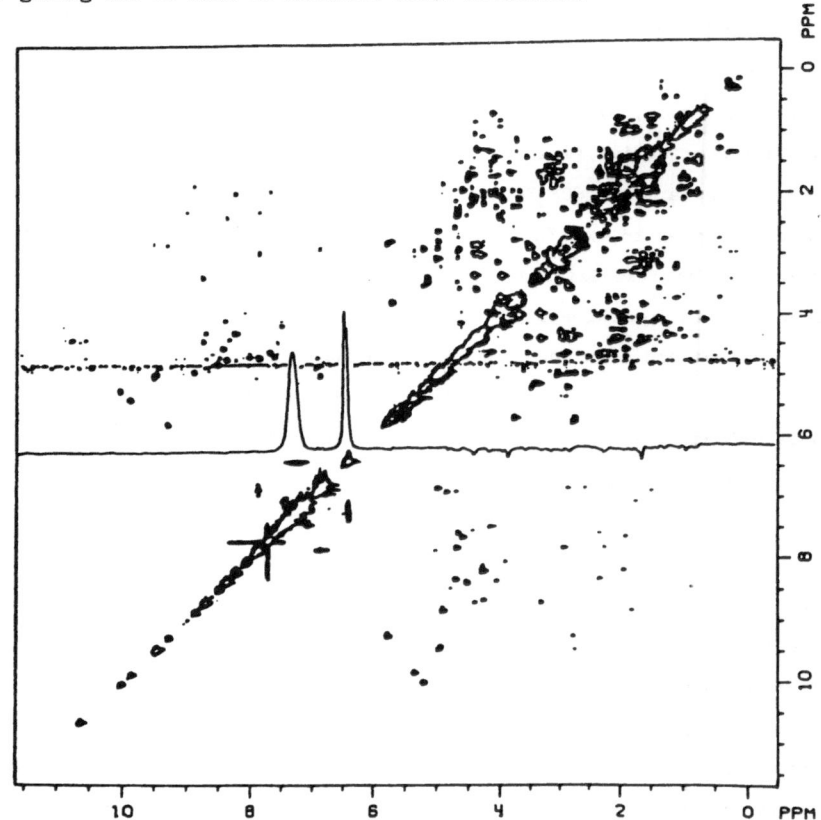

Figure 6 – 500 MHz 2D HOHAHA spectrum of 8 mM bovine trypsin inhibitor in D_2O, pH 4.6, at 37°C, using a 34 ms mixing period and τ' durations of 1.5 ms. The total measuring time was 3 h. The inset shows a F_1 section taken at the F_2 frequency of the Tyr-23 C_ϵ protons and displays the connectivity to the C_δ protons, despite the broadening introduced by intermediately slow ring flipping

MIXING WITH A MLEV-16 COMPOSITE PULSE CYCLE

The major limitation in the application of homonuclear Hartmann Hahn mixing is the decay of the magnetization during the spin lock, with decay
constant $T_{1\varrho}$ which is approximately equal to the transverse relaxation
time T_2. However, mixing schemes can be developed that prolong this
apparent decay constant, the theoretical limit being T_1. The best we
have been able to achieve, so far, is a "spin lock" period using a MLEV-
16 composite pulse decoupling cycle (31), instead of the phase alternated spin lock field (Fig. 3d). A single MLEV-16 cycle consists of a
repetitive series of 16 composite pulses, A and B, applied in the order:
AABB ABBA BBAA BAAB. The composite pulse A consists of $90°_x\ 180°_y\ 90°_x$,
and B is the inverse of this: $90°_{-x}\ 180°_{-y}\ 90°_{-x}$. The net effect of A or B
is a rotation of 180° about the +y (A) or -y (B) axis. Thus, the MLEV
sequence can be considered as a rapidly phase alternating spin lock
sequence along the +y axis. However, composite pulses A and B rotate
magnetization vectors, initially parallel to the y axis, through space
in such a way that effectively one half of the time the magnetization is
aligned along the static magnetic field. During this time, the relaxation is determined by T_1, which for macromolecules is usually much
longer than $T_{1\varrho}$ and T_2. In the limit, $T_1 \gg T_{1\varrho}$ one can therefore expect
an apparent lengthening of the decay constant by up to ~100%. Experi-
mentally, we have observed a lengthening of 96% for the non-exchangeable
amide protons in ribonuclease. Unfortunately, the working of the MLEV
sequence for obtaining efficient magnetization transfer between homonuclear coupled spins is very sensitive to pulse imperfectione (imperfect 90° phase shifts and unbalanced rf amplitudes) and therefore requires properly adjusted hardware. As shown elsewhere (31), these pulse
imperfections during MLEV mixing can slow down the transfer rate and can
also cause difficulties in obtaining absorptive spectra. However, the

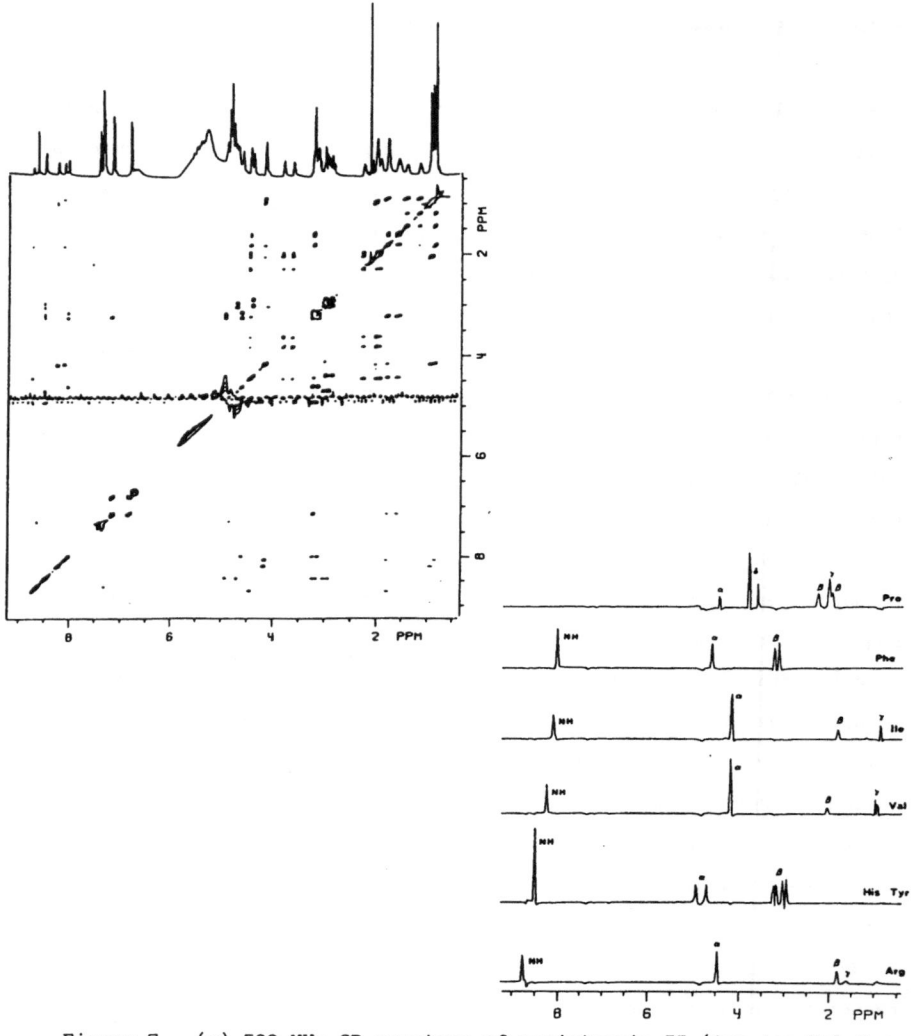

Figure 7 - (a) 500 MHz 2D spectrum of angiotensin-II (Asp-Arg-Val-Tyr-
Ile-His-Pro-Phe) obtained with an MLEV mixing period of 54
ms, preceded and followed by a 3 ms trim pulse. The sample
was dissolved in 80%H_2O/20%D_2O, pH 3, at 30°C. The total
data acquisition time was 2h.
(b) Sections parallel to the F_1 axis through the 2D
spectrum, taken at the F_2 frequencies of the various amide
protons and the high-field Pro C_δ proton

advantage of lengthening the decay constant during the mixing period may be crucial in the study of macromolecules. Improvements to the MLEV sequence that will make it less sensitive to pulse imperfection are currently under investigation.

An example of a 2D spectrum of angiotensin-II, a linear octapeptide, dissolved in 90% H_2O/10% D_2O, obtained with a MLEV mixing period of 54 ms is shown in Figure 7a. Intense relay peaks between NH and C_β and $C\gamma$ protons are observed. Cross sections parallel to the F_1 axis taken at the F_2 frequencies of the various amide resonances provide partial sub-spectra of the various amino acids (Figure 7b).

GENERATION OF ONE-DIMENSIONAL SUBSPECTRA

The spin propagation idea can easily be used to generate subspectra of parts of a molecule that are not scalar coupled to other parts. For example, it is possible to generate subspectra of individual sugar units in an oligosaccharide, or individual amino acids in a peptide. The idea is very similar to 1D NOE difference spectroscopy: a selective 180° pulse is used to invert the ^{1}H multiplet of one particular hydrogen, and a 90° pulse followed by a propagation period of variable length is employed. This spectrum is then substracted from a spectrum obtained in an identical way, but without the selective 180° pulse (Fig. 8). Only magnetization that originates from the inverted ^{1}H multiplet will appear in the difference spectrum. For short propagation periods (< 25 ms), magnetization that has not propagated much further than to geminal or vicinal protons will appear in the difference spectrum. For longer propagation delays (100-200 ms), one can obtain entire subspectra of individual parts of the molecule. As an example, Fig. 9 shows how the spectrum of sucrose can be separated into a glucose and a fructose subspectrum. Note that in the fructose subspectrum tha C1 protons are absent because coupling to the C3 proton is vanishingly small.

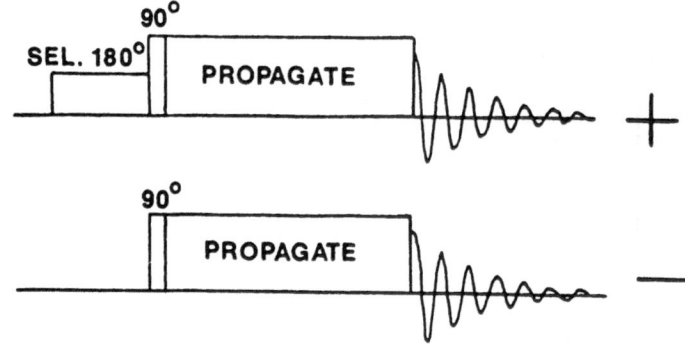

Figure 8 – Pulse scheme for the generation of one-dimensional subspectra. Any of the mixing schemes used in Fig.3 can be used to induce spin propagation. A difference experiment, with and without a selective 180° pulse, generates a subspectrum of all protons directly or indirectly coupled to the selectively inverted proton

We find the one-dimensional propagation method very useful for ^{1}H resonance assignment in complex overlapping spectra of rather small molecules ($M_r < 1500$) that often have a relatively long $T_{1\varrho}$ and permit the use of 1D spin propagation without excessive loss in sensitivity.

DISCUSSION AND CONCLUSIONS

Above, the idea of spin propagation has been discussed in very quali-tative terms and a attempt has been made to convey some of the ideas that are helpful in understanding the propagation mechanism. A more rigorous treatment requires diagonalization of the propagation Hamilto-nian and generally requires a substantial amount of computer arithmetic. An advantage of the computer analysis of the problem is that resonance

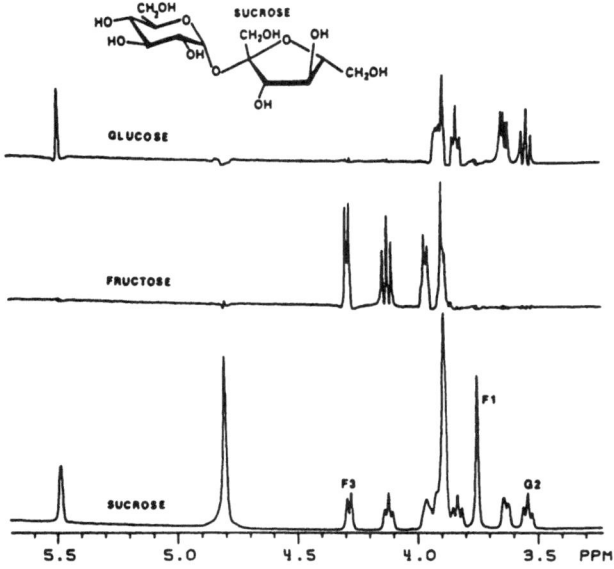

Figure 9 - Spectra of sucrose.(a) Regular ^{1}H spectrum and (b) propagation spectrum of fructose, obtained by selective inversion of H3 and (c) propagation spectrum of glucose, obtained by inverting glucose H2. Propagation periods of 100 ms
were used for both spectra (b) and (c), and the compensated
alternated spin lock mixing (Fig.3c) was used. Note that
fructose protons H1 do not appear in (b) due to the absence
of a scalar coupling pathway to the rest of the fructose
protons

offset effects do not provide major additional problems. A second simple

computer solution to describe the behaviour of the spin system during

propagation is to use a numerical evaluation of the Liouville equation:

$$\dot{\sigma} = -i\,[\,\sigma,H\,]$$

(10)

where no diagonalization of H is required. Using such computer calcula-

tions one can estimate the size of unresolved scalar couplings from the buildup rate of cross peaks in the 2D spectra recorded for a series of propagation delays. This application is currently under investigation.

We do not expect that the HOHAHA method will replace the COSY technique because the latter has the advantage of showing exclusively direct scalar connectivity, whereas the HOHAHA method can show weak relay peaks for mixing times as short as 15 ms. In this case it may not always be obvious whether such a peak originates from a direct coupling or whether it concerns a relay effect. However, it is expected that HOHAHA spectroscopy will be used as an additional device for cases where the COSY method does not provide an unambiguous answer. In our experience, the HOHAHA method appears more effective, both in sensitivity and resolution, than the homonuclear RELAY experiment (32-34) and we expect this RELAY method to become obsolete for most common cases of interest.

A large number of other, more complicated pulse sequences, based upon the homonuclear Hartmann Hahn effect can be envisioned. For example, all current techniques that rely on COSY type relay of magnetization can be modified in such a way that magnetization relay occurs _via_ the Hartmann Hahn mechanism. For the heteronuclear $^1H-^1H-^{13}C$ relay experiment (35) such an experimental modification has shown to be very useful, especially when looking at saturated aliphatic ring systems (36).

ACKNOWLEDGMENTS

The authors are indebted to Ted Becker, Gary Drobny, Al Garroway, Luciano Muller and Attila Szabo for stimulating discussion and to Laura Lerner for critical comments during the preparation of the manuscript. The angiotensin sample was kindly provided by James Ferretti and technical support was provided by Rolf Tschudin.

REFERENCES

1) J.Jeener, Ampere International Summer School, Basko Polje, Yugoslavia, 1971.

2) W.P.Aue,E.Bartholdi and R.R.Ernst, J.Chem.Phys. **64**, 2229 (1976).

3) A.Bax and R.Freeman, J.Magn.Reson. **44**, 542 (1981).

4) K.Nagayama,A.Kumar,K.Wüthrich and R.R.Ernst, J.Magn.Reson. **40**, 321 (1980).

5) U.Piantini,O.W.Sorensen and R.R.Ernst, J.Amer.Chem.Soc. **104**, 6800 (1982).

6) M.Rance,O.W.Sorensen,G.Bodenhausen,G.Wagner,R.R.Ernst and K.Wütrich Biochem.Biophys.Res.Commun. **117**, 479 (1983).

7) A.Bax, "Two-Dimensional Nuclear Magnetic Resonance", Reidel, Boston, 1982, Ch.2.

8) L.Braunschweiler and R.R.Ernst, J.Magn.Reson. **53**, 521 (1983).

9) D.P.Weitekamp,J.R.Garbow and A.Pines, J.Chem.Phys. **77**, 2870 (1982).

10) P.Carravatti,L.Braunschweiler and R.R.Ernst, Chem.Phys.Letters **100**, 305 (1983).

11) S.R.Hartmann and E.L.Hahn, Phys.Rev. **128**, 2042 (1962).

12) A.Pines,M.G.Gibby and J.S.Waugh, J.Chem.Phys. **59**, 569 (1973).

13) L.Muller and R.R.Ernst, Mol.Phys. **38**, 963 (1979).

14) G.C.Chingas,A.N.Garroway,R.D.Bertrand and W.B.Moniz, J.Chem.Phys. **74**, 127 (1981) and references therein.

15) J.A.Pople,W.G.Schneider and H.J.Bernstein, "High Resolution Nuclear Magnetic Resonance", McGraw-Hill, New York, 1959.

16) D.J.States,R.A.Haberkorn and D.J.Ruben, J.Magn.Reson. **48**, 286 (1982).

17) A.Bax, Bull.Magn.Reson., in press.

18) N.Chandrakumar and S.Subramanian, J.Magn.Reson. **62**, 346 (1985).

19) A.A.Bothner-By,R.L.Stephens,J.T.Lee,C.D.Warren and R.W.Jeanloz, J.Amer.Chem.Soc. **106**, 811 (1984).

20) A.Bax and D.G.Davis, J.Magn.Reson. **63**, 207 (1985).

21) D.G.Davis and A.Bax, J.Amer.Chem.Soc. **107**, 2821 (1985).

22) J.Hennig and H.H.Limbach, J.Magn.Reson. **49**, 322 (1982).

23) H.Bleich and J.Wilde, J.Magn.Reson. **56**, 149 (1984).

24) D.G.Davis and A.Bax, J.Magn.Reson. submitted for publication.

25) E.J.Wells and H.S.Gutowsky, J.Chem.Phys. **43**, 3414 (1965).

26) A.Allerhand, J.Chem.Phys. **44**, 1 (1966).

27) I.Solomon, Phys.Rev.Letters 2, 301 (1959).

28) G.Wagner and K.Wüthrich, J.Mol.Biol. **155**, 347 (1982).

29) G.Wagner, Quart.Rev.Biophys. **16**, 1 (1985).

30) M.H.Levitt,R.Freeman and T.A.Frenkiel, J.Magn.Reson. **47**, 328 (1982).

31) A.Bax and D.G.Davis, J.Magn.Reson. submitted for publication.

32) G.Eich,G.Bodenhausen and R.R.Ernst, J.Amer.Chem.Soc. **104**, 3732 (1982).

33) G.Wagner, J.Magn.Reson. **55**, 151 (1983).

34) A.Bax and G.Drobny, J.Magn.Reson. **61**, 306 (1985).

35) P.H.Bolton, J.Magn.Reson. **48**, 336 (1982).

36) A.Bax,D.G.Davis and S.K.Sarkar, J.Magn.Reson. **63**, 230 (1985).

PBB, Vol. 2
Advanced Magnetic Resonance Techniques
in Systems of High Molecular Complexity
© 1986 Birkhäuser Boston, Inc.

SIMPLIFICATION OF 2D SPECTRA BY (a) TOPOLOGY-SELECTIVE

MULTIPLE-QUANTUM FILTRATION OR (b) BY BILINEAR MIXING

M.H.Levitt*, C.Radloff and R.R.Ernst

Laboratorium für Physikalische Chemie

ETH-Zentrum, Zürich, Switzerland

*Present address:Francis Bitter Magnet Lab

MIT Bldg.NW14, Cambridge, Mass 02139, USA

Two very different ways of simplifying homonuclear 2D correlation spectra are presented.

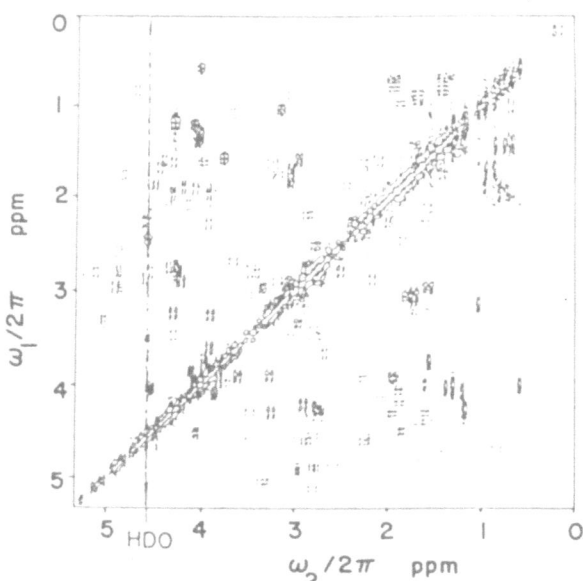

Figure 1 - Double-quantum filtered COSY of BPTI

In the first case the simplification is achieved by using a pulse sequence designed to excite intense high-order multiple-quantum coherences
providing the spin system conforms to a given coupling topology (1). In
the case demonstrated, AX_3-selective (±4)-quantum excitation is used. By
using a phase cycle to suppress lower orders, a COSY spectrum containing
only responses from AX_3 systems can be obtained. The method was
demonstrated for selection of responses of alanine residues in the COSY
spectrum of the small protein BPTI (2). (Figures 1 and 2).

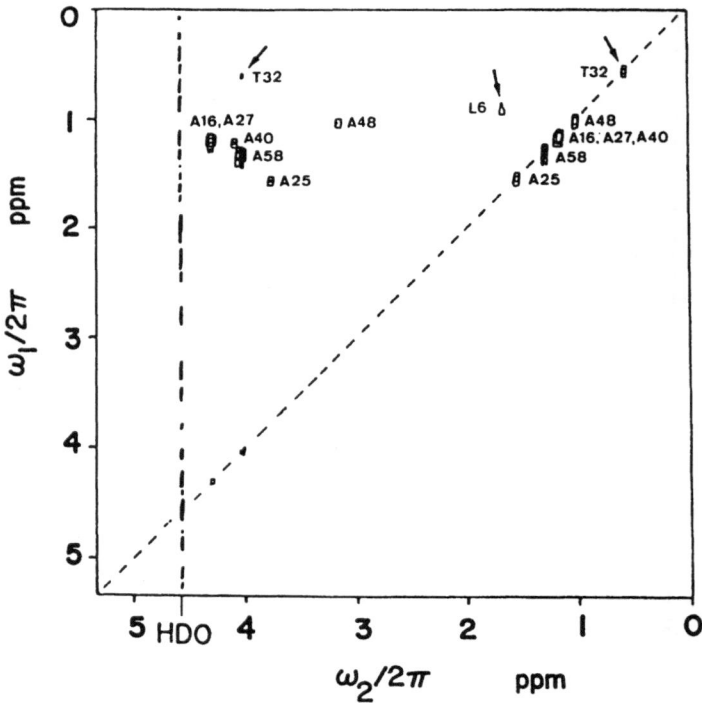

Figure 2 - AX-filtered COSY of BPTI

In the second case, the desired reduction in the number of signals in
the 2D spectrum is achieved by substituting a sequence of pulses with
appropriate commutation properties for the usual 90° mixing pulse (3). A

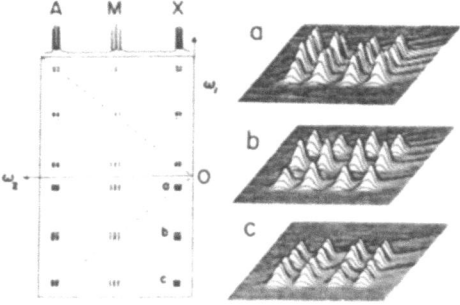

Figure 3 - Conventional COSY of 1,3-dibromobutane

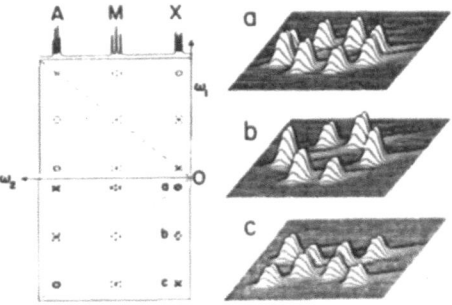

Figure 4 - Same with bilinear mixing of duration 50 ms

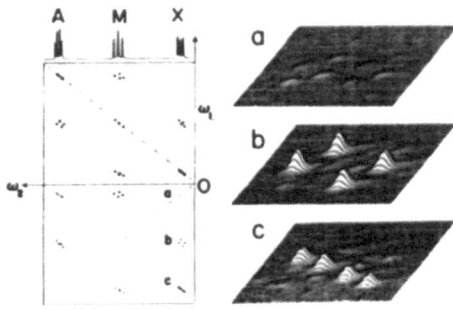

Figure 5 - Same with bilinear mixing of duration 5 ms

bilinear mixing sequence 90°-τ/2-180°-τ/2-90° has the desired effect in a weakly-coupled system. Figures 3, 4 and 5 show the progressive sim-plification of the COSY spectrum of 1,3-dibromobutane obtained using (Fig.3) simple 90° mixing, (Fig.4) bilinear mixing τ = 50 ms, (Fig.5) bilinear mixing τ = 5 ms. In all cases (+1)-quantum peaks ("N-type") are shown at the bottom and (-1)-quantum peaks ("P-type") are shown at the top. A notable feature of the last spectrum is the absence of the (+1)-quantum diagonal, making this method potentially useful for larger molecules. The unusual form of the 2D multiplets in Figs.4 and 5 can be explained purely by symmetry arguments (conservation rules for "π-numbers") as explained in ref.3.

REFERENCES

1) M.H.Levitt and R.R.Ernst, Chem.Phys.Letters **100**, 119 (1983).
2) M.H.Levitt and R.R.Ernst, J.Chem.Phys., in press.
3) M.H.Levitt,C.Radloff and R.R.Ernst,Chem.Phys.Letters **114**,435 (1985).

PBB, Vol. 2
Advanced Magnetic Resonance Techniques
in Systems of High Molecular Complexity
© 1986 Birkhäuser Boston, Inc.

^{17}O NMR AS A PROBE TO STUDY HYDRATION AND HYDROGEN

BONDING OF AMINO ACIDS AND POLYPEPTIDES

A.Spisni,E.D.Gotsis,E.Ponnusamy and D.Fiat

Istituto di Chimica Biologica, Università di Parma

43100 Parma, Italy

Dept. of Physiology and Biophysics

University of Illinois at Chicago

College of Medicine

P.O. Box 6998, Chicago IL 60680, U.S.A.

INTRODUCTION

The conformation of amino acids and polypeptides is known to play a crucial role in the modulation of their biological activity. Several inves-
tigators have indicated that solute-solvent and solute-solute interactions are responsible for the modifications as well as the stabilization
of the solute conformation (1,2) and that intra- and intermolecular hydrogen bonding is the principal route through which these interactions
take place (3,4).

It is well recognized that carboxyl and carbonyl groups are strongly involved in dimerization and binding processes (5-9). As previous studies
have shown (10-11), ^{17}O NMR is a very sensitive probe in assessing hydrogen bonding effects, thus the present study has focused on the carboxyl group of glycine, alanine and proline and also on the carbonyl
oxygens of TRH, in order to probe the hydration state of these groups as
well as the effect of intra- and intermolecular hydrogen bonding on the

conformation of these molecules.

RESULTS AND DISCUSSION

AMINO ACIDS

The ^{17}O chemical shift and linewidth of enriched glycine, alanine and proline (12) were measured in H_2O/DMSO solvent mixtures.

Because of the equivalence of the two carboxyl oxygens a single resonance signal is observed and therefore any change of the chemical shift is an average value of the two oxygens. Figure 1a is representative of the data obtained for the three amino acids and it depicts the dependence of the ^{17}O chemical shift of glycine as a function of the molar fraction of DMSO, at pH's where it is in the zwitterionic and anionic form. The linear dependence of the chemical shift on the DMSO molar fraction allows one to extrapolate to 100% DMSO, a point otherwise not achievable experimentally, due to the low solubilty of the zwitterionic and anionic amino acids in DMSO. Taking into consideration the non-proton-donor character of DMSO (13), the carboxylate group is expected to be free from hydrogen bonding, thus it is reasonable to assume that the downfield shift of the carboxylate group in going from H_2O to DMSO is the result of the rupturing of hydrogen bonds to water molecules. We will refer to this downfield shift as $\Delta\delta$, which for glycine, alanine and proline was found to be of 19.9, 17.5 and 15 ppm, respectively (14,15). The comparison of these values with the reported values of 13-35 ppm obtained for other carboxyl and carbonyl groups upon breaking of one hydrogen bond, suggests that in the case of the amino acids discussed in this work, two water molecules are hydrogen bonded to the carboxylic group, one per each oxygen.

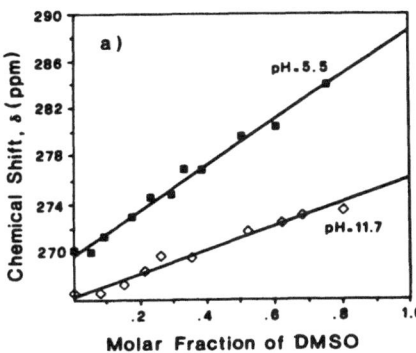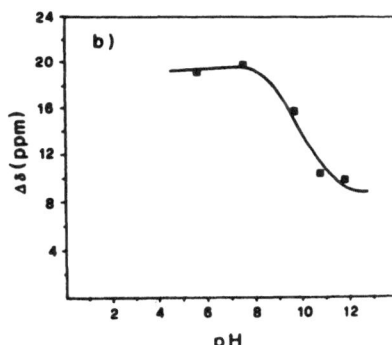

Figure 1 - a)Dependence of the ^{17}O chemical shift of Gly on the molar
fraction of DMSO, at selected pH values of the initial
aqueous solution.
b)pH dependence of the difference in chemical shift of Gly,
$\Delta\delta$, between 100% DMSO and 100% water

The analysis of the behavior of $\Delta\delta$ with respect to pH, figure 1b, shows
that deprotonation of the amino group produces a drastic decrease of $\Delta\delta$.
Such a drastic reduction of $\Delta\delta$ is interpreted as the result of the ability of the lone pair of electrons of the amino group to perturb, at variable extent, the electron distribution on the carboxylic group not
only via inductive effect but also through a direct electrostatic interaction. However, the number of water molecules hydrogen bonded per carboxyl group remains two also at basic pH.

We can therefore conclude that the ionization of the amino group although capable of inducing a perturbation of the carboxyl electron density
does not seem to alter the hydration number of the carboxyl group. Figure 2 summarizes the hydration of the carboxyl group for an amino acid in
its two limiting states: zwitterionic and anionic.

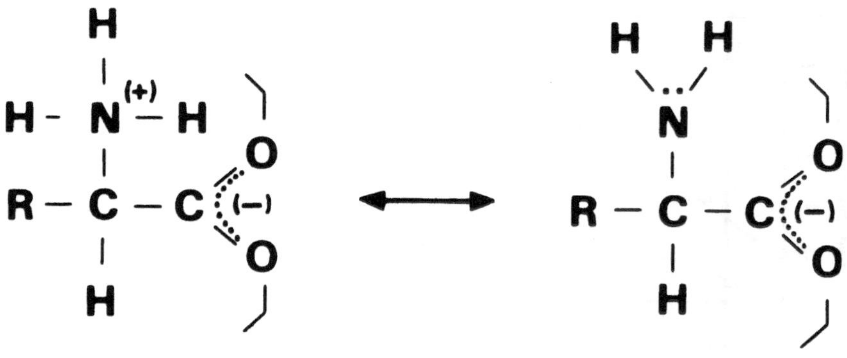

Figure 2 - Model of the hydration of an amino acid in its zwitterionic
and anionic form

The analysis of the more complex non-linear dependence of the chemical
shift of the amino acids in their cationic form will be the subject of a
forthcoming paper.

POLYPEPTIDES: TRH

TRH is a tripeptide, pGlu-His-Pro-NH$_2$, that stimulates the release of
thyrotropin and prolactin besides being an important neurotransmitter.
In spite of the numerous works carried out so far (16-19), a precise
description of its conformation has yet to be reached. In particular,
there are controversial reports on the existence of intramolecular
hydrogen bonds (20,21).

The preliminary data presented here have been obtained with TRH
selectively enriched in ^{17}O at the proline and p-glutamic carbonyl
sites. Table I shows the chemical shifts measured in water for the two
cases. While the Pro-enriched TRH shows only one resonance, figure 3a,
the pGlu-enriched TRH yielded two signals, figure 3b, tentatively assigned, on going from high to low field, to the peptide carbonyl and to
the pGlu residue, respectively.

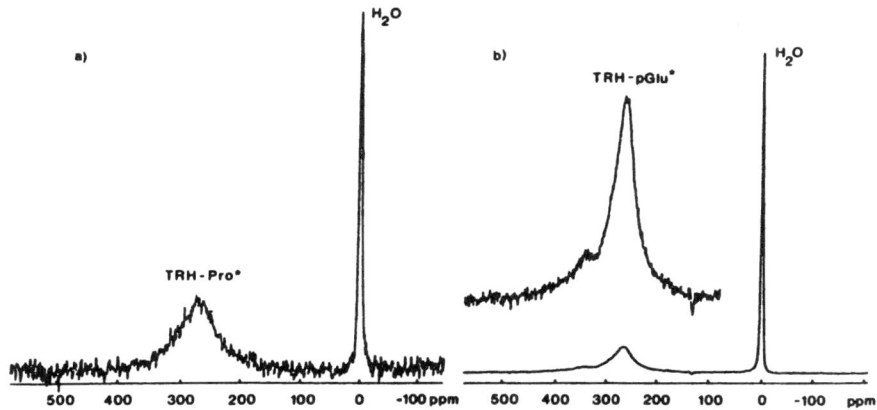

Figure 3 - ^{17}O spectra at 40.67 MHz of TRH enriched in ^{17}O at selected
amino acid residues: a)Pro*-TRH; b) pGlu*-TRH

In an attempt to determine the existence in aqueous solution of intra or
inter molecular hydrogen bonds, the behavior of the ^{17}O chemical shift
has been studied in H_2O/CH_3CN solvent mixture. Preliminary results (not

shown) indicate that, while the changes of the chemical shift for the proline residue are linearly proportional to the CH_3CN concentration, in the case of the pGlu residue the ^{17}O chemical shift shows an S-shaped dependence on the molar fraction of CH_3CN for both resonances, suggesting the existence of two possible stable states that could be due either to two distinct hydration forms or to the combination of an intra and an intermolecular hydrogen bond, as was found for MIF (21).

Table I

Chemical shifts and linewidths of the ^{17}O enriched carbonyls of the TRH's Pro and pGlu residues in 100% H_2O at acidic pH

Amino acids Residues	Chemical shift* (ppm)	Linewidth (Hz)
Pro C=O	268±2	1800±200
pGlu C=O(1)	266±1	2000±200
C=O(2)	320±5	2300±250

*Chemical shifts are downfield from external water

Studies are now in progress to clarify this point and to also better define the structural and functional importance of the histidine residue in the TRH molecule.

ACKNOWLEDGMENT

We would like to thank the "Servizio di Spettroscopia NMR ad alto campo del CNR di Bologna" for the use of Bruker CXP 300.

REFERENCES

1) J.A.Glasel, J.Amer.Chem.Soc. **92**, 375 (1970).

2) A.K.Covington and K.E Newman, Pure Appl.Chem. **51**, 2041 (1979).

3) R.Taylor,O.Kennard and W.Verichel, J.Amer.Chem.Soc. **105**, 5761 (1983).

4) M.Meot-Ner, J.Amer.Chem.Soc. **106**, 278 (1984).

5) S.Patai, in "The Chemistry of Carboxylic Acids and Ethers", Interscience Publ., 1969.

6) K.C.Chang and E.Grunwald, J.Phys.Chem. **80**, 1422 (1976).

7) M.Onishi and D.W.Urry, Science **168**, 1091 (1970).

8) E.D.Gotsis and D.Fiat, in preparation.

9) G.Formicka-Kozlowska,M.Bezer and L.D.Pettit, J.Inorg.Biochem. **18**, 335 (1983).

10) M.I.Burgar,T.E.St.Amour and D.Fiat, J.Phys.Chem. **85**, 502 (1981).

11) T.E.St.Amour,M.I.Burgar,B.Valentine and D.Fiat, J.Amer.Chem.Soc. **103**, 1128 (1981).

12) A.Steinshneider,M.I.Burgar,A.Buku and D.Fiat, Int.J.Pept.Prot.Res. **18**, 324 (1981).

13) L.Baltzer and E.D.Becker, J.Amer.Chem.Soc. **105**, 5730 (1983).

14) A.Spisni,E.D.Gotsis and D.Fiat, J.Chem.Soc.Chem.Comm. (1985), submitted for publication.

15) A.Spisni,E.D.Gotsis and D.Fiat, in preparation.

16) M.Montagut,B.Lemanceau and A.M.Belocq, Biopolymers **13**, 2615 (1974).

17) J.Feeney,G.R.Bedford and P.L.Wessels, FEBS Lett. **42**, 347 (1974).

18) J.Vicar,E.Abillon,F.Toma,F.Pirion,K.Lintner,K.Blaha,P.Fromageot and S. Fermandjian, FEBS Lett. **97**, 275 (1979).

19) P.Mavalan and F.A.Momany, Biopolymers **19**, 1943 (1980).

20) K.Kamiya,M.Takamoto,Y.Wada,M.Fujino and M.Nishikawa, J.Chem.Soc.

Chem.Comm. 438 (1980).

21) H.Gilboa,A.Steinshneider,B.Valentine,D.Dhawan and D.Fiat, Biochim. Biophys.Acta **800**, 251 (1984).

PBB, Vol. 2
Advanced Magnetic Resonance Techniques
in Systems of High Molecular Complexity
© 1986 Birkhäuser Boston, Inc.

INDIRECT, NEGATIVE HETERONUCLEAR OVERHAUSER EFFECT

DETECTED IN A STEADY-STATE, SELECTIVE $^{13}C-\left\{^{1}H\right\}$ NOE

EXPERIMENT AT NATURAL ABUNDANCE*

K.E.Kövér[a] and G.Batta[b]

[a]Biogal Pharmaceutical Works,

H-4032 Debrecen, Hungary

[b]Department of Organic Chemistry,

L. Kossuth University,

H-4010 Debrecen, Hungary

*We dedicate this manuscript to the memory of Peter Kerekes

It has recently been shown that both selective $^{13}C-\left\{^{1}H\right\}$ NOE measurements (1-10) and non-selective two-dimensional (2D) NOE (^{1}1-13) (cross-relaxation) spectroscopy are promising techniques for the determination of carbon-proton distances in organic molecules in solution.

In the present communication we show that the quasi-simultaneous saturation (14) of all ^{13}C satellite lines of a proton in a ^{13}C isotopomer is an efficient new technique for selective $^{13}C-\left\{^{1}H\right\}$ NOE measurements (15). More importantly, we have unambiguously detected a negative, indirect $^{13}C-^{1}H$ NOE at a protonated carbon when the lines of a not directly bound proton were saturated.

As a model system, /-/α-Hydrastine (1) (Figure 1) has been investigated. Assignments of proton and carbon resonances are supported by the basic chemical shift correlation methods (16-18). Non-protonated carbons have been assigned via long-range $^{13}C-^{1}H$ spin-spin couplings in a 2D DEPT experiment (19-22). The preferred conformation of **1** has been verified by [1]H-^{1}H NOE experiments (23). Finally, a heteronuclear 2D NOE (cross-relaxation) (11,12) spectrum (Figure 2) has corroborated the carbon assignments. In Fig.2 two cross-peaks (indicated by small frames) may arise in theory from either direct or indirect effects. We demonstrate that they arise from the latter.

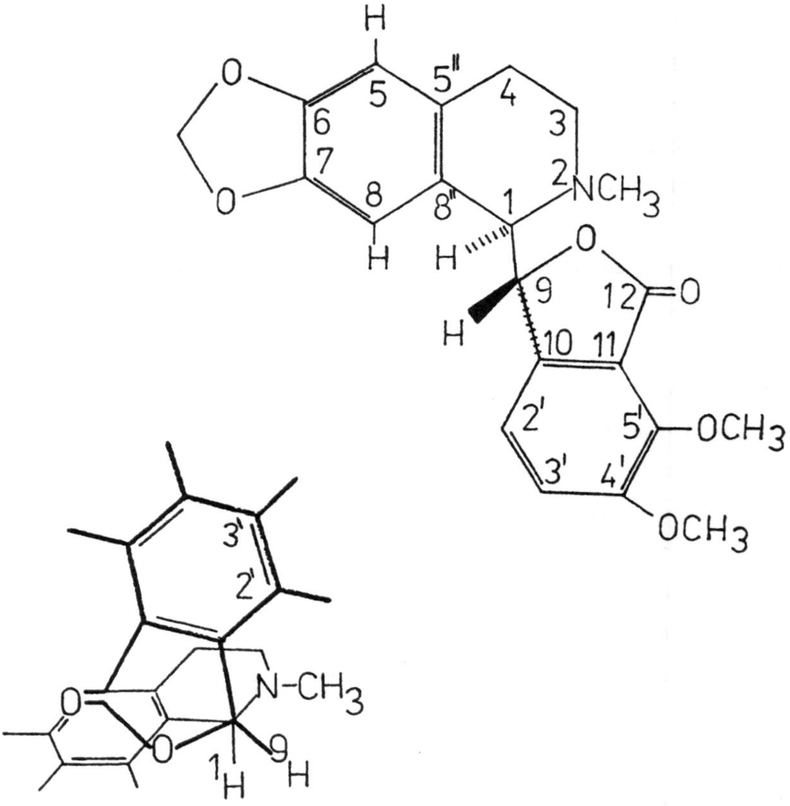

1

Figure 1

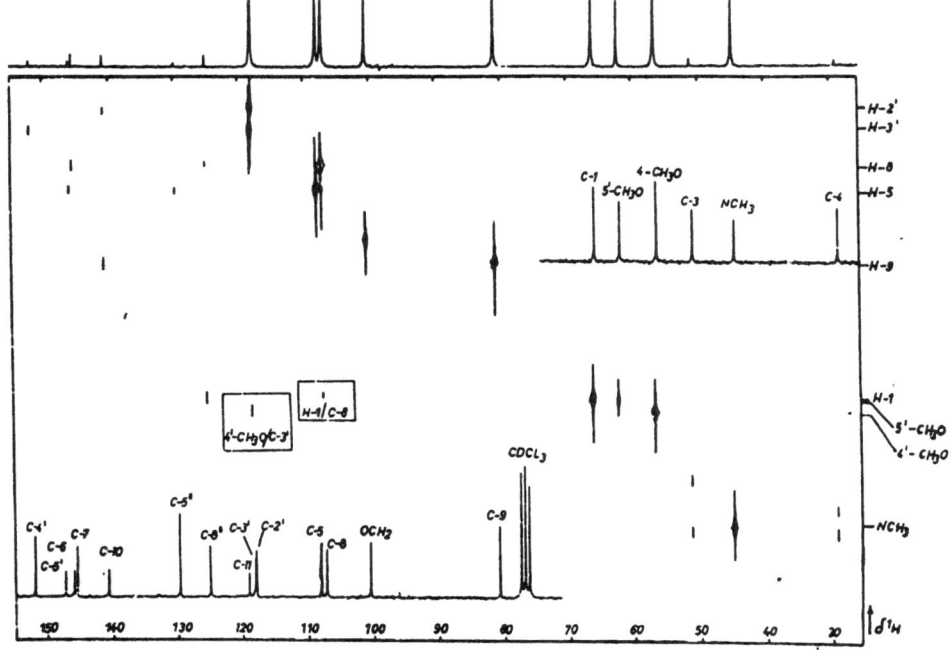

Figure 2 — Contour plot of an absolute-value 50/200 MHz 2D heteronuclear NOE (11) spectrum (BRUKER WP-200 SY) of 0.5 M **1** in
CDCl$_3$, T = 298 K. F2 projection of signals is shown on the
top, while normal ^{13}C spectrum is inside the frame. Indirect NOE effects at protonated carbons are indicated by
small frames. The 2D map is composed of 128x8 K data points. A 2.5 s waiting time was allowed between each pulse
sequence. A fixed mixing time of 1.5 s was used before the
final ^{13}C pulse and the 0.5 s acquisition time. Before the
2D Fourier transform data were multiplied with suitable
Gaussian type weighting functions

Indirect effects are well known (24-27) for spin systems containing more
than two protons. In a spin system containing two protons and one ^{13}C
nucleus, similar effects may occur. When the three spins are collinear,
for example, and the ^{13}C nucleus is situated at one end while the proton

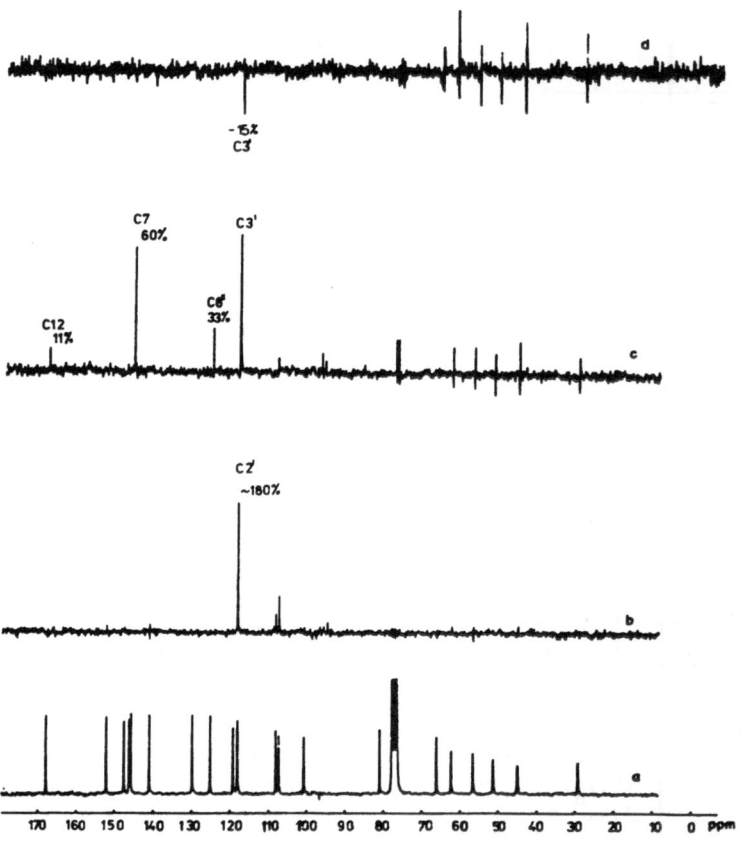

Figure 3 - One-dimensional $^{13}C-\{^{1}H\}$ NOE difference spectra of **1** using the quasi-simultaneous transition-selective saturation technique. a) Normal spectrum. b) Selective measurement of heteronuclear NOE at protonated carbon C-2' after quasi-simultaneous saturation of the one-bond ^{13}C satellites of H-2'. Residual positive signals below 110 ppm arise from accidental saturation of other satellites. c) Heteronuclear NOE of non-protonated carbons after saturation (1000x π/2/x//44ms/ pulses) of the H-8 parent proton lines. C-3' is also enhanced because the H-8 proton lines overlap with one of the one-bond H-3' satellites. d) Measurement of an indirect, negative heteronuclear NOE at protonated carbon C-3', after saturation of the 4'-CH$_3$O singlet

being irradiated is at the other end, the increased Boltzmann population at the proton in the middle decreases the population difference at the carbon. This is the negative, indirect effect, which may exceed the positive direct effect, and results theoretically into an overall negative "enhancement" up to ca.-90% at the carbon nucleus (28).

Such an effect has been observed after selective irradiation of the 4'-CH_3O methyl singlet. Figure 3d shows the NOE difference spectrum, which results into a -15% "enhancement" at C-3'. The negative sign verifies that this is an indirect, three-spin effect, which can be explained by considering the strong dipole-dipole interaction between the 4'-CH_3O and H-3' protons ($f_{H3'}/4'-CH_3O/=30\%$) and the nearly complete $^{13}C-\{^1H\}$ NOE between C-3' and H-3'.

The use of the quasi-simultaneous line-selective preirradiation technique is demonstrated in figures 3c and 3d for the measurement of $^{13}C-^1H$ NOE at non-protonated and protonated carbons.

The unprecedented negative, heteronuclear NOE reported here may be thought as a $^1H \longrightarrow {}^1H \longrightarrow {}^{13}C$ relay NOE, and it can be used for homo and/or heteronuclear distance determination in accordance with theory (24). Indirect effects are probably not restricted to protonated carbons, and should be taken into account to improve the reliability of heteronuclear distances calculated from $^{13}C-^1H$ NOE data.

ACKNOWLEDGEMENT

We are indebted to Drs. L. Radics and L. Szilàgyi for discussion concerning the manuscript.

REFERENCES

1) J.Uzawa and S.Takeuchi, Org.Magn.Reson. **11**, 502 (1978).

2) H.Seto,T.Sasaki,H.Yonehara and J.Uzawa, Tetr.Lett. 923, (1978).

3) K.Kakinuma, N.Imamura, N.Ikekawa, H.Tanaka, S.Minami and S.Ómura, J.Amer.Chem.Soc. **102**, 7493 (1980).

4) J.J.Ford,W.A.Gibbons and N.Niccolai, J.Magn.Reson. **47**, 522 (1982).

5) M.F.Aldersley,F.M.Dean and B.E.Mann, J.Chem.Soc.Chem.Commun. 107, (1983).

6) V.Leon,R.A.Bolivar,M.L.Tasayco,R.Gonzalez and C.Rivas, Org.Magn. Reson. **21**, 470 (1983).

7) M.A.Khaled and C.L.Watkins, J.Amer.Chem.Soc. **105**, 3363 (1983).

8) N.Niccolai, C.Rossi, V.Brizzi and W.A.Gibbons, J.Amer.Chem.Soc. **106**, 5732 (1984).

9) N.Niccolai, C.Rossi, P.Mascagni, P.Neri and W.A.Gibbons, Biochem. Biophys.Res.Commun. **124**, 739 (1984).

10) M.J.Shapiro,M.X.Kolpak and T.L.Lemke, J.Org.Chem. **49**, 187 (1984).

11) P.L.Rinaldi, J.Amer.Chem.Soc. **105**, 5167 (1983).

12) C.Yu and G.C.Levy, J.Amer.Chem.Soc. **105**, 6994 (1983).

13) C.Yu and G.C.Levy, J.Amer.Chem.Soc. **106**, 6533 (1984).

14) K.E.Kövér, J.Magn.Reson. **59**, 485 (1984).

15) In the transition-selective saturation technique, the total duration of the consecutive 90° pulses should be set to ca. 5-10 times the generally longer ^{13}C T_1 relaxation time. A heteronuclear NOE at a protonated carbon may be obtained by quasi-simultaneous irradiation of the one-bond satellites. In the case of non-protonated carbons (if the long-range satellites cannot be resolved), effective saturation may be achieved by non-coherent irradiation of the parent lines. It is our experience that the saturation technique presented here offers a better saturation/selectivity ratio than the conventional methods; however, our technique is not a prere-

quisite for any measurements when the selectivity is not critical.

16) W.P.Aue,E.Bartholdi and R.R.Ernst, J.Chem.Pys. **64**, 2229 (1976).

A.Bax,R.Freeman and G.J.Morris, J.Magn.Reson. **42**, 164 (1981).

A.Bax and R.Freeman, J.Magn.Reson. **44**, 542 (1981).

17) A.A.Maudsley,L.Müller and R.R.Ernst, J.Magn.Reson. **28**, 463 (1977).

18) A.Bax, J.Magn.Reson. **53**, 517 (1983).

19) D.M.Doddrell, D.T.Pegg and M.R.Bendall, J.Magn.Reson. **48**, 323 (1982).

20) M.R.Bendall and D.T.Pegg, J.Magn.Reson. **53**, 144 (1983).

21) M.H.Levitt,O.W.Sorensen and R.R.Ernst, Chem.Phys.Letters **94**, 504 (1983).

22) G.Batta and A.Lipták, J.Chem.Soc. Chem.Commun. 368 (1985).

23) K.E.Kövér and P.Kerekes, Magn.Reson. in Chem., submitted.

24) J.H.Noggle and R.E.Schirmer, "The Nuclear Overhauser Effect" Academic Press, New York, 1971.

25) J.K.M.Sanders and J.D.Mersh, Progr.NMR Spectrosc. **15**, 353 (1982).

26) J.D.Mersh and J.K.M.Sanders, Org.Magn.Reson. **18**, 122 (1982).

27) J.D.Mersh and J.K.M.Sanders, J.Chem.Soc. Chem.Commun. 306 (1983).

28) All the NOE data mentioned in the text mean fractional enhancements; $(I_S - I_0)/I_0$.

PBB, Vol. 2
Advanced Magnetic Resonance Techniques
in Systems of High Molecular Complexity
© 1986 Birkhäuser Boston, Inc.

DETERMINATION OF ZERO- AND DOUBLE-QUANTUM RELAXATION TRANSITION

PROBABILITIES BY MULTIPLE-SELECTIVE IRRADIATION METHODS

G.Valensin, G.Sabatini and E.Tiezzi

Department of Chemistry, University of Siena

Pian dei Mantellini 44, 53100 Siena, Italy

Any multispin system can be reasonably approximated by a sum of pairwise spin-spin interactions. As a consequence spin-lattice relaxation of any spin i coupled to a group of spins j can be approximated by summing the relaxation effects of all the spin pairs, yielding (1):

$$d<I_{zi}>/dt = -R_i (<I_{zi}> - I_{oi}) - \sum_{i \neq j} \sigma_{ij} (<I_{zj}> - I_{oj}) \qquad (1)$$

where

$$R_i = \sum_{i \neq j} \varrho_{ij} + \varrho_i^* \qquad (2)$$

I_{oi} is the equilibrium value of $<I_{zi}>$ which represents the expectation value of the nuclear spin operator and it is proportional to the total integrated intensity of the NMR signal. σ_{ij} is the cross relaxation term which is usually contributed by dipole-dipole interactions only. ϱ_{ij} is the dipole-dipole direct relaxation term, while ϱ_i^* accounts for direct relaxation of spin i due to mechanisms other than the dipole-dipole.

It follows that the proton relaxation rate, measured in these conditions, is in practice determined only by single-(W_1) and double-quantum

(W_2) relaxation transition probabilities summed over all the effective spin pairs and over all the possible relaxation mechanisms. Non selective proton relaxation rates thus provide poor information unless predominance of a single spin-spin interaction and of a single relaxation mechanism can be assumed.

Selective irradiation of the proton resonance of any spin i leaves all the energy levels of coupled spins j unperturbed, whereby eq.(1) can be simplified as follows(2):

$$d <I_{zi}> /dt = -R_i (<I_{zi}> - I_{oi}) \qquad (3)$$

The longitudinal magnetization decay following such selective irradiation cannot be described by a single exponential; however a spin-lattice relaxation rate can still be measured in the initial rate approximation (3). Such selective relaxation rate is contributed by W_1, W_2, as before, but also by the zero-quantum (W_o) relaxation transition probabilities, summed over all the spin pairs and all the effective relaxation mechanisms. Derived information is still poor unless the same assumptions as in the case of nonselective relaxation are possible.

Double-selective irradiation of any two coupled spins i and k, however, yields the following relaxation equation from eq.(1):

$$d <I_{zi}> /dt = -R_i (<I_{zi}> - I_{oi}) - \sigma_{ik} (<I_{zk}> - I_{ok}) \qquad (4)$$

The spin-lattice relaxation rate of spin i, measured under conditions of double-selective irradiation of spins i and k, contains, as compared to the selective spin-lattice relaxation rate of the same spin i, the additional term σ_{ik} which, as precedently stated, is the cross relaxation term originating from the mutual dipole-dipole interaction between the two irradiated spins.

If we indicate by R_i and by R_i^{ik} the selective and double-selective spin-lattice relaxation rates respectively of any spin i, as measured under the above-stated conditions, it is consequent that:

$$\sigma_{ik} = R_i^{ik} - R_i \qquad (5)$$

The cross relaxation term σ_{ik} is contributed only by W_0 and W_2 between the energy levels of the two irradiated spins:

$$\sigma_{ik} = W_{2(ik)} - W_{0(ik)} \qquad (6)$$

Such term is originated by the mutual dipole-dipole interaction between the two magnetic moments μ_i and μ_k separated by the distance r_{ik}. The relaxation transition probabilities W_2 and W_0 are determined by modulation of such interaction at frequencies $\omega_i + \omega_k = 2\omega_H$ and $\omega_i - \omega_k = 0$ respectively. The explicit expressions are given by(1):

$$W_{2(ik)} = 0.6 \; \gamma^4 \hbar^2 \, r_{ik}^{-6} \cdot \tau_c / (1 + 4\omega_H^2 \tau_c^2) \qquad (7)$$

$$W_{0(ik)} = 0.1 \; \gamma^4 \hbar^2 \, r_{ik}^{-6} \cdot \tau_c \qquad (8)$$

It follows that

$$\sigma_{ik} = 0.1 \; \gamma^4 \hbar^2 \, r_{ik}^{-6} \left\{ 6\tau_c / (1 + 4\omega_H^2 \tau_c^2) - \tau_c \right\} \qquad (9)$$

Measuring the difference between double-selective and selective proton spin-lattice relaxation rates yields evaluation of σ_{ik} for any proton pair, that is to say a parameter that is a relatively simple function of internuclear separation and of the rate of modulation of the relaxation vector.

We emphasize herein that such measurements can be very suitably applied for evaluation of rotational correlation times of molecular segments. This seems to be particularly relevant when dealing with small molecules irrotationally bound to macromolecules.

As an example we report the proton relaxation data of sulfisomidine (N^1-(2,6-dimethyl-4-pyrimidinil)sulfanilamide) (see figure 1) having a well documented biological activity (4,5). The H_{10}, $H_{10'}$, H_{11}, $H_{11'}$ protons give rise to a AA'BB' spin system multiplet where the distance $r_{10,11}$ is fixed at 2.43 Å. The whole proton relaxation data within this spin system are reported in Table I. The measured proton relaxation rates can be used for calculating $\sigma_{10,11}$ in the following way:

$$\sigma_{10,11} = R_{10}^{10,11} - R_{10} = 0.084 \text{ s}^{-1} \qquad (10)$$

$$\sigma_{11,10} = R_{11}^{10,11} - R_{11} = 0.090 \text{ s}^{-1} \qquad (11)$$

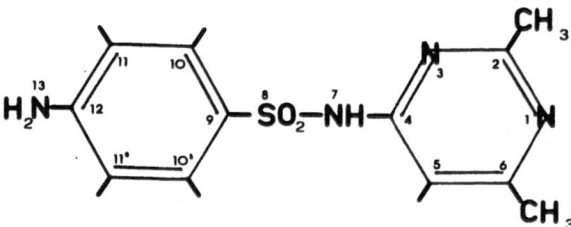

Figure 1

TABLE I

Proton relaxation rate measurements for sulfisomidine

2.78 mM in D_2O at pH = 5.0 and T = 298 K

Resonance	R_{all} (s^{-1})	R_i (s^{-1})	R_i^{ik} (s^{-1})
$H_{10,10'}$	0.417	0.286	0.370
$H_{11,11'}$	0.377	0.292	0.382

$\sigma_{10,11}$ can be therefore taken as the average 0.087 s^{-1}. The fact that it is a positive quantity means that $W_2 > W_o$ (see eq.(9)) and therefore that $\omega_H \tau_c \ll 1$. If the values of the constants are substituted in eq.(9) one obtains:

$$\sigma_{10,11} = 5.96 \times 10^{-38} \, r^{-6} \, f(\tau_c) \qquad (12)$$

wherefrom one calculates $\tau_c = 7.5 \times 10^{-11}$ s at 298 K by substituting r = 2.43 Å.

With the aim of assessing the method of double-selective irradiation, the interaction of sulfisomidine with bovine serum albumin (BSA) was then investigated. It was previously known (6) that at pH = 5.0 two molecules of drug are bound per protein molecule with an overall binding constant log K_b = 2.49. The proton spin-lattice relaxation rates were therefore measured also in the presence of BSA, as reported in Table II.

TABLE II

Proton relaxation rate measurements for sulfisomidine 2.78 mM

in the presence of BSA 0.21 mM at pH=5.0 and T=298 K

Resonance	R_{all} (s^{-1})	R_i (s^{-1})	R_i^{ik} (s^{-1})
$H_{10,10'}$	0.627	2.299	1.470
$H_{11,11'}$	0.575	1.923	1.282

It is evident that, while the nonselective rate R_{all} is only poorly affected by the presence of the macromolecule, as observed elsewhere (2, 7,8), the selective R_i and the double-selective R_i^{ik} rates undergo large enhancements in the presence of the protein. The fractions of bound (p_b) and free (p_f) sulfisomidine were calculated according to the known binding constant (6) as $p_f = 0.961$ and $p_b = 0.039$ and fast exchange between free and bound environments was assumed. As a consequence the following equations could be given:

$$\sigma_b = (\sigma_{obs} - p_f \sigma_f)/p_b \qquad (13)$$

$$\sigma_{obs} = R_i^{ik}(obs) - R_i(obs) \qquad (14)$$

If the experimental values are substituted in eqs.(13) and (14), one obtains:

$$\sigma_{10,11}(bound) = -23.3 \ s^{-1} \qquad (15)$$

$$\sigma_{11,10}(\text{bound}) = -18.6 \text{ s}^{-1} \tag{16}$$

The disagreement between the two values can be possibly due to the fact that a proton resonance from protein protons partially overlaps the spin multiplet of protons 11,11'. Anyway it is apparent that $\sigma_{10,11}$ is a negative quantity, wherefrom it is inferred that, as expected, $W_o < W_2$, being $\omega_H \tau_c \gg 1$.

As it was done for the free molecule, the correlation time of the bound ligand, directly at the binding site, can now be calculated at $\tau_c = 1.15 \times 10^{-9}$ s at 298 K, suggesting irrotational binding of the aromatic moiety, which the relaxation vector belongs to, to the serum protein.

It may be concluded that multiple-irradiation methods make it possible to extract the dipole-dipole relaxation contributions of single proton spin pairs from the overall relaxation rate of any proton spin. Such contributions are expressed by means of cross relaxation terms σ's that depend upon interproton distances and dynamics of relaxation vectors. It is therefore possible to determine the rotational correlatione time of any molecular segment provided the relative interproton distance is known, independently of the chemical environment in which the investigated molecule is located.

The method seems to have the potential of delineating receptor sites within macromolecules. In fact in any ligand, diverse relaxation vectors can be isolated (e.g. the proton vectors 10-11, 10-7, 7-5, 5-6 can be possibly investigated in sulfisomidine) wherefrom molecular dynamics at the binding site might be inferred yielding a good piece of information on the conformational features of the receptor site.

REFERENCES

1) J.H.Noggle and R.E.Schirmer, "The Nuclear Overhauser Effect", Academic Press, New York, 1971.

2) G.Valensin,T.Kushnir and G.Navon, J.Magn.Reson. **46**, 23 (1982).

3) R.Freeman,H.D.W.Hill,B.L.Tomlinson and L.D.Hall, J.Chem.Phys. **61**, 4466 (1974).

4) I.Moriguchi,S.Wada and T.Nishizawa, Chem.Pharm.Bull. **16**, 601 (1968)

5) T.Fujita, J.Med.Chem. **15**, 1049 (1972).

6) A.Agren,R.Elofsson and S.-O.Nilsson, Acta Pharmacol.Toxicol. **29**, 48 (1971).

7) G.Valensin,A.Casini,A.Lepri and E.Gaggelli, Biophys.Chem. **17**, 297 (1983).

8) G.Valensin,A.Lepri and E.Gaggelli, Biophys.Chem. **22**, 83 (1985).

PBB, Vol. 2
Advanced Magnetic Resonance Techniques
in Systems of High Molecular Complexity
© 1986 Birkhäuser Boston, Inc.

THE ROLE OF CONFORMATION AND CONFORMATIONAL DYNAMICS

IN BIOLOGICAL INFORMATION TRANSFER

W.A.Gibbons,P.Mascagni,N.Zhou,A.E.Aulabaugh,

A.Prugnola,M.Kuo and N.Niccolai

Department of Pharmaceutical Chemistry,

School of Pharmacy, University of London

Department of Chemistry, University of Siena, Italy

INTRODUCTION

Despite the advances in NMR theory and the introduction of modern pulsed
1D and 2D techniques, our knowledge of the conformation and dynamics of
biopolymers in solution is still in its infancy. Although we can study
dynamics by a variety of techniques it is true to say that for no single
polypeptide or protein do we know the full structure and dynamics, especially the latter; we need to know the modes, frequencies, phases and
amplitudes of the different motions and how they are coupled.

This coupling of motions plays a critical role in biological recogni-
tion, in the nature of the interface between the components of a biocom-
plex and in the transfer of information across the interfaces of the
complex. Conformational moieties, e.g. helices, sheets and reverse
turns, not only play a structural role in determining molecular archi-
tecture but they participate in the transfer of information (say) across
a biological macromolecule.

In this manuscript we report our NMR studies of the conformation and dynamics of several polypeptides. The knowledge of structure gained so far of these systems is far in advance of that for many biological macromolecules in solution. It remains to progress deeper into our knowledge of structure, to relate structure to dynamics and both to biological functions.

DERIVATION OF CARBON–PROTON DISTANCES IN BIOMOLECULAR SYSTEMS FROM ^{13}C RELAXATION RATES AND HETERONUCLEAR OVERHAUSER EFFECTS

It has been demonstrated that homonuclear NOE measurements yield proton-proton distances (1) as well as conformational information. Conformational dynamics can also be elucidated from such measurements and relaxation rates. One of the major remaining problems in studying solution conformation is the elucidation of the nature of each of the component moieties involved in intermolecular hydrogen bonding and of the proton-carbon and proton–nitrogen distances characteristic of each hydrogen bond in the molecule. In this section, heteronuclear Overhauser effects (HOE) and spin-lattice ^{13}C relaxation studies are reported, in order to obtain structural information on hydrogen bonding patterns and, hence, conformation by using the experimental approach previously reported for small organic molecules (2),natural products (3) and other peptides (4).

A combination of HOE and ^{13}C relaxation rate measurements yielded information on single proton-carbon distances since

$$HOE_{Cm}(Hn)R_{Cm} = (\gamma_H/\gamma_C)\ \sigma_{mn} \qquad (1)$$

The internuclear distance (r_{mn}) between the carbon atom (Cm) and the Hn proton can be obtained by two independent methods.

Method A: When the saturation of Hn gives Overhauser effects on two or

more carbon resonances, internuclear distances can be calculated from the following type of relationship:

$$\text{HOE}_{C1}(\text{Hn})R_{C1}/\text{HOE}_{C2}(\text{Hn})R_{C2} = r_{2n}^6/r_{1n}^6 \quad (2)$$

In order to evaluate R's from eq(2), a knowledge of correlation times is not required, but one of the two distances has
to be used as a calibration one.

Method B: If both the correlation time and the cross-relaxation term are
known, an absolute determination of r_{mn} is possible with use
of eq 3.

$$r_{mn}^6 = \hbar^2\gamma_H^2\gamma_C^3/10\text{HOE}_{Cm}(\text{Hn})R_{Cm} \quad 6\tau_c/1+(\omega_H+\omega_C)^2\tau_c^2 - /1+(\omega_H-\omega_C)^2\tau_c^2 \quad (3)$$

The use of either of the methods depends on the particular system being
investigated, but it seems reasonable that, as in the present work, they
can be used simultaneously, thus allowing a double check on calculated r
values and hence the assumptions behind calculated data.

In Table I the experimental relaxation parameters for the cyclic tetrapeptide $\left[\text{Val}^3\right]$-HC toxin are reported. Experimental details are given in
our previous papers (2-4).

A feature common to all the protonated carbons is a large non-selective
HOE (~ 1.63) which indicated that the $^1\text{H}-^{13}\text{C}$ dipolar interactions are
the main source for the spin-lattice relaxation of these nuclei. It also
suggested that all the dipolar couplings are modulated by molecular
motions which satisfy (or are very close to) the narrowing limit

Correlation times for each class of C-H vectors of the molecule were
calculated using the experimental data from Table I, the fractional

effectiveness of the dipolar relaxation mechanism (X^{DD}) and the dipolar contribution to R (R^{DD}). These two latter quantities are expressed as:

$$X^{DD} = HOE(BB)/HOE_o(BB) \qquad (4)$$

and

$$R^{DD} = R_{exp} * X^{DD} \qquad (5)$$

where $HOE_o(BB)$ is the maximum non-selective HOE at a given correlation time, assuming a 100% dipolar relaxation mechanism.

TABLE I

^{13}C relaxation parameters and ^{13}C chemical shifts for [Val^3]-HC toxin in CDCl$_3$

	ppm	R**	nOe(BB)*	X^{DD}***	R^{DD}
Ala$_2$C=O	185.78	0.27	0.96	0.5	0.13
Ala$_1$C=O	185.49	0.42	0.73	0.38	0.16
Val$_3$C=O	185.43	0.38	0.63	0.33	0.13
Pro$_3$C=O	183.09	0.39	0.81	0.42	0.13
Val$_4$C	58.33	2.42	1.70	0.88	2.15
Pro$_1$C	57.14	2.49	1.73	0.90	2.23
Ala$_2$C	47.77	2.65	1.67	0.86	2.29
Ala$_4$C	46.90	2.53	1.63	0.84	2.13
Pro$_3$C	46.37	1.76	1.71	0.88	1.55
Val$_4$Cβ	27.40	2.22	1.67	0.86	1.91
Pro$_4$Cβ	24.54	1.75	1.65	0.87	1.49
Pro$_3$C	24.41	1.40	1.68	0.87	1.22
Val$_3$C^D_u	18.81	0.52	1.86	0.93	0.42
Val$_1$C	18.02	0.51	1.89	0.95	0.48
Ala$_2$Cβ	14.63	0.61	1.93	0.97	0.59
Ala^{2}Cβ	13.54	0.54	1.84	0.92	0.50

* Non selective nOe's calculated as $(I_z - I_o)/I_o$
** ^{13}C relaxation rates in sec^{-1}. R rates for CH$_3$ and CH$_2$ groups have been devided by the number of attached protons
*** See text for explanations

For the C-H vectors a τ_c = 1.1x10^{-10} could be derived that fits the experimental data. Thus this proposed correlation time and the standard C-H distance of 1.09 Å yielded R = 2.24 s^{-1} and HOE$_0$ = 1.94, in the hypothesis of a mechanism entirely dipolar for the alpha carbons.

For these carbon nuclei eqs.(4) and (5) then gave R^{DD} values that on the average were consistent with R = 2.24 previously calculated and hence with τc = 1.1x10^{-10}. The methyl carbons exhibited longer spin-lattice relaxation, presumably because of faster reorientation of their C-H vectors. Using the theoretical HOE$_0$(BB) = 1.99 and the appropriate R^{DD} rates from table I, correlation times of 2.3x10^{-11} were obtained for the CH$_3$ groups of Ala^2, Ala^1 and Val^3 respectively.

For the ß, γ and δ methene carbons a behavior intermediate between that of the alpha carbons and that of methyl groups should be expected and in accordance with the experimental results, 1.94 $\leq$ HOE$_0$(BB)$\leq$ 1.99. If the inferior limit was used, R^{DD} rates slower than the alpha carbon ones were found showing that indeed an additional mobility characterises the dipolar interactions between protons and these side-chain carbons.

As expected, the carbonyl carbons, owing to their unsuitable magnetic environment, had inefficient relaxation mechanisms. Nevertheless, long range proton-carbon dipolar couplings contributed with a minimum of 33% to their nuclear dipolar relaxation, under the reasonable assumption that these couplings are modulated by the molecular correlation time τ_c = 1.1x10^{-10} s. This allowed the use of eqs.(2) and (3) as structural approach for obtaining the proton-carbon distances reported in Table II.

CONFORMATION OF $\left[\text{Val}^3\right]$-HC TOXIN FROM INTERNUCLEAR DISTANCES

The experimentally derived proton-carbon distances elucidated the hydro-

gen bonding pattern of this tetrapeptide. Thus two $N^1H-^{13}CO$ dipolar couplings, that between Val^3-CO and Ala^1-NH and that between Val^3-NH and
Ala^1-CO were ascribed to hydrogen bonds across the cyclic backbone; the
corresponding distances (see table II) were in agreement with those
generated by γ-turns in other similar cyclic tetrapeptides.

TABLE II

Proton-carbon distances for $[Val^3]$-HC toxin from relaxation parameters*

	Ala^1NH	Ala^2NH	Val^3NH
Ala^1CO	2.6_5 (2.5_7)	2.0_9 (1.7_7)	2.5_7 (2.6_4)
Ala^2CO	------	---- (2.9_8)	2.2_1 (1.9_3)
Val^3CO	2.5_5 (2.6_3)	------	2.4_9 (2.6_0)
Pro^4CO	2.1_6 (1.8_8)	------	------

* Data in parentheses are from X-ray of dihydrochlamydocin

Consistent with this conclusion was also the extent of the NH-CO dipolar
interactions between adjacent residues. Thus the experimentally measured
distances indicated that the NH and CO moieties of Ala^1 and Val^3 residues have a cis conformation that can exist only in the presence of the
two hydrogen bonds above mentioned (see figure 1).

Molecular models incorporating these features yielded interproton and
C-H distances fully in agreement with those calculated using the relaxation parameters. Furthermore, these distances were consistent within
0.1 A with those previously calculated for the peptide desdimethyl-

chlamydocin (5) and, more remarkably, with those from x-ray data for dihydrochlamydocin (6). These two peptides were found to possess the same conformation proposed for $[Val^3]$-HC toxin.

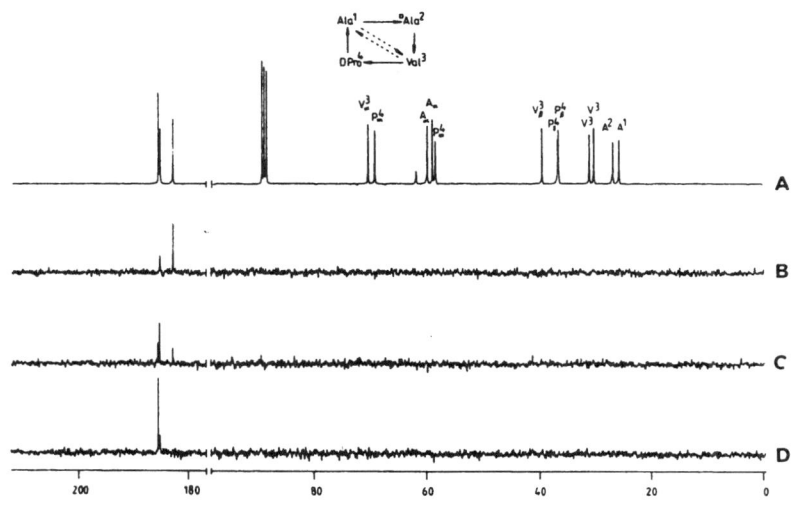

Figure 1 - Selective heteronuclear NOEs for $[Val^3]$-HC toxin. A) ^{13}C spectrum at 75 MHz of the peptide in $CDCl_3$ solution (0.8 M, T = 298 K). B) Difference spectrum after selective irra-diation of tha Ala^1-NH proton; C) Difference spectrum after irradiation of the Ala^2-NH proton. D) Difference spectrum after irradiation of the Val^3-NH proton.

CONFORMATIONAL DYNAMICS AND BIOLOGICAL INFORMATION TRANSFER: THE TYROCIDINE PEPTIDES

The three processes commonly involved in biology are molecular recogni-tion, complexation and information transfer between the components of the complex. The structural information alone is insufficient to explain these events in physico-chemical terms; a knowledge of the dynamics is also necessary. Examples of such events are enzyme-substrate, antigen-antibody, drug-receptor interactions.

a) STRUCTURAL STUDIES BY NMR

In order to elucidate these processes for proteins and peptides we need to have information at several levels. The latter includes the role of (i) the different regular conformations such as helices, extended sheets and reverse turns; (ii) motions of side chains; (iii) and motions of the backbone; (iiii) coordinated or geared motions. One major role that NMR spectroscopy can play is to elucidate the characteristics of these types of motion from the NMR parameters of the individual atoms. Despite numerous experimental and theoretical studies there is still a gulf between physico-chemical data and the explanation of biological phenomena.

The tyrocidin family of cyclic decapeptides are the most detailed model system yet available. The proton, carbon and nitrogen spectra have been assigned; the secondary conformation has been established by proton-proton distance measurements from NOEs and cross relaxation rates; the hydrogen bonds have been defined; and an almost rotamer analysis of the side chains has been achieved. These polypeptides were the first biopolymers to be so studied and have served as models for other studies using both 1D and 2D NMR methods (7-11).

Crystallographic studies subsequently confirmed these structural conclusions from NMR in the case of gramicidin S, the simplest tyrocidin.

b) CONFORMATIONAL DYNAMICS OF THE TYROCIDINS

NMR spectroscopy provides many parameters for the elucidation of poly-peptide dynamics and each will be discussed. These include (a) chemical shift anisotropy and the temperature dependence of anomalously shifted resonances; (b) carbon-13 relaxation rates; (c) proton relaxation rates and cross-relaxation rates; (d) line broadening, T_2 and T_1 measure-ments; (e) motional averaging of scalar coupling constants; (f) transi-ent NOEs.

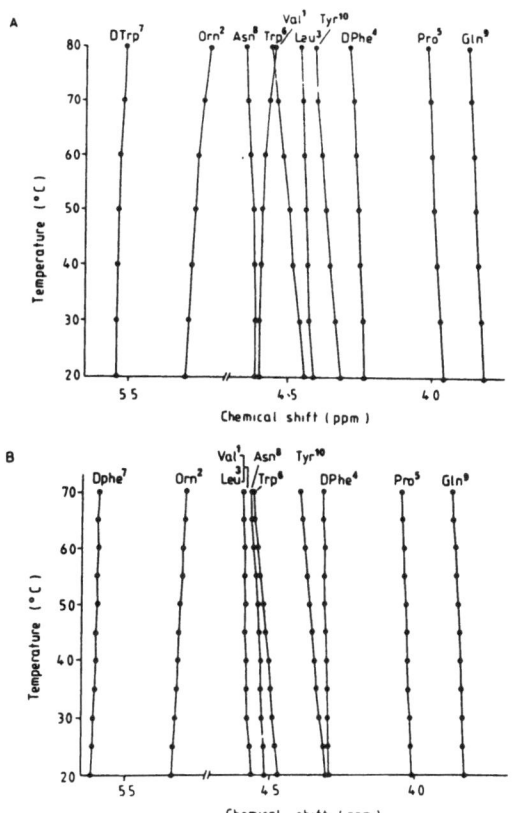

Figure 2 – The temperature dependence of the H_α chemical shifts of tyrocidine B (fig.2A) and tyrocidine C (fig.2B) in DMSO-d$_6$

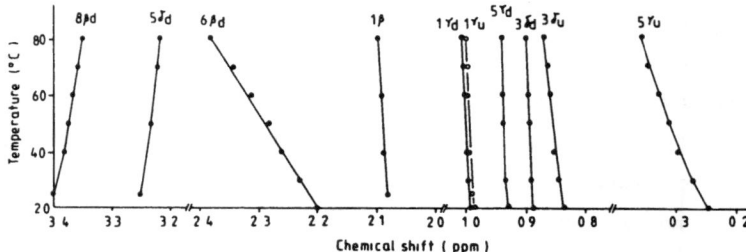

Figure 3 - The temperature dependence of the aliphatic side chain protons of tyrocidine C in DMSO-d$_6$

Anisotropically shifted proton resonances for Orn^2, Pro^5, Phe^6 and Phe^7 have been identified previously and used to define proton-aromatic ring chromophore distances and hence the $(\chi_1\chi_2)$ conformations of the aromatic residues causing the anisotropy. Generally only the temperature dependence of amide protons is used to elucidate structure and motion. As shown in Figures 2 and 3, in the case of the tyrocidins the anomalously shifted resonances exhibit two phenomena as a function of temperature: case (1) the D-Phe^{7}H and Orn^{2}H are downfield shifted by 1 ppm from their random coil positions but their chemical shifts are essentially temperature independent. Case (2) the Pro^{5}H$_\delta$, Pro^{5}H$_\gamma$ and Pro^{5}H$_\beta$ are anomalously shifted upfield but their chemical shifts are highly sensitive. In case (2) the Pro proton shifts are caused by one specific $(\chi_1\chi_2)$ conformer of the D-Phe^4 and the LPhe^6 residues - each residue having six possible classical side chain conformations. The temperature sensitivity is due to changing statistical weights of the $(\chi_1\chi_2)$ conformation due to temperature-induced χ_1 and/or χ_2 rotation of the aromatic residues. In case (1), by contrast, the lack of temperature effects on H$_\alpha$ shifts is attributed to a completely frozen $(\chi_1\chi_2)$ rotamer of D-Phe^7.

We have not completed quantitative pulsed T$_2$ or line width measurements as a function of temperature but a great deal of information was qualitatively available in CH$_3$OH solution down to -50°. In this case the Tyr^{10}

ring aromatic protons were severely and selectively broadened due to slowing of the χ_2 motions of the Tyr[10] ring. Since the Tyr[10] residue is 90% populated in one χ_1 rotamer at 25°C, it is probable that frozen χ_1 and χ_2 motions takes place at -70°C.

Although scalar $^3J_{\alpha\beta}$ values directly reflect conformation equilibria as distinct from dynamic effects some information on the latter can be inferred from rotamer populations. In the tyrocidins all aromatic residues clearly populate one χ_1 rotamer to a statistical weight of greater than 90%. We need to perform $^3J_{\alpha\beta}$ versus temperature to definitively establish accurate χ_1 rotation. On the other hand we have performed ^{13}C NT_1 measurements at 200, 300 and 400 MHz. For all aromatics NT_1 for both the α and β protons are the same within the experimental error of $\approx 10\%$. This data confirms the inferences from $^3J_{\alpha\beta}$ equilibria studies.

A great deal remains to be done before we can understand fully the backbone φ and ψ motions as well as the χ side chain motions of even a simple a polypeptide as the tyrocidine A, B and C and gramicidin S. Even more is needed to relate our knowledge of the structure and dynamics to specific and general biological functions. As described above, our data clearly delineate the motions of some tyrocidin side chains and point to a more detailed interpretation in the near future.

REFERENCES

1) C.R.Jones,C.T.Sikakana,M.C.Kuo and W.A.Gibbons, Biophys.J. **24**, 815 (1978).

2) N.Niccolai,A.Sega,M.Scotton and C.Rossi, Gazz. Chim. It. **115**, 149 (1985).

3) N.Niccolai,C.Rossi,V.Brizzi and W.A.Gibbons, J.Amer.Chem.Soc. **106**,

5732 (1984).

4) N.Niccolai,C.Rossi,P.Mascagni,P.Neri and W.A.Gibbons, Biochem. Biophys.Res.Commun. **124,** 739 (1984).

5) P.Mascagni,W.A.Gibbons,D.H.Rich and N.Niccolai, J.Chem.Soc. Perkin Trans.I, 245 (1985).

6) J.Flippen and I.Karle, Biopolymers **15,** 1081 (1976).

7) C.R.Jones,C.T.Sikakana,M.Kuo and W.A.Gibbons. J.Amer.Chem.Soc. **100,** 5960 (1978).

8) C.R.Jones,M.Kuo and W.A.Gibbons. J.Biol.Chem. **254,** 10307 (1979).

9) M.Kuo and W.A.Gibbons.Biophys.J. **32,** 807 (1980).

10) M.Kuo,T.Drakenberg and W.A.Gibbons. J.Amer.Chem.Soc. **102,** 520 (1980).

11) N. Zhou, Ph.D. Thesis, University of Wisconsin (1985).

PBB, Vol. 2
Advanced Magnetic Resonance Techniques
in Systems of High Molecular Complexity
© 1986 Birkhäuser Boston, Inc.

STRUCTURE AND DYNAMICS STUDIED VIA TWO-DIMENSIONAL NMR.

ILLUSTRATIONS WITH DNA FRAGMENTS

Thomas L. James, Michelle S. Broido[a] Nadege Jamin,

Gregory B.Young and Gerald Zon[b]

Departments of Pharmaceutical Chemistry and Radiology

Schools of Pharmacy and Medicine,University of California

San Francisco, CA 94143

[a]Present address: Department of Chemistry

Hunter College and Graduate Center of the City University of New York

New York, NY 10021

[b]National Center for Drugs and Biologics

Food and Drug Administration,Bethesda, MD 20205

INTRODUCTION

A major objective of scientists for many years has been the
determination of molecular structures in non-crystalline environments.In
general, researchers were forced to accept limited structural
information in these cases due to the techniques available; certainly
high-resolution structures such as those derived from x-ray diffraction
on crystals could not be remotely attained. Two recent developments
together provide a direct route to molecular structure in
non-crystalline phases: two-dimensional NMR (2D NMR) (1) and the
distance geometry algorithm (2). The essential ideas can be described
quite simply. The structure of any molecule can be determined if one can
obtain a sufficient number of structural constants, e.g., intranuclear

distances and bond torsion angles, to use in conjunction with holonomic constraints of bond lengths, bond angles and steric limitations. A high-resolution structure requires a large number of such constraints. Although it will probably not be possible to match the achievements of x-ray crystallography, in principle, the new two-dimensional NMR procedures enable us to obtain many structural constraints, and could lead to high-resolution structures. The distance geometry algorithm provides an efficient way to convert the structural measurements (internuclear distances and torsion angles), with their experimental errors, into three-dimensional structures. As the number and accuracy of structural constraints increases, the structures become better defined.

A serious attempt to determine high-resolution structures in non-crystalline environments must be cognizant of the possibility that more than one conformation may exist with rapid exchange between con-formations occurring. High frequency, small amplitude bond vibrations and hindered librations do not pose as much a problem as low frequency, large amplitude conformational fluctuations. As this is very possible, we must maintain the perspective that any structure determined is a time- and ensemble-averaged structure. Details of the conformational flexibility may be explored for example by independent molecular mecha-nics calculations (3) and NMR relaxation studies (4), with some insight being evident from the 2D NMR results (vide infra).

A 2D NMR technique, the two-dimensional nuclear Overhauser effect (2D NOE) experiment, has been introduced in recent years (5); it has the potential for providing numerous internuclear distances. Elegant studies in Wuthrich's lab have semiquantitatively utilized 2D NOE spectra to examine small protein structure in solution (6,7). We recently inve-stigated the 2D NOE technique for determination of solution structure using a complete relaxation matrix approach theoretically (8) and in a rigid molecule test case (9). The method of calculation is similar to that previously used to examine spin diffusion effects in one--dimensional NMR (10). Unlike previous 2D NOE studies, this method is not limited by assumptions regarding the number or geometry of proton spins, spin diffusion, the details of molecular motions, or the range of mixing times employed in the experiment.

Sequence-dependent structural variations in DNA evidently are important for the protein-DNA interactions necessary for life. Recent progress in synthesis techniques have enabled synthetic double-stranded DNA fragments to be prepared and crystal structures determined via x-ray crystallography (11-14). The sequence-dependent structural variations appear to be in accord with the structural rules of Calladine (15). With the availability of the synthetic oligonucleotide duplexes, studies of DNA solution structure have been initiated by exploiting 2D NOE (16-21).

The present article describes studies on two related DNA fragments: $[d-(GGAATTCC)]_2$ and $[d-(GGTATACC)]_2$. These octamer duplexes are sufficiently short that nearly all non-exchangeable proton resonances are resolved in the two-dimensional nuclear Overhauser effect (2D NOE) spectra obtained at 500 MHz, but they are sufficiently long that interesting structural features are to be found in the interior base pairs. The $[d-(GGAATTCC)]_2$ octamer contains the sequence recognized by Eco Rl endonuclease for cleavage. Furthermore, the central six base pairs are homologous to the dodecamer duplex $[d-(CGCGAATTCGCG)]_2$ which has been the subject of x-ray crystal structure studies by Dickerson's group (12) and theoretical structure calculations by Kollman and colleagues (e.g., 3). The other octamer variant we have been studying contains the "TATA box" which is in the promoter region of DNA and binds to RNA polymerase preparatory to transcription.

This paper assesses progress on the project to determine the molecular structures in aqueous solution for the two DNA fragments. Before examining conformational features, proton resonances are assigned. Conformational details are explored by comparing the pure absorption 2D NOE spectra (500 MHz) experimentally obtained at four mixing times with theoretical spectra calculated with either a standard B-form structure or an energy-minimized structure for DNA.

THEORY

The theory for 2D NOE was developed in Ernst's lab (5). Subsequently, we (8) described a matrix method relating all interproton distances in a molecule to peak intensities in a theoretical 2D NOE spectrum which can be iteratively compared with the experimental pure absorption 2D NOE spectrum. The 2D NOE pulse sequence $\left[(\pi/2)-t_1-(\pi/2)-\tau_m-(\pi/2)-t_2\right]$ yields a 2D spectrum following Fourier transforms over t_2 and t_1. We have used a modification of the phase cycling route of States et al. (22) to obtain pure absorption phase spectra.

The intensity of an auto- (diagonal) or cross-peak for any value of the experimental parameter τ_m, the mixing time, may be calculated from

$$a(\tau_m) = \chi \exp(-\lambda_m) \chi^{-1} \tag{1}$$

where element a_{ij} of matrix **a** gives the 2D NOE cross-peak intensity for nuclei i and j, χ is the matrix of eigenvectors of the relaxation rate matrix **R**, and λ is a diagonal matrix of eigenvalues, i.e., the solution to the system of equations comprising the rate matrix. Relaxation rate matrix **R** has diagonal elements

$$R_{ii} = \sum_j (W_0^{ij} + 2W_1^{ij} + W_2^{ij}) + R_{1i} \tag{2}$$

representing the direct relaxation contributions from all sources to spin i, and off-diagonal elements

$$R_{ij} = W_2^{ij} - W_0^{ij}$$ (3)

representing the cross-relaxation rate between spins i and j. R_{1i} comprises all sources of relaxation other than proton-proton dipolar interactions. The zero-, single-, and double-quantum transition probabilities W_0^{ij}, W_1^{ij}, and W_2^{ij} are proportional to $J(\omega)/r_{ij}^6$ where $J(\omega)$ is the spectral density for the molecular motion modulating the ij interaction and r_{ij} is the distance between protons i and j.

A pure absorption 2D NOE spectrum will depend on choice of the mixing time τ_m used, molecular motions as manifest in the spectral densities, and internuclear distances between all protons, not just the two giving rise to a cross-peak. Macura and Ernst (5) showed that a pair of spins i and j can be considered as isolated if τ_m is sufficiently short, thereby enabling use of only a couple terms in a series expansion and obviating the need for the complete matrix solution outlined above. In practice it is difficult to isolate spins i and j completely from spin k if $r_{ik} \leqslant r_{ij}$. Although an ij cross-peak can still be observed and used qualitatively, the peak intensity will be modified, thus preventing quantitative internuclear distance determinations.

A theoretical investigation of the potential for 2D NOE spectral determinations of internuclear distances lead to the following conclusions (8): (a) several pure absorption 2D NOE spectra at a series of mixing times should be obtained; (b) a practical upper limit on determination of distance is $\simeq 5$ Å; (c) distances should be determined to an accuracy of ~10% if a good fit to all data (non-overlapping peaks assumed) is obtained; and (d) isotropic motion with a single effective correlation time can be assumed for 10% distance accuracy even though the actual motion may be more complicated.

There are a few ways of handling the 2D NOE dependence on molecular motions (spectral densities). At the present, we calculate an initial spectral density, assuming effective isotropic motion, from proton spin-lattice relaxation time (T_1) measurements. Small adjustments are permitted in the spectral densities to obtain a good fit of experimental and theoretical spectra. We have had extensive experience calculating spectral densities from considerably more sophisticated (and realistic) motional models (e.g., 4), which will be utilized in future calculations.

The complete matrix analysis of 2D NOE results had not been employed to determine the solution structure of a molecule with the accuracy we intend. Consequently, we chose to study a rigid molecule in solution that had a crystal structure already determined (9). Proflavine was selected as it is rigid with known crystal structure (23). The number of interacting protons is limited to CH protons by deuteron exchange of the amino protons.

With six cross-peak intensities and four diagonal peak intensities measured from each 240 MHz pure absorption 2D NOE spectrum at four different mixing times (240, 480, 640 and 850 msec), forty pieces of experimental data were acquired, from which the six interproton distances in proflavine can be determined without knowledge of the crystal structure. Interproton distances for the crystal structure protons, located by a Fourier difference synthesis (23), were compared to those from the 2D NOE analysis of proflavine in solution. The difference in internuclear distances was less than 10%, thus justifying the 2D NOE method. Justifications for the differences up to 10% were presented (9).

RESULTS AND DISCUSSION

Our intent is to obtain detailed structures in solution for DNA fragments, via internuclear distances, from the 2D NOE spectra. To obtain a
large number of these distances accurately, it is necessary to resolve
most 2D NOE cross-peaks and to assign most proton resonances.

RESONANCE ASSIGNMENTS

With our interest in DNA structural details, we have studied the octa-
nucleotide duplex $[d(GGAATTCC)]_2$, which has a weight $\sim$5000 daltons. We
have used 2D NOE to assign all seventy non-exchangeable proton resonances (20,21). More recently, we have assigned all seventy non-
exchangeable proton resonances in the related octamer $[d-(GGTATACC)]_2$.

Figure 1 contains the proton pure absorption 2D NOE spectrum (500 MHz)
of $[d-(GGTATACC)]_2$ acquired with a 250 msec mixing time. As described by
Scheek et al. (19), a sequential assignment procedure can be used to
identify the non-exchangeable base 1', 2', and 2" proton resonances
assuming that the DNA duplex structure is in the B family of structures,
i.e., right-handed helix. Any problems in this assumption are immediately obvious when the 2D NOE spectrum is examined since A-, B- and
Z-forms of DNA have very different values for internuclear distances and
consequently for cross-peak patterns.

The resonance assignments are based on the dipolar connectivities evident in the 2D NOE spectrum. This is illustrated in figure 2 where a
segment of B-DNA is shown. Close physical proximity of two protons in
the structure will lead to a strong dipole-dipole interaction between

$$[d_-(GGTATACC)]_2$$

pure absorption 2D NOE

$$t_m = 250\,msec.$$

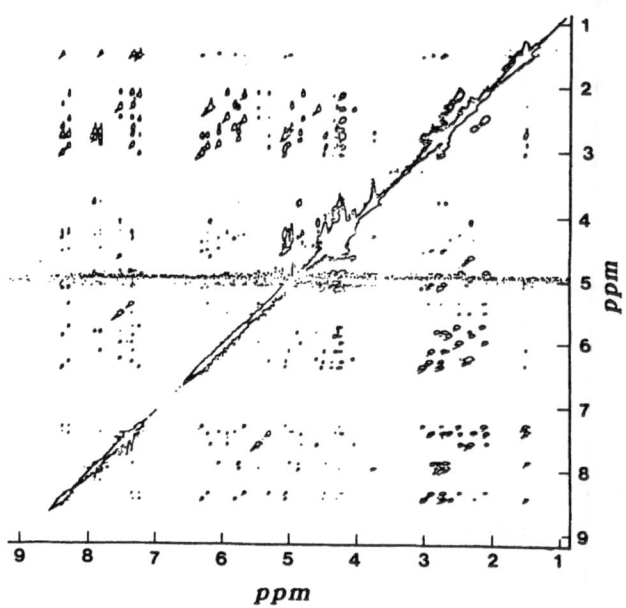

Figure 1 – Proton homonuclear pure absorption 2D NOE spectrum (500 MHz) of $[d-(5'GGTATACC3')]_2$ acquired with a mixing time of 250 msec. The pure absorption phase spectrum was obtained via a modification of the States et al.(22) procedure and required a 32-step phase cycling, with 12 sec delay between each step, for each of the 400 free induction decays obtained by incrementing t_1. In the t_2 dimension,4K data points were acquired. The final data matrix was 1Kx1K of real points following zero filling, Gaussian multiplication, and Fourier transformation. Zero quantum coherences were suppressed using a slight random variation in the mixing time (24). T = 15°C.

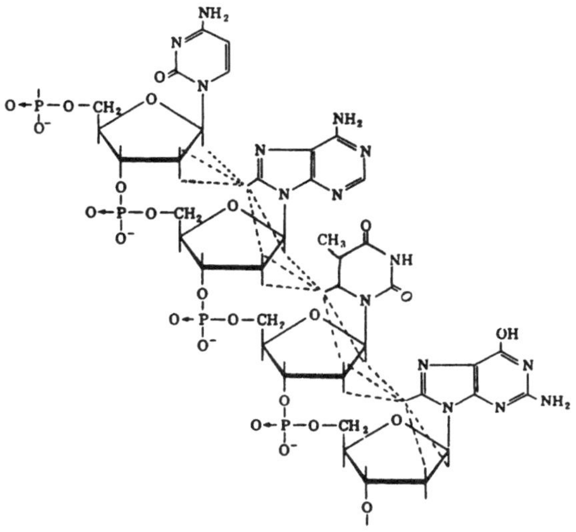

Figure 2 - B-DNA illustrating the dipolar connectivities between the base 1', 2' and 2" protons.

the two and a relatively intense cross-peak in the 2D NOE spectrum located at (ω_1, ω_2) where ω_1 and ω_2 are the resonance frequencies for the two interacting protons. With knowledge of some proton resonance assignments, the others can be obtained by following the dipolar connectivities in a sequential manner.

Although assignments are often more difficult due to the diminished frequency range of the 3', 4', 5', and 5" protons on the sugar moieties, they too can often be assigned using the 2D NOE dipolar connectivities. Figure. 3 summarizes the dipolar connectivities used for the assignments of the non-exchangeable proton resonances of the "TATA octamer". The as-

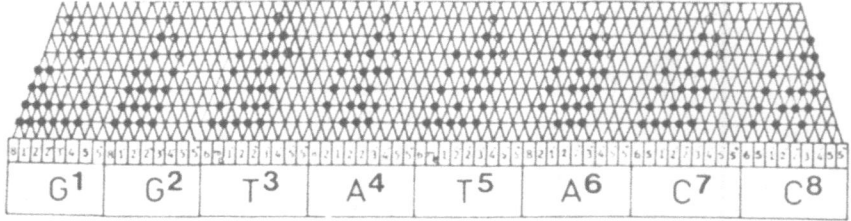

Figure 3 - Summary of the dipolar connectivities manifest in the 2D
NOE cross-peaks which were used for assignment of the proton resonances of $[d-(GGTATACC)]_2$. Closed symbols denote
intranucleotide interactions and open symbols denote internucleotide interactions.

signments for all seventy of the proton resonances are given in Table I.

STRUCTURE AND DYNAMICS

Some structural features of the octamer duplexes can be discerned in
spite of the fact the analysis is not yet complete. This is based on a
comparison of the experimental 2D NOE spectra (pure absorption), obtained at four mixing times, with theoretical 2D NOE spectra calculated
as described in the THEORY AND RATIONALE section. As indicated there, we
must incorporate spectral densities in the calculations. Since octamer
length ($\sim$27 Å) is about equal to the helix diameter ($\sim$26 Å) and since
assumption of simple isotropic motion introduces little error (8), we
estimate the effective isotropic correlation time of the two octamer
duplexes to be a couple nanoseconds on the basis of non-selective proton
T_1 values. Initial analysis of the 2D NOE data has allowed some
variability in the correlation time (vide infra).

TABLE I

Chemical shifts[a] (ppm) of the non-exchangeable protons in

$$[d-(5'GGTATACC3')]_2$$

	G^1	G^2	T^3	A^4	T^{5c}	A^6	C^7	C^8
H8	7.87	7.79		8.35		8.26		
H6			7.31		7.21		7.31	7.49
H5							5.30	5.45
H2[b]				7.40		7.40		
				7.31		7.31		
CH$_3$			1.45		1.47			
H1'	5.74	6.04	5.80	6.29	5.66	6.20	5.91	6.16
H2'	2.64	2.66	2.24	2.71	2.08	2.70	2.04	2.27
H2"	2.75	2.84	2.58	2.99	2.46	2.88	2.43	2.27
H3'	4.82	4.98	4.90	5.05	5.05	5.04	4.78	4.54
H4'	4.23	4.43	4.27	4.47	⎛4.22	4.45	4.29	4.16
H5'	3.73	4.29	4.24	4.33	⎝4.18	4.31	4.19	4.04
H5"	3.73	4.18	4.19	4.20	4.18	4.19	4.16	4.00

[a] $\pm$0.03 ppm; chemical shifts relative to HDO resonating at 4.90 ppm from TSP; 15°C.

[b] Distinction between A^4-H2 and A^6-H2 not resolved.

[c] Distinction between H4' and H5' for the T5 nucleotide is not possible.

Theoretical 2D NOE spectra for both octamer duplexes were calculated using a standard B-DNA structure (25) and an energy-minimized structure for each, generated using the energy refinement program AMBER (26). For each octamer, theoretical 2D NOE spectra were generated using an isotropic correlation time covering a few values from 1.7 to 2.9 nsec. All non-exchangeable protons in the octamer duplex were included in the calculations with the location of each proton specified by the structure assumed; one exception to this is that the thymine methyl protons were treated as a three-proton centroid.

For either of the octamer duplexes, all experimental 2D NOE data are not completely reproduced by any single correlation time with either of the two theoretical structures considered in this initial analysis. But many aspects of the experimental spectra are reproduced by both the B-DNA and energy-minimized structure. Furthermore, sequence-dependent structural features manifest in the 2D NOE spectra are mimicked in the theoretical spectra calculated from the energy-minimized strucutre, in particular, at the purine-pyrimidine junction as observed previously in the crystal structure of the partially homologous dodecamer (11).

For $[d-(GGAATTCC)]_2$, the energy-minimized structure does yield theoretical spectra which reproduce many, of not all, aspects of the experimental 2D NOE spectra (21). The structural features of the energy-minimized form are essentially those of the homologous moiety of the DNA dodecamer previously published (3,27). The intensities of cross-peaks arising from interactions of base protons with nearby protons are among the most sensitive to sequence-dependent structural variations. Figure 4 shows a composite experimental and theoretical relaxation matrix at a single mixing time as an example. The upper triangle contains the theoretical predictions, based on the energy-minimized B-DNA structure, for the relative magnitudes of the intensities of the base proton-to-neighboring proton cross-peaks. The lower triangle contains the experimentally observed ordering of the relative magnitudes of the same

$$[(\text{d-GGAATTCC})]_2$$

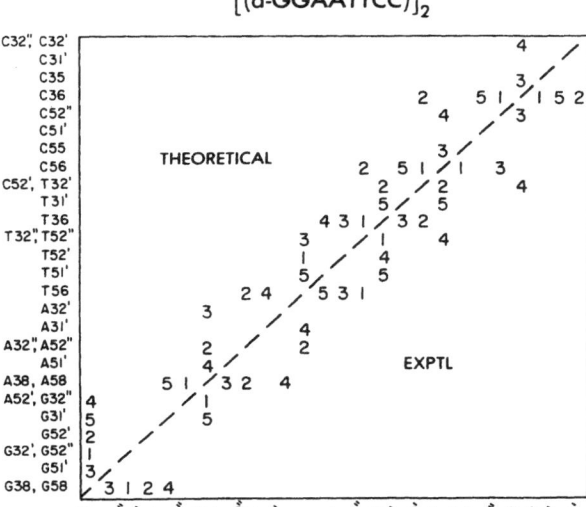

Figure 4 – Theoretical (upper triangle) and experimental (lower triangle) 2D NOE matrices (500 MHz, 250 msec mixing time) of the
base protons interactions in d-(GGAATTCC)$_2$. The five most
intense cross-peaks are given for each base proton with the
most intense being "1" and the least intense "5". The
theoretical matrix assumed a 2.7 nsec isotropic correlation
time and the energy-minimized B-DNA structure.

cross-peaks. For clarity, cross-peaks involving other protons are exclu-
ded. Where experimental resonances overlap, intensities of the individual peaks have been added (in both theoretical and experimental) as,
for example, H8s on the two G residues. See figure captions for further
explanation. If the minimized structure and the average structure are
identical and spectral densities were accurately determined, this matrix
should be completely symmetric about the diagonal. Considering the
sensitivity of cross-peak intensities to slight changes in internuclear
distances (9), the symmetry is fairly good. A change in ordering of
experimental intensities relative to theoretical intensities of one or
two units is not surprising, since many of the predicted intensities are
within a few percent of each other. The largest discrepancy between the

theoretical and experimental 2D NOE spectra of fig. 4 emanates from the terminal C residue. That discrepancy may reflect "fraying effects" as the terminal residue could be subject to transient single-stranded character with consequent conformational variations. Note that fig. 4 was obtained for data with the relatively long mixing time of 250 msec. This fit can only be obtained with consideration of the complete relaxation matrix (8) and not with the initial rate approximations commonly used.

In the case of $[d-(GGTATACC)]_2$, an energy-minimized structure has not previously been presented in the literature, so the structural characteristics of the energy-minimized structure generated by AMBER are listed in Table II along with those of standard B-DNA.

For the "TATA octamer", the base proton, anomeric proton 1', 2' and 2" proton peak intensities have been examined in more detail to date, but their axial and cross-peak intensities are affected by dipolar coupling to the other nearby sugar protons. Even with the limited analysis so far, a very large amount of data is still available to compare with an even larger amount of theoretical cross-peak and axial peak intensities generated with the different structural models and correlation times. As a sample of the comparison of experimental with theoretical axial peak intensities, Table III gives that comparison for the aromatic base protons. Figure 5 gives a few examples comparing experimental and theoretical cross-peak intensities. In summary, most axial and cross-peak intensities for the G and C nucleotides could be modeled fairly well with the standard B-form structure. But the "TATA box" could not be fit (with some exceptions) by either the regular B-form or the energy-minimized form.

An interaction between the A^6- H8 and T^3- CH_3 protons is detected in the

TABLE II

Torsion, base pair twist and tilt angles for the

energy-minimized and regular B-DNA structures

Glycosyl angle	Main chain torsion angles (°)						Base pair twist(°)	Tilt angle(°)
χ	α	β	γ	δ	ε	ξ		

I- ENERGY-MINIMIZED STRUCTURE

	χ	α	β	γ	δ	ε	ξ	twist	tilt
G^1	47	-81	-180	62	150	-171	- 85	14	15
G^2	60	-68	-173	56	137	176	-107	19	17
T^3	60	-73	176	60	137	-178	- 97	14	15
A^4	56	-69	178	59	125	-179	-110	9	10
T^5	65	-74	178	61	141	-173	-120	9	2
A^6	66	-70	175	60	151	-173	-127	14	4
C^7	61	-68	-174	62	150	177	-104	19	2
C^8	53	---	----	60	149	----	----	14	7
G^9	47	-81	-179	62	150	-171	- 85	14	15
G^{10}	60	-68	-174	56	136	176	-107	19	17
T^{11}	60	-73	176	60	137	-178	- 97	14	15
A^{12}	56	-69	178	59	125	-178	-109	9	10
T^{13}	65	-74	178	61	141	-173	-120	9	2
A^{14}	66	-70	176	60	151	-173	-128	14	4
C^{15}	61	-68	-175	62	150	177	-104	19	2
C^{16}	53	---	----	59	149	----	----	14	7

II- REGULAR B-DNA STRUCTURE

	χ	α	β	γ	δ	ε	ξ	twist	tilt
	82	-47	-146	36	156	155	- 95	4	5 to 8

TABLE III

Experimental and theoretical axial peak intensities

for six of the aromatic base protons of   d-(GGTATACC)$_2$[a]

	Exp[b]	MBF[c]	MBM[d]	MBS[e]	RBF[f]	RBM[g]	RBS[h]
G8[1]							
100 msec	0.85	0.97	0.95	0.94	0.91	0.87	0.85
175 msec	0.80	0.94	0.91	0.90	0.86	0.79	0.74
250 msec	0.63	0.92	0.88	0.86	0.81	0.73	0.69
400 msec	0.53	0.88	0.82	0.80	0.73	0.62	0.57
G8[1]							
100 msec	0.72	0.80	0.71	0.67	0.83	0.76	0.72
175 msec	0.65	0.68	0.57	0.53	0.73	0.63	0.58
250 msec	0.38	0.59	0.47	0.42	0.65	0.52	0.47
400 msec	0.29	0.48	0.34	0.30	0.52	0.38	0.33
A8[4]							
100 msec	0.72	0.87	0.81	0.78	0.82	0.75	0.71
175 msec	0.64	0.78	0.69	0.65	0.72	0.61	0.57
250 msec	0.39	0.71	0.60	0.56	0.64	0.51	0.46
400 msec	0.28	0.60	0.47	0.41	0.51	0.37	0.32
T6[5]							
100 msec	0.61	0.79	0.71	0.67	0.73	0.63	0.61
175 msec	0.56	0.68	0.57	0.51	0.60	0.48	0.45
250 msec	0.31	0.58	0.45	0.40	0.50	0.37	0.35
400 msec	0.25	0.47	0.31	0.27	0.36	0.25	0.23

TABLE III (continued)

A8[6]

100 msec	0.74	0.73	0.63	0.59	0.82	0.75	0.72
175 msec	0.67	0.60	0.49	0.44	0.72	0.62	0.57
250 msec	0.42	0.51	0.39	0.35	0.64	0.51	0.46
400 msec	0.20	0.38	0.27	0.24	0.51	0.37	0.32

C6[8]

100 msec	0.72	0.68	0.57	0.52	0.72	0.62	0.57
175 msec	0.61	0.53	0.41	0.36	0.58	0.46	0.41
250 msec	0.42	0.42	0.31	0.27	0.47	0.35	0.31
400 msec	0.32	0.29	0.20	0.18	0.34	0.23	0.20

---

[a] Boxed values indicate best fit to experimental data. Since T6[3] and C6[7] axial peaks overlap, their intensities are not listed.

[b] Experimental values.

[c] Energy-minimized structure, 1.7 nsec correlation time.

[d] Energy-minimized structure, 2.5 nsec correlation time.

[e] Energy-minimized structure, 2.9 nsec correlation time.

[f] Regular Arnott B-DNA, 1.7 nsec correlation time.

[g] Regular Arnott B-DNA, 2.5 nsec correlation time.

[h] Regular Arnott B-DNA, 2.9 nsec correlation time.

2D NOE results acquired at the longer mixing times, 250 and 400 msec. The distances between these protons (inter- or intrastrand) are too large in both model structures to account for the observation. Feigon et al. (16) have reported cross-strand A-H2....A-H2 interactions in the TpA, but not the ApT, sequence, Similarly, it seems that cross-strand interactions are also manifest in the TATA octamer with A[6]- H8 being closer to its base-paired T[3]-CH_3 than in either of the models examined.

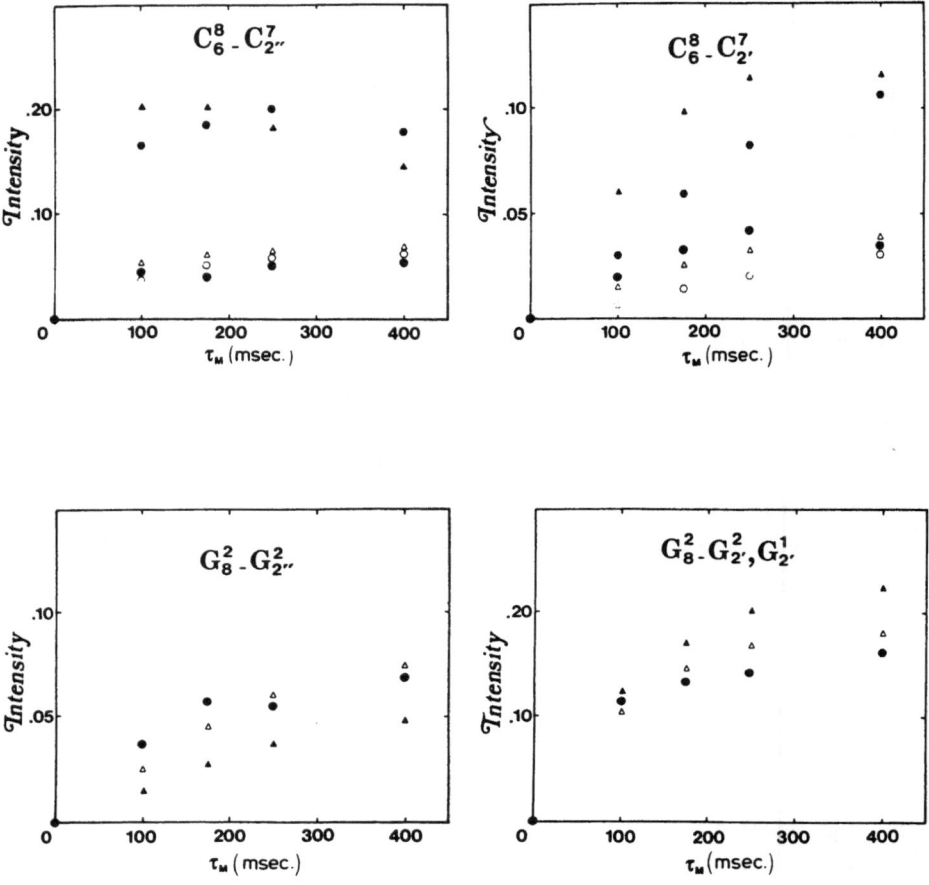

Figure 5 – Cross-peak intensity (relative to a total normalized intensity of 1) as a function of mixing time τ_m for $[$d-(5'GGTATA
CC3')$]_2$. The superscript in the nucleotide symbol gives its
position in the octamer sequence (numbering beginning at
the 5' terminus), and the subscript designates the hydrogen
in that nucleotide. Experimental data (●); calculated with
the regular B-DNA structure, 2.9 nsec correlation time (△)
or 1.7 nsec correlation time (○); or with the energy-minimized DNA structure, 2.9 nsec correlation time (▲) or 1.7
correlation time (◐)

CONCLUSION

The 2D NOE experiment is extremely useful for assigning proton resonances. In conjuntion with a complete relaxation matrix analysis of the diagonal and cross-peak intensities in pure absorption phase 2D NOE spectra obtained as a function of mixing time, it is possible to determine several internuclear distances in relatively small molecules. There is a potential for determining more than one hundred internuclear distances similarly for medium-sized molecules ($\sim$ 5000 daltons). That potential has not yet been realized, but initial work with two DNA fragments is promising.

REFERENCES

1) A.Bax, "Two-Dimensional Nuclear Magnetic Resonance in Liquids", Reidel Publishing, Dordrecht, Netherlands, 1982.

2) T.F.Havel, I.D.Kuntz and G.M.Crippen, Bull. Math. Biol. **45**, 665 (1983).

3) J.W.Keepers,P.A.Kollman,P.K.Weiner and T.L.James, Proc. Natl. Acad. Sci. USA **79**, 5537 (1982).

4) T.L.James,in "Phosphorus-31 NMR: Principles and Applications", D. Gorenstein (ed.), Academic Press, New York, 1984, pp. 349-400.

5) S.Macura and R.R.Ernst, Mol. Phys. **41**, 95 (1980).

6) M.Billeter,W.Braun, and K.Wüthrich, J. Mol. Biol. **155**, 321 (1982).

7) A.Kumar,G.Wagner,R.R.Ernst and K.Wüthrich, J. Amer. Chem. Soc. **103**, 3654 (1981).

8) J.W.Keepers and T.L.James, J. Magn. Reson. **57**, 404 (1984).

9) G.B.Young and T.L.James, J. Amer. Chem Soc. **106**, 7986 (1984).

10) A.A.Bothner-By and J.H.Noggle, J.Amer. Chem. Soc. **101**, 5152 (1979).

11) R.E.Dickerson and H.R.Drew,, J.Mol. Biol. **149**, 761 (1981).

12) R.E.Dickerson, J. Mol. Biol. **166**, 416 (1983).

13) A.H.J.Wang,S.Fujii,J.H.van Boom and A.Rich, Proc. Natl. Acad. Sci USA **79**, 3968 (1982).

14) Z.Shakked,D.Rabinovich,O.Kennard,W.B.T.Cruse,S.A.Salisbury, and M.A. Viswamitra, J. Mol. Biol. **166**, 183 (1983).

15) C.R.Calladine, J. Mol. Biol. **161**, 343 (1982).

16) J.Feigon,W.Leupin,W.A.Denny and D.R.Kearns, Biochemistry **22**, 5930 and 5943 (1983).

17) D.R.Hare,D.E.Wemmer,S.H.Chou,G.Drobny and B.R.Reid, J. Mol. Biol. **171**, 755 (1983).

18) R.M.Scheek,R.Boelens,N.Russo,J.H.van Boom and R.Kaptein, Biochemistry **23**, 1371 (1984).

19) R.M.Scheek,N.Russo,R.Boelens,R.Kaptein and J.H.van Boom, J. Amer. Chem. Soc. **105**,2914 (1983).

20) M.S.Broido,G.Zon and T.L.James, Biochem. Biophys. Res. Commun. **119**, 663 (1984).

21) M.S.Broido,T.L.James,G.Zon and J.W.Keepers, Eur. J. Biochem., in press.

22) D.J.States,R.A.Haderkorn and D.J.Ruben, J. Magn. Reson. **48**, 286 (1982).

23) A.Achari and S.Neidle, Acta Crystallogr. **B32**, 2537 (1976).

24) S.Macura,Y.Huang,D.Suter and R.R.Ernst, J. Magn. Reson. **43**, 259 (1981).

25) S.Arnott,P.Campbell-Smith and P.Chandrasekharan, in "CRC Handbook of Biochemistry", Vol. 2, (G.D. Fasman ed.), CRC Press, Cleveland, 1976, pp. 411-422.

26) P.K.Weiner and P.A.Kollman, J. Comp. Chem. **2**, 287 (1981).

27) P.A.Kollman,J.W.Keepers and P.K.Weiner, Biopolymers **21**, 2345 (1982).

PBB, Vol. 2
Advanced Magnetic Resonance Techniques
in Systems of High Molecular Complexity
© 1986 Birkhäuser Boston, Inc.

SELECTIVE ^{1}H NMR RELAXATION DELINEATION OF RECEPTOR BINDING EQUILIBRIA

E.Gaggelli,A.Lepri,N.Marchettini and S.Ulgiati

Department of Chemistry, University of Siena

Pian dei Mantellini 44, 53100 Siena, Italy

The NMR spectrum of a macromolecule is usually a broad envelope which seldom allows one to distinguish any individual resonance line. Moreover it often happens that a low amount of the macromolecule, insufficient to give a strong resonance adsorption, can be isolated. As a consequence, the problem of studying interactions between a small ligand and its macromolecular receptor has then to be approached by observation of any change in the NMR parameters of the ligand caused by the presence of the macromolecule. Exchange between at least two environments, free (f) and bound (b) to the macromolecule must then be considered in developing theoretical equations for the NMR parameters of the ligand; namely either chemical shift or relaxation rates have been shown (1,2) to depend on the rate of chemical exchange. Occasionally it may be possible to detect the resonances of the bound ligand directly, but generally the concentration of the macromolecule, and hence also that of the bound ligand, will be small and this detection will be difficult. If, however,the rate of chemical exchange between free and bound environments is fast in respect of either the difference in chemical shift or the nuclear relaxation rate, then the observed NMR parameters will be a weighted average of those in each environment and thus information on the bound resonance signals can be gained in the bulk.

The applicability of the NMR approach suffers however of the following drawbacks:

(a) The amount of bound ligand may be too small to be detected such that under conditions of slow exchange, non effect on bulk resonances will be observed.

(b) Conditions of fast exchange may not apply to all the resonances, since they may have different relaxation rates and chemical shifts at the bound site.

(c) The observed parameter may be intrinsically unsensitive to the presence of the macromolecule.

The third point needs further discussion since it introduces the reasons why diamagnetic systems have been up to now neglected and why the method of selective irradiation has been suggested (3-7) to allow a very suitable approach to diamagnetic ligand-receptor pairs. In fact, if two environments are assumed, a fast exchange rate yields the following equation:

$$M_{obs} = p_f M_f + p_b M_b \tag{1}$$

where M is a generalized NMR parameter, f and b refer to free and bound environments respectively and the p's are fractions of ligand molecules in the two environments ($p_f + p_b = 1$). Since p_b must be usually kept very small ($p_b \ll 1$, $p_f \simeq 1$), the NMR approach is feasible only where M_b is quite different from M_f. If M is the chemical shift, observation of any effect practically occurs only in the presence of nearby paramagnetic centres within the molecule. If M is the spin-lattice relaxation rate (the spin-spin relaxation rate will not be considered herein as many complications arise in the measurements), the relaxation mechanism as well as the modulation of the relaxing field must be considered. In the case of protons relaxed by mutual intramolecular ^{1}H-^{1}H dipole-dipole interactions, the spin-lattice relaxation rate can be expressed as a sum of pairwise interactions extended over all the actual spin pairs (8):

$$R_{1i} = \sum_{i \neq j} \varrho_{ij} + \sum_{i \neq j} \sigma_{ij} \qquad (2)$$

ϱ_{ij} and σ_{ij}, called the direct- and cross-relaxation terms of any spin i dipolarly coupled to any spin j, depend upon the relaxation transition probabilities among the four energy levels ($\alpha\alpha, \alpha\beta, \beta\alpha, \beta\beta$):

$$\varrho_{ij} = 2W_1 + W_2 + W_0 \qquad (3)$$

$$\sigma_{ij} = W_2 - W_0 \qquad (4)$$

where W_1 is the single-quantum transition probability ($\alpha\alpha \leftrightarrow \alpha\beta, \alpha\beta \leftrightarrow \beta\beta$), W_0 the zero-quantum transition probability ($\alpha\beta \leftrightarrow \beta\alpha$) and W_2 the double-quantum transition probability ($\alpha\alpha \leftrightarrow \beta\beta$). The explicit forms of ϱ_{ij} and σ_{ij}, in the case of intramolecular dipolar interactions, are:

$$\varrho_{ij} = 0.1 \, \gamma_H^4 \hbar^2 r_{ij}^{-6} \left\{ 3\tau_c/(1+\omega_H^2 \tau_c^2) + 6\tau_c/(1+4\omega_H^2 \tau_c^2) + \tau_c \right\} \qquad (5)$$

and

$$\sigma_{ij} = 0.1 \, \gamma_H^4 \hbar^2 r_{ij}^{-6} \left\{ 6\tau_c/(1+4\omega_H^2 \tau_c^2) - \tau_c \right\} \qquad (6)$$

where r_{ij} is the interproton distance, ω_H is the proton Larmor frequency and τ_c is the motional correlation time. Substitution of eqs.(5) and (6) into eq.(2) yields, for a given spin pair:

$$R_{1i} = 0.3 \, \gamma_H^4 \hbar^2 r_{ij}^{-6} \left\{ \tau_c/(1+\omega_H^2 \tau_c^2) + 4\tau_c/(1+4\omega_H^2 \tau_c^2) \right\} \qquad (7)$$

Since binding of a ligand to a macromolecular receptor slows down the reorientational motions to the $\omega_H \tau_c \gg 1$ region, where $f(\tau_c)$ is very small, the spin-lattice relaxation rate of the bound ligand will be large only in cases where the relaxation mechanism has changed, which, as a matter of fact, has limited the applicability of relaxation studies to paramagnetic systems.

However, when selective irradiation is used to excite spins i such that spins j are not perturbed, eq.(2) modifies as follows (9):

$$R_{1i}^S = \sum_{i \neq j} \varrho_{ij} \tag{8}$$

and, as a consequence, the selective spin-lattice relaxation rate becomes (10-13):

$$R_{1i}^S = 0.1 \, \gamma_H^4 \hbar^2 r_{ij}^{-6} \left| 3\tau_c/(1+\omega_H^2\tau_c^2) + 6\tau_c/(1+4\omega_H^2\tau_c^2) + \tau_c \right| \tag{9}$$

In such a case, even if the relaxation mechanism does not change, entering the slow motion region does not alter the direct dependence of R_{1i} on τ_c and, therefore, the selective relaxation rate of bound ligand is expected to be quite different from that of the free ligand in solution as shown in figure 1. It follows that selective irradiation of properly

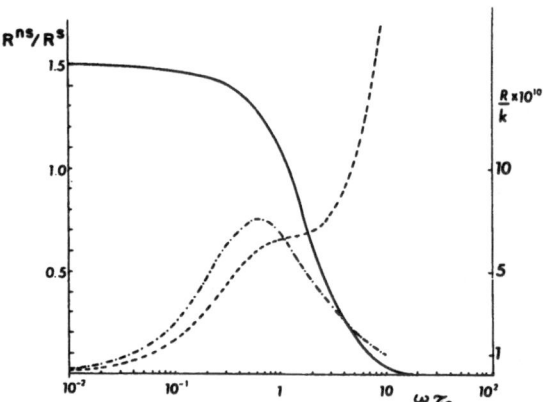

Figure 1 - Ratio R^{NS}/R^S (solid line), R^{NS} (dotted line), and R^S (dashed line) for a proton pair vs $\omega\tau_c$. In the figure k = $\hbar^2\gamma_H^4/r_{ij}^6$

chosen spin systems within relatively small "NMR visible" biomolecules allows detection of binding to macromolecular receptors without requiring insertion of paramagnetic spin probes.

As an example, binding of the substrate glycyl-tyrosine to the Zn-enzyme carboxypeptidase can be tested as shown in Table I. Selective irradiation of whichever doublet within the AA'BB' spin system of the aromatic ring (figure 2) yields changes in R^S in the range 40-100 % depending upon the Larmor frequency; whereas changes in R^{NS} (=non-selective) are negligible, especially at high frequency (3).

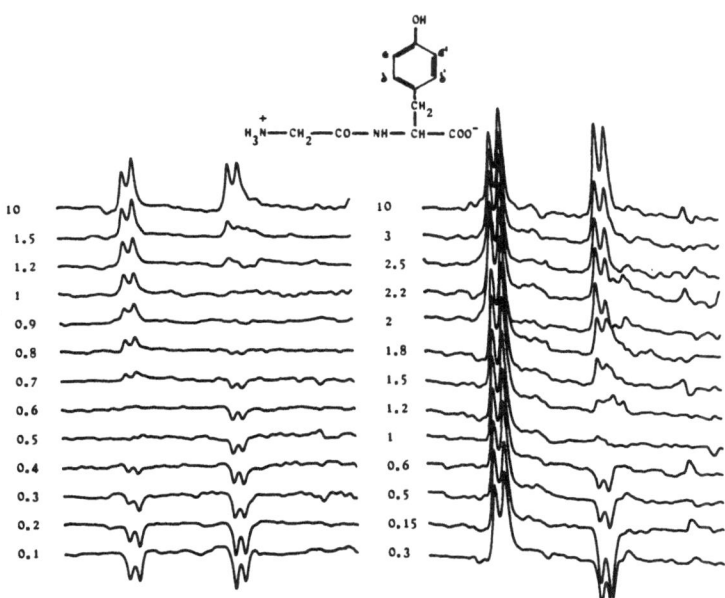

Figure 2 - Nonselective (left) and selective (right) partially relaxed proton spectra of glycyl-L-tyrosine 0.043 mol/dm^3 at 270 MHz. T=274 K. The low-field doublet is attributable to $H_{\varepsilon\varepsilon'}$ and the high-field doublet to $H_{\delta\delta'}$

Table I

Non-selective (R^{NS}) and selective (R^{S}) longitudinal relaxation rates
of phenyl protons of Glycyl-L-Tyrosine bound to Zn-Carboxypeptidase A

ν_H (MHz)	[E] (mM)	[S] (mM)	R^{NS} (s^{-1})	Tyr H$_,$ R^{S} (s^{-1})	R^{NS}/R^{S}	R^{NS} (s^{-1})	Tyr H$_,$ R^{S} (s^{-1})	R^{NS}/R^{S}
90	0	43	1.19	0.76	1.59	1.56	1.11	1.41
90	0.66	43	1.28	1.39	0.17	1.72	2.00	0.20
90	0	22	1.14	0.84	1.36	1.51	1.06	1.42
90	0.83	22	1.35	2.04	0.21	1.78	2.50	0.22
270	0	43	0.65	0.51	1.27	1.19	1.06	1.12
270	0.66	43	0.78	0.85	0.92	1.19	1.43	0.83
270	0	10	0.64	0.53	1.21	1.13	0.94	1.20
270	0.16	10	0.69	0.77	0.90	1.13	1.26	0.90

The method of selective irradiation can then be exploited to detect binding to ill-defined receptors within biological preparations, such as
blood or whole cell samples. In Table II the effect of adding different
samples to a solution of colchicine (figure 3) (an anti-mytotic and
tumor-inhibiting alkaloid) are reported. It is evident that true binding
only can be responsible for the observed R^{S} enhancement, since no change
in viscosity, such that molecular motions are slowed down of more than
one order of magnitude, can be figured out.

Once the ligand-receptor binding has been isolated, as in the case of
colchicine interacting with erythrocytes, details of the interaction can
be gained by proper changes in experimental variables, such as pH, temperature, concentration etc. In particular, titration of the R^{S} enhancement vs. ligand concentration allows evaluation of the apparent asso-

ciation constant K_{ass}. For an interaction 1:1 the binding equilibrium can be schematized as:

$$C + RBC \rightleftharpoons RBC\text{--}C \qquad\qquad (10)$$

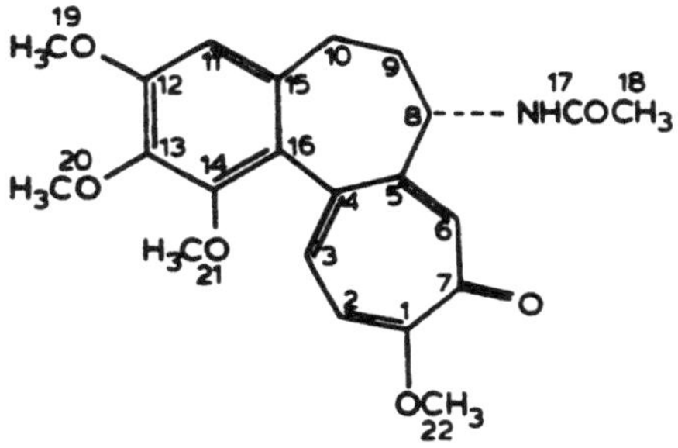

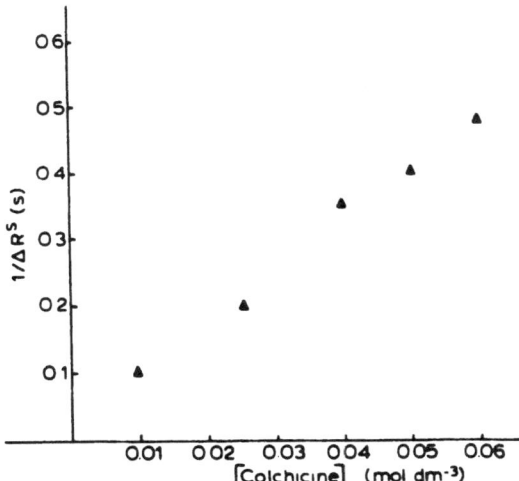

Figure 3

Figure 4 – $1/\Delta R^S$ vs [colchicine] plot for the H_2 protons in the presence of red blood cells

Table II

Non-selective and selective longitudinal relaxation rates
(s^{-1}) of colchicine 0.1 M at pH = 7 after addition
of 0.1 ml of different biological preparations.

Preparation added	H_{11} R^{NS}	R^{S}	H_3 R^{NS}	R^{S}	H_2 R^{NS}	R^{S}	H_6 R^{NS}	R^{S}
------	1.29	0.89	2.96	2.07	3.95	2.72	4.35	3.02
DPPC vesicles	1.31	1.18	3.08	2.78	3.97	3.65	4.31	3.97
RBC 4% (*)	1.42	1.51	3.01	3.48	4.07	4.56	4.36	5.13
RBC ghost 4% (*)	1.34	1.21	3.06	2.81	4.01	3.72	4.30	3.89

(*) RBC = red blood cells. T = 300 K.

The apparent association constant is given by:

$$K_{ass} = [RBC-C]/[C][RBC] = [RBC-C]/[C]\{[RBC]_o - [RBC-C]\} \quad (11)$$

where $[C]$ is the concentration of free colchicine and $[RBC]_o$ represents
the initial concentration of red blood cells.

The fraction of bound colchicine, P_b, is given by:

$$P_b = [RBC-C]/([C] + [RBC-C\ )] \approx [RBC-C]/[C] \quad (12)$$

Eq.(11) can be rearranged in the following way:

$$[RBC-C] = K_{ass}[C][RBC]_o/(1 + K_{ass}[C]) \quad (13)$$

Substituting eq.(13) into eqs.(1) and (12) and considering R^S instead of M, the following equation can be derived:

$$1/\Delta R^S = (K_{ass}^{-1} + [c]).1/R_b^S[RBC]_o \qquad (14)$$

Eq.(14) states that extrapolating the plot of $1/\Delta R^S$ vs $[c]$ to zero allows evaluation of K_{ass} ($1/\Delta R^S$ is zero when $[c] = K^{-1}$). A typical plot is shown in figure 4 for the H_2 proton of colchicine: the linear relationship is evident allowing straightforward extrapolation to $1/\Delta R^S = 0$.

REFERENCES

1) T.J.Swift and R.E.Connick, J.Chem.Phys. **37**, 307 (1962).

2) Z.Luz and S.Meiboom, J.Chem.Phys. **40**, 2686 (1964).

3) G.Valensin,T.Kushnir and G.Navon, J.Magn.Reson. **46**, 23 (1982).

4) G.Valensin,A.Casini,A.Lepri and E.Gaggelli, Biophys.Chem. **17**, 297 (1983).

5) G.Valensin and P.E.Valensin, Magn.Reson.Med. **2**, 501 (1985).

6) I.Barni Comparini, E.Gaggelli, N.Marchettini and G.Valensin, Biophys.J. **48**, 247 (1985).

7) G.Valensin,A.Lepri and E.Gaggelli, Biophys.Chem. **22**, 83 (1985).

8) J.H.Noggle and R.E.Schirmer, "The Nuclear Overhauser Effect", Academic Press, New York, 1971.

9) R.Freeman,H.D.W.Hill,B.L.Tomlinson and L.D.Hall, J.Chem.Phys. **61**, 4466 (1974).

PBB, Vol. 2
Advanced Magnetic Resonance Techniques
in Systems of High Molecular Complexity
© 1986 Birkhäuser Boston, Inc.

NUCLEAR SPIN RELAXATION STUDY OF WEAK MOLECULAR

INTERACTIONS IN BENZENE SOLUTIONS OF QUINUCLIDINE

A.Maliniak and J.Kowalewski

Division of Physical Chemistry,

Arrhenius Laboratory,

University of Stockolm,

S-106 91 Stockholm, SWEDEN.

INTRODUCTION

The question whether weak interactions between two molecules lead to the formation of a molecular complex or should only be seen as an effect of the preferential solvation is complicated and has been the subject of many investigations (1). To give a precise definition of the term "molecular complex" is a difficult task. One explanation is given by Mulliken and Person (2) who describe molecular complex as a type of an association, stronger than van der Waals interactions, but with the properties of the original molecules mostly preserved. On the other hand, according to Laszlo (3), the concept of "molecular complex" should have a more precise meaning and be used with more care. The problem is not only the strength of the interactions between the components, but also the lifetime of the complex. If the species exists during a period which is considerably shorter than the rotation correlation time, the question arises whether it is meaningful to describe the species as a complex.

In the present work the interactions of 1-azabicyclo[2.2.2]octane or quinuclidine with benzene in cyclohexane are investigated, by measuring the effects on molecular dynamics. In an earlier work (4) we have used quinuclidine as a model for studies of hydrogen bonds formed with different solvents. In absence of specific intermolecular interactions it can be expected that quinuclidine reorients almost isotropically, whereas the anisotropy increases when interactions, stronger than van der Waals forces, are present.

This work presents results of measurements of relaxation rates of ^{13}C and ^{14}N nuclei. Dipole-dipole (DD) mechanism with directly bonded hydrogen nuclei is usually most efficient for relaxation of carbon-13. The contribution of the DD- mechanism to the observed longitudinal relaxation time, T_1 can be determined by measurements of the nuclear Overhauser enhancement (NOE) factors. Standard equations for evaluation of correlations times from the relaxations rates in the limit of extreme narrowing are used and Woessner's model for dipole relaxation is applied (5). The geometrical parameters (angles and bond lengths) are obtained from electron diffraction studies (6). Since quinuclidine is a symmetric rotor, its reorientation in solution can be described in terms of two rotational diffusion constants, for molecular rotation parallel and perpendicular to the symmetry axis which are denoted $R_{\parallel}$ and $R_{\perp}$ respectively. The rotation, parallel to the symmetry axis, is mainly determined by the friction due to viscosity of the solvent and is not sensitive to specific interactions. On the other hand, the motion perpendicular to the molecular axis is restricted by the specific interactions. Anisotropy expressed as a ratio between the two rotational diffusion constants $R_{\parallel}$ and $R_{\perp}$, can therefore be taken as a measure of the interactions between quinuclidine and benzene.

The dominant mechanism in relaxation of nitrogen-14 nuclei is the coupling between nuclear electric quadrupole moment and the electric field gradients, due to the molecular environment. The quadrupole relaxation time, T_2 is determined from the spectral width at half-height. The molecular reorientation determining the nitrogen relaxation rates is the rotation perpendicular to the symmetry axis. The variable temperature study is here performed on nitrogen-14 nucleus.

EXPERIMENTAL

The preparation of the samples as well as techniques for measuring of the longitudinal relaxation times and NOE factors follow the procedures described previously (4). All the experiments were performed on JEOL GX 400 NMR instrument. The carbon-13 measurements were made at 30°C. The mole fraction of quinuclidine was constant 0.05, in all samples, which corresponds to a concentration of about 0.5M. 10% of deuterated cyclohexane or benzene was added for field-frequency locking.

RESULTS AND DISCUSSION

In figure 1 the ratio between the rotational diffusion constants, $R_\parallel$ and $R_\perp$ is plotted versus the concentration of benzene. The anisotropy of reorientation of quinuclidine is small ($R_\parallel / R_\perp = 1.3$) in cyclohexane and increases asymptotically with increased concentration of benzene.

The variation of nitrogen-14 relaxation times with temperature has been studied for quinuclidine in pure benzene and cyclohexane; the temperature dependence can be described by an Arrhenius equation of the form $T_2 = T_2^\circ \exp(E_a / RT)$. From figure 2 it can be seen that the plots are linear giving activation energies, $E_a = 8.5$ kJ/mole and $E_a = 6.3$ kJ/mole for benzene and cyclohexane as the solvents, respectively. The corresponding activation energies for the viscosities of the pure solvents (7), are $E_a = 4.4$ kJ/mole for benzene and $E_a = 5.4$ kJ/mole for cyclohexane. The large change in the activation energy in benzene solutions can be interpreted in terms of restricted motion of quinuclidine in benzene, implying the

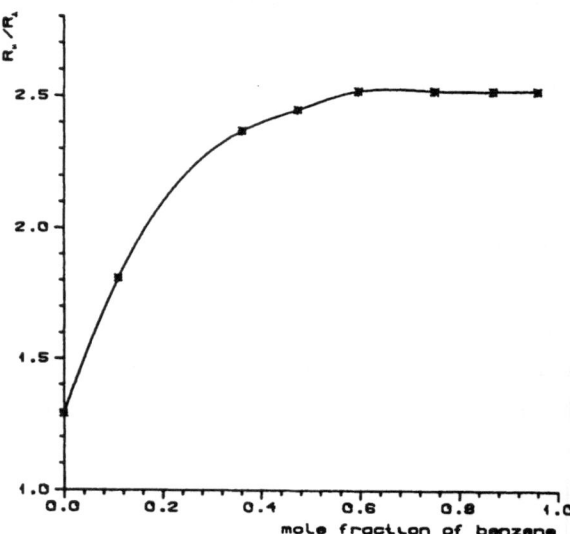

Figure 1 – Anisotropy in reorientations of quinuclidine in benzene expressed as the ratio between the rotational diffusion constants $(R_\parallel/R_\perp)$

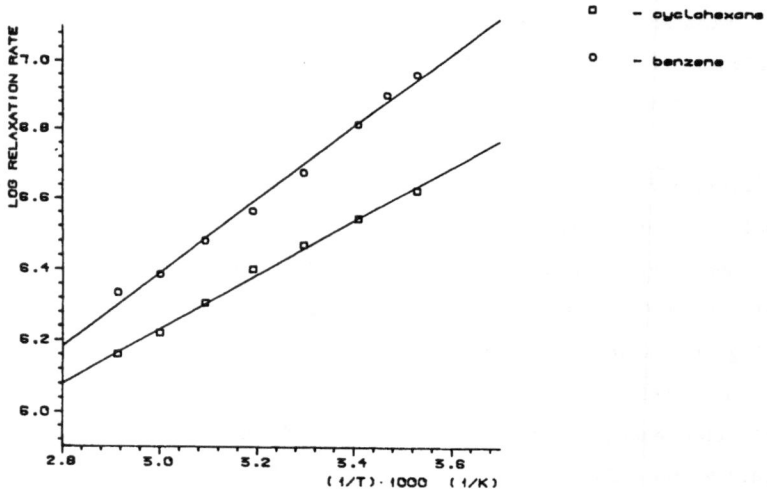

Figure 2 – Temperature dependence of nitrogen-14 relaxation rates in cyclohexane and benzene solutions

energy barrier of rotation due to specific interactions. The study of chemical shift shows that the equilibrium constant for the formation of a complex between quinuclidine and benzene is very low, which is also expected from thermodynamical studies of triethylamine/benzene system (8). The conclusion of the present study is that weak interactions between quinuclidine and benzene are clearly showed by a change in molecular reorientations; however the more specific information about those interactions remains still to be obtained.

REFERENCES

1) E.M.Engler and P.Laszlo, J.Amer.Chem.Soc. **93**, 1317 (1971).

A.K.Covington and J.M.Thain, J.Chem.Soc. Faraday.Trans.I 1879 (1974).

H.Grahn,U.Edlund and G.C.Levy, J.Magn.Reson. **56**, 61 (1984).

2) R.S.Mulliken and W.B.Person, "Molecular Complexes", John Wiley and Sons, New York, 1969.

3) P.Laszlo, Progr.NMR Spectr. **13**, 257 (1979).

4) A.Maliniak,J.Kowalewski and I.Panas, J.Phys.Chem. **88**, 5628 (1984).

5) D.E.Woessner, J.Chem.Phys. **37**, 647 (1962).

6) H.Shei,Q.Shen and R.L.Hilderbrandt, J.Mol.Struct. **65**, 297 (1980).

7) Landolt-Börnstein "Zahlenwerte und Funktionen", Springer-Verlag, West Berlin, 1969; Aufl., Bd II/5a.

8) B.I.Mattingley and D.V.Fenby, J.Chem.Therm. **7**, 307 (1975).

PBB, Vol. 2
Advanced Magnetic Resonance Techniques
in Systems of High Molecular Complexity
© 1986 Birkhäuser Boston, Inc.

PROTON NMR STUDIES OF THE ACTIVE CENTRE OF ACTH

F.Toma[a],V.Dive[a],M.Löw[b] and L.Kisfaludy[b]

[a]Service de Biochimie, Département de Biologie, CEN-Saclay

91191 Gif sur Yvette, France

[b]Chemical Works of Gedeon Richter

Budapest, Hungary

INTRODUCTION

Adrenocorticotropic hormone (ACTH), a linear polypeptide (figure 1) of
the anterior lobe of the pituitary, is responsible for a variety of important biologic functions in mammalians. Among these, steroidogenesis
at the adrenal cortex and activity at the central nervous system have
been most extensively studied (1-3). The 4-10 part of the ACTH sequence,
Met-Glu-His-Phe-Arg-Trp-Gly, is present also in the primary structure of
the biogenetically related hormones αMSH, βMSH, ß- and γ-lipotropins,
with which ACTH features in common melanotropic and lipolytic activities
(1,2).

Ser[1]-Tyr-Ser-Met-Glu[5]-His-Phe-Arg-Trp-Gly[10]-Lys-Pro-Val-Gly-Lys[15]-
Lys-Arg-Arg-Pro-Val[20]-Lys-Val-Tyr-Pro-Asn[25]-Gly-Ala-Glu-Asp-Glu[30]-
Ser-Ala-Glu-Ala-Phe[35]-Pro-Leu-Glu-Phe[39]

Figure 1 - Primary structure of ACTH of human origin. The 1-24 part
of the sequence is invariant with the species

Structure-activity studies have indicated that residues 5-10 are involved in the stimulation of the cell receptors and that the 15-18 sequence

is responsible for the binding of the hormone to the cell membrane (1, 2). N-terminal peptide fragments may have full-ACTH(1-32)- or high-residual-ACTH(1-24)-in vivo steroidogenic activity; C-terminal fragments are devoid of biological activity and may have antagonistic character (2).

Physicochemical studies have shown that, although flexible, the ACTH molecule may adopt preferred conformations in both aqueous and organic solutions and that a helical structure is stabilized in trifluoroethanol solution in the N-terminal part (4,6,9,10).

The present study deals with the experimental evidence that ACTH(1-14) undergoes a conformational transition in DMSO solution including a helical structure at positions 4-9. No similar behaviour is observed for ACTH(1-10). The possible consequent conformation-steroidogenic activity relationships are discussed.

RESULTS AND DISCUSSION

The tetradecapeptide ACTH(1-14) has been studied in DMSO solution together with the other shorter fragments (1-10), (5-14), (4-10), (5-10) in two ionic states. These are: (i) the cationic form (pH ~2 in aqueous solution) characterized by the protonated carboxylic function at the C-terminus and at Glu5 side chain, by the positively charged groups at the N-terminus and His6,Arg8,Lys11 side chains; (ii) the zwitterionic form (pH ~6 in aqueous solution) corresponding to negatively charged carboxyl groups at positions 5 and 10, to positively charged groups at positions 1,8,11 and ca. 20% of neutral form of the imidazole group of His6 side chain.

The analysis of the proton spin systems could be carried out completely at 500 MHz for the different peptides with the aid of the two-dimensional techniques (Cosy45,relay-Cosy,Noesy). Assignment of the signals to protons corresponding to individual residues could also be achieved fully by comparison of the fragments and the pH dependence of chemical shifts, the search of long range coupling connectivities by Cosy (ß protons-aromatic protons for the identification of Tyr2, His6, Phe7, Trp9 resonances), intra- and interresidue NOE's.

Line broadening occurs on the amide protons of ACTH(1-14), and to a lesser extent in ACTH(5-14), in the zwitterion form. This does not depend on the concentration in the 1-10 mM range. The proton-proton vicinal coupling constants in the backbone $^3J_{HNC\alpha H}$ could not therefore be measured accurately enough by direct reading on the corresponding signals in the normal or in 2D-J resolved spectra (short T_2 values). Accurate values were obtained (namely for Glu5, His6 and Phe7) by monitoring the intensity I of the amide resonances as observed by spin-echo (90°-τ-180°-Acq) under decoupling of the corresponding proton multiplet. Under these conditions, I=0 for $\tau=1/4J$ (11).

Chemical shifts and their temperature dependence, backbone and side chain vicinal coupling constants could be measured for each residue in this series of peptides.

In the following discussion we will deal essentially with $^3J_{HNC\alpha H}$ coupling constants, which account for the possible backbone conformations (through the Karplus-type relationship to the φ angles (12)), and respectively with the temperature dependence of the amide proton chemical shifts, which give indications about the formation of hydrogen bonds (backbone coupling constants $J \sim 5\text{-}6$ Hz are typical of helical structures; $J \sim 7$ Hz characterize a random conformation; $J \sim 7\text{-}9$ Hz indicate ß structures. Temperature coefficients, S, more negative than -3ppb/degree are believed to belong to "free" amide protons).

Table I

Backbone coupling constants (J,in Hz) and temperature
dependence (S,in ppb/degree) of the amide proton
chemical shifts of ACTH peptides in DMSO

ACTH	5–10		4–10		5–14		1–10		1–14	
	J	S	J	S	J	S	J	S	J	S
Tyr2 c									7.8	−2.4
z							6.5	−7.1	no	no
Ser3 c									7.2	−6.4
z							7.5	−7.1	7.2	−5.4
Met4 c									7.8	−4.0
z							7.7	−5.0	6.8	−1.8
Glu5 c			7.3	−1.3					7.2	−4.0
z			no	no			7.5	−5.1	5.1	−1.2
His6 c	8.2	−3.8	8.0	−3.4	8.0	−3.2			7.2	−4.2
z	7.9	−3.2	7.5	−3.7	7.7	−4.2	7.4	−3.6	5.9	−2.6
Phe7 c	7.7	−2.4	7.9	−2.1	7.3	−2.4			7.6	−4.6
z	8.1	−3.4	8.2	−4.2	7.6	−3.4	8.1	−3.8	7.0	−0.6
Arg8 c	7.8	−6.0	8.0	−4.1	7.8	−5.8			7.6	−5.6
z	7.7	−6.6	7.4	−5.7	7.5	−4.6	7.8	−4.2	no	no
Trp9 c	8.0	−4.6	8.0	−3.4	7.7	−4.0			7.5	−4.0
z	8.7	−3.8	8.6	−5.5	8.2	−4.2	8.7	−5.9	7.0	−3.2
Gly10c			5.9						4.8	
	5.9	−5.6	5.6	−6.1	5.9	−5.8			5.9	−5.8
z	5.0		5.2		6.0				6.0	
	4.2	−2.8	4.4	−1.6	5.4	−4.6	nm	−0.7	5.5	−3.9
Lys11c					7.8	−6.2			7.8	−6.0
z					7.2	−3.4			7.5	−3.4
Val13c					8.9	−6.2			8.9	−5.4
z					8.6	−5.2			8.9	−3.6
Gly14c					5.6				5.6	
					5.9	−5.6			6.1	−5.4
z					5.5				5.0	
					4.4	−2.4			4.0	−2.7

(c) cation form; (z) zwitterion form ; (no) signal not observed because
of line broadening; (nm) not measured.

The J and S values reported in Table I suggest the absence of any regular structure for all the peptides in both ionic forms and for ACTH(1--14) only in the cationic form. This does not preclude completely the existence of some local preferred conformations, namely in the two shorter fragments (5-10) and (4-10), under the influence of intermolecular and intramolecular interactions. This is so for ACTH(4-10) in the zwitterion for which the following main changes were observed in the concentration range from 4 mM to 0.1 mM: the amide proton of Gly10 shifts from 7.52 ppm at 4 mM to 7.35 ppm at 0.1 mM, while simultaneously S varies from -1.6 to -2.4 ppb/degree; the $N^\varepsilon H$ proton of Arg8 side chain shifts from 9.28 to 9.70 ppm and shows a parallel change of S from -4.6 to -0.3 ppb/degree; the rotamer distributions of Arg8 and Trp9 side chains, as observed by the corresponding $^3J_{\alpha\beta}$ and $^3J_{\alpha\beta'}$, vary at the same time; and finally the α proton chemical shift of Met4 changes from 3.59 to 3.28 ppm. Clearly, as found in the solid state, ionic interactions in this peptide favour intermolecular associations in solution at higher concentrations. The crystal structure of ACTH(4-10), although not resolved in detail, shows the existence of antiparallel β-sheets in infinitely long strands in which successive molecules are arranged head to tail by H bonds between COO^- and $NH3^+$ (13). The above changes of the NMR parameters of ACTH(4-10) as a function of concentration in DMSO solution can be interpreted in terms of a concentration dependent equilibrium between inter- and intramolecular interactions: at higher concentrations, the Gly10 amide proton,is H bonded to a second solute molecule; the Arg8 and Trp9 side chains are in close contact. Upon dilution, the former proton becomes more exposed to the solvent, Arg8 and Trp9 reorient their side chains far apart from each other and the $N^\varepsilon H$ proton of Arg8 is H bonded to the COO^- of Gly10. The same interaction has been described for the (6-10)-pentapeptide, His-Phe-Arg-Trp-Gly, in DMSO solution (14). Similar backbone-side chain ionic interactions have been observed for the ACTH (4-7) tetrapeptide, Met-Glu-His-Phe, in the solid state (15).

The data in Table I show that a pH dependent conformational transition occurs in the ACTH(1-14) molecule. The backbone coupling constants, J, of residues 4-7 decrease from the cationic to the zwitterionic state and have values close to those expected for a helical structure. Correspondingly, the temperature coefficients, S, of the same residues become less negative indicating the presence of intramolecular H bonds. J and S could not be measured for Arg8 because of extensive line broadening of the corresponding amide signal. Trp9 presents in this case its lowest values of J (7.0 Hz versus 8.2-8.7 Hz in the (5-14) and (1-10) peptides) and of S (-3.2 ppb/degree versus -5.9 and -4.2 ppb/degree). These data can reflect a conformational equilibrium including a helix around Arg8-Trp9. In fact, semiempirical conformational energy calculations carried out on the blocked ACTH(4-10) have shown that the very stable α-helix is in equilibrium with a C_7 axial conformation at Trp9 and a C_7 conformation at Gly10. This entails a reorientation of the Arg8 side chain (χ_1 changes from -60° to 180°) (16). Under the same experimental conditions (pH, concentration, temperature), ACTH(1-10) adopts an average conformation more close to a random coil in the 1-6 part and to an extended conformation in the C-terminal part (with the Gly 10 amide proton strongly H bonded). Analogous conclusions were obtained for ACTH(1-10) in trifluoroethanol solution by proton NMR (17) and circular dichroism (18).

These findings, together with similar observations for ACTH(5-14), allow some conclusions. (i) It is now well established that the helical conformation (either of α- or 3_{10}-type) which ACTH adopts in DMSO solution (as in trifluoroethanol) involves the 4-9 amino acid sequence essential for the biological activity of the hormone. (ii) The induction and stabilization of the helical structure require the presence in the peptide of both the 1-3 segment, Ser-Tyr-Ser, and of the 11-14 segment, Lys-Pro-Val-Gly. (iii) In contrast, the shorter peptides have average, random or extended conformations. (iv) Glu5 is a structurally important residue (the side chain must bear a negative charge COO^-).

All the ACTH fragments studied in this work are capable of stimulating the steroidogenesis and have relative maximum effect (with respect to the parent hormone) $a=1$, but they differ from each other by their molar potencies (2). Thus, while ACTH(1-10) is only twice as potent as ACTH(4-10), which in turn is 3 times more active than ACTH(5-10), the Ac.ACTH (1-13)-amide (i.e. a-MSH) is 15 times more potent than ACTH(4-10). This ratio is expected to be higher for ACTH(1-14), for which pharmacologic data are lacking, in view of the important role played by the charged primary amino group at position 1 on the biological activity (1). These data hint at the part taken by the 1-3 and the 11-14 segments on the affinity of these peptides for the receptor (with a larger influence of the latter segment). The data of this work indicate that the same amino acid sequences are also responsible for the differences of conformational features of the active centre 5-10 as observed in ACTH(1-14) and, respectively, in the shorter fragments (1-10), (5-14), (4-10).

The fact that part of the ACTH molecule may adopt a helical conformation is not unique, since other biologically relevant linear polypeptides of similar size, as e.g. the S-peptide of ribonuclease A (20 residues) (19) or the glucagon hormone (29 residues) (20), show pH-concentration-solvent-dependent a-helix formation in a limited part of the backbone.

The presence of a helical structure in equilibrium with other possible conformational states is nevertheless biologically significant in ACTH as proved by the relationship between the steroidogenic activity and the helical content which we found in a series of D-substituted analogs of ACTH(1-19) in which residues 2 to 18 have been replaced by the corresponding D-amino acid (10). In these, both the activity and the helix content (as measured by circular dichroism in trifluoroethanol) either simultaneously decrease with respect to the parent hormone or increase as a function of D-substitution. The curves display two minima respectively for D-Arg8 (absolute minimum) and for D-Pro12 substitution. Again, these observations can be interpreted in terms of the present conformational data on ACTH(1-14) (a similar investigation now in progress on ACTH(1-19) confirms that the 1-14 part of this potent molecule has, as expected, the same conformational behaviour as in ACTH(1-14)). A large perturbation of the helical structure is in fact expected by the change of the side chain orientation, and consequently of side chain side chain interactions, caused by D-substitution of residues belonging to the helix. A similar, although less pronounced, effect on the helix stability must occur by the stereochemical modification of the nearby residues 11-13. This changes the orientations of the different groups fitting the best both the interaction of the hormone with the receptor and its stimulation and therefore entails a decrease of the activity.

REFERENCES

1) J.Ramachandran, in "Hormonal Proteins and Polypeptides", (C.H.Li ed.), vol.2, pp.1-28, Academic Press, New York (1973).

2) R.Schwyzer, Ann.New York Acad.Sci. **297**, 3 (1977).

3) D.De Wied, Ann.New York Acad.Sci. **297**, 263 (1977).

4) D.Greff, F.Toma, S.Fermandjian, M.Löw and L.Kisfaludy, Biochim. Biophys.Acta **439**, 219 (1976).

5) F.Toma, D.Greff, S.Fermandjian, M.Löw and L.Kisfaludy, in "Peptides 1976" (A.Loffet ed.), pp. 625-631, Editions de l'Universite de Bruxelles (1976).

6) E.Nabedryk-Viala, C.Thiery, P.Calvet, S.Fermandjian, L.Kisfaludy and J.Thiery, Biochim.Biophys.Acta **536**, 252 (1978).

7) F.Toma,S.Fermandjian,M.Löw and L.Kisfaludy, Biochim.Biophys.Acta **534**, 112 (1978).

8) F.Toma,S.Fermandjian,M.Löw and L.Kisfaludy, Biopolymers **20**, 901 (1981).

9) M.Löw, L.Kisfaludy, Gy.Hajos, L.Szporny, K.Mihaly, G.B.Makara, F.Toma,V. Dive and S.Fermandjian, in "Peptides 1980" (K.Brunefeldt ed.), pp. 513-519, Scriptor, Copenhagen (1981).

10) F.Toma, V.Dive, H.Lam-Thanh, F.Piriou, K.Lintner, S.Fermandjian, M.Löw and L. Kisfaludy, Biochimie **63**, 907 (1981).

11) I.D.Campbell,C.M.Dobson,R.G.Ratcliffe and R.J.P.Williams, J.Magn. Reson. **31**, 341 (1978).

12) V.F.Bystrov, Progr.NMR Spectrosc. **10**, 41 (1976).

13) G.Admiraal and A.Vos, Int.J.Peptide Protein Res. **23**, 151 (1984).

14) N.Higuchi,Y.Kyogoku and H.Yajima, Biopolymers **20**, 2203 (1981).

15) G.Admiraal and A.Vos, Acta Cryst. **C39**, 82 (1983).

16) J.L.De Coen,D.Tourwe and G.Van Binst, Proc. "Peptide Forum", Cap d'Agde, France, Sept. 24-28 (1984).

17) B.J.Rawson,J.Feeney,B.J.Kimber and H.M.Greven, J.Chem.Soc. Perkin Trans.II 1471 (1982).

18) F.Toma, unpublished results.

19) P.S.Kim and R.L.Baldwin, Nature **307**, 329 (1984).

20) P.Y.Chou and G.D.Fasman, Biochemistry **14**, 2536 (1975).

PBB, Vol. 2
Advanced Magnetic Resonance Techniques
in Systems of High Molecular Complexity
© 1986 Birkhäuser Boston, Inc.

^{13}C AND ^{1}H NMR CHARACTERIZATION OF A $Na^{+}ClO_{4}^{-}$ COMPLEX

OF A NEW LIPOPHILIC CAGE LIGAND

P.L.Anelli[a],T.Beringhelli[b],H.Molinari[a],F.Montanari[a] and S.Quici[a]

[a]Centro CNR and Dipartimento Chimica Organica e Industriale

Università Milano, Via Golgi 19, 20133 Milano

[b]Dipartimento di Chimica Inorganica e Metallorganica

Università Milano, Via Venezian 21, 20133 Milano

The importance of macropolycyclic ligands arises from their ability to form inclusion complexes, to selectively bind substrates and to perform transport or reactions on the bound substrate (1).

The new lipophilic sodium complex (I) (Figure 1) is a very efficient anion activator in non-polar media due to its great stability constant (2). Herein we report the results of a preliminar NMR investigation aimed to fully characterize the title compound, in view of studies on conformational changes due to the electronegativity and/or polarizability of the anionic counterpart.

The aliphatic region of the ^{13}C NMR spectrum of (I) (1M in $CDCl_3$, Me_4Si as internal standard) exhibits twelve carbon resonances (Table 1), thus indicating that the molecule has an apparent C_2 symmetry. ^{13}C NMR evidence for complexation is derived from the upfield shifts of CH_2 signals of (I) with respect to those of the uncomplexed ligand.

^{13}C resonances have been assigned on the basis of $^{13}C-^{1}H$ heteronuclear correlated (3) 2D experiments (Figure 2).

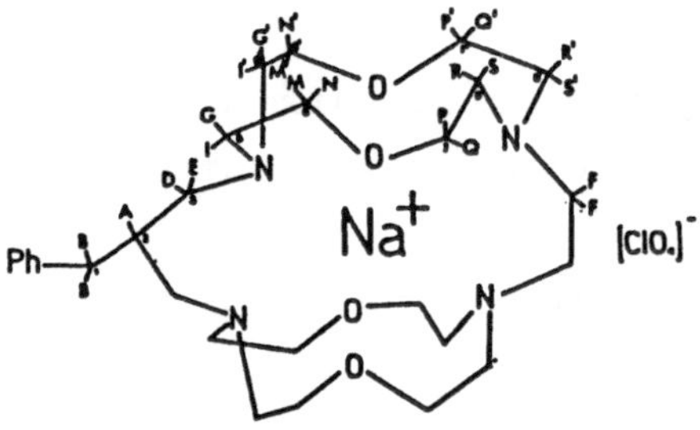

(I)

Figure 1

Table 1

$\delta^{13}C$ chemical shifts (ppm from Me_4Si) within the aliphatic region of ^{13}C NMR spectra of the ligand and of its sodium complex (I)

Ligand		Complex
38.00	C_2	37.92
42.24	C_1	41.41
54.06	$C_{8'}$	51.56
56.90	C_8	52.69
57.30	C_4	52.90
57.65	$C_{5'}$	53.74
58.70	C_5	55.00
68.77	C_3	66.65
69.71	$C_{6'}$	66.83
70.00	C_7	66.85
70.36	$C_{7'}$	67.05
72.72	C_6	68.11

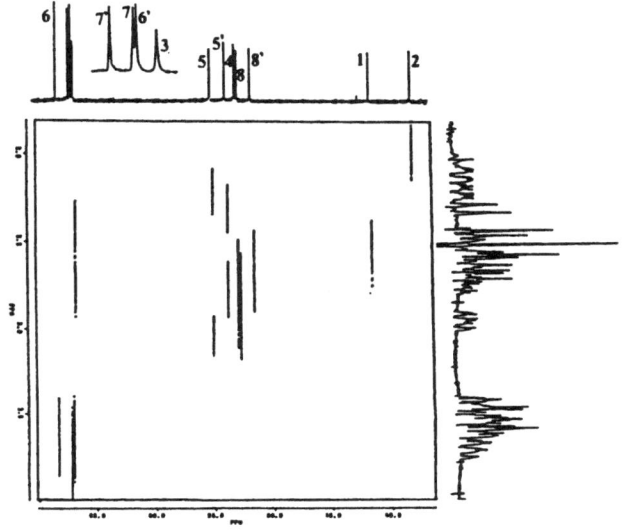

Figure 2 - 67.8 MHz ^{13}C-^{1}H shift correlated 2D NMR of (I) in CDCl$_3$
(0.1 M) (aliphatic region). Parameters: X_1=512, X_2=4K, NS=
64, N_1=1K, N_2=8K, W_1=365 Hz, W_2=3000 Hz. Window functions:
sine-bell in T_2 and sine-bell squared shifted /10 in T_1

It is worth noting the unusual downfield shift of C_3 (Fig.1) which might
be ascribed to conformational strains. In order to elucidate this point
molecular mechanics calculations are in progress.

^{1}H Homonuclear correlated (4) (Figure 3) and ^{1}H J resolved (5) 2D experiments allowed the recognition of the resonances due to different types
of O-CH$_2$-CH$_2$-N fragments and of those related to the CH$_2$ of the bridging
chains (Fig.3).

^{13}C relaxation times together with NOE measurements showed that dipole-
dipole is the unique relaxation mechanism. The calculated rotational

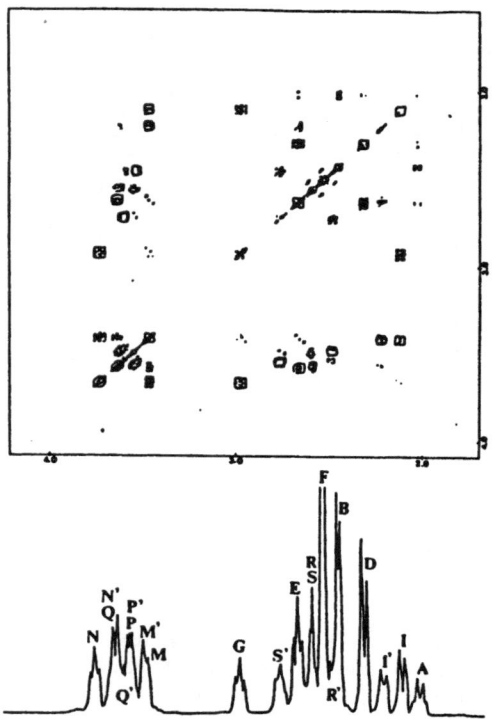

Figure 3 - 500 MHz ^{1}H shift correlated 2D NMR of (I) in CDCl$_3$ (0.1 M) (aliphatic region). Parameters: X$_1$=512, X$_2$=2K, NS=16, N$_1$= 1K, N$_2$=2K, W$_1$=1582.28 Hz, W$_2$=3164.56 Hz, recycle delay 2 sec. Window functions: sine-bell in both dimensions

correlation times (τ_c = 0.77x10^{-10} s) are the same for all the carbons, thus indicating that molecular reorientation is isotropic.

REFERENCES

1)   J.M.Lehn, Acc.Chem.Res. **11**, 49 (1978).

2)   P.L.Anelli,F.Montanari and S.Quici, JCS,Chem.Commun. 132 (1985).

3)   A.A.Maudsley,L.Müller and R.R.Ernst, J.Magn.Reson. **28**, 463 (1977);

G.Bodenhausen and R.Freeman, J.Magn.Reson. **28**, 471 (1977);

G.Bodenhausen and R.Freeman, J.Amer.Chem.Soc. **100**, 320 (1978).

4)   A.Bax and R.Freeman, J.Magn.Reson. **42**, 164 (1981);

A.Bax and R.Freeman, J.Magn.Reson. **44**, 542 (1981).

5)   W.P.Aue,J.Karhan and R.R.Ernst, J.Chem.Phys. **64**, 4226 (1976).

PBB, Vol. 2
Advanced Magnetic Resonance Techniques
in Systems of High Molecular Complexity
© 1986 Birkhäuser Boston, Inc.

B AND Z FORMS OF A MODIFIED d(C-G) SEQUENCE IN SOLUTION

L.P.M.Orbons,G.A.van der Marel,J.H.van Boom and C.Altona

Gorlaeus Laboratories, State University of Leiden

P.O. Box 9502, NL 2300 Leiden, The Netherlands

Since the discovery of the "high salt form" of DNA (1,2) (which was baptized Z DNA) there has been an increasing interest in the structural and dynamical properties of this Z form (3-6) and its possible biological role (7,8). Extensive studies on the polymorphism of DNA (9,10) have revealed that, for instance $d(C-G)_n$ sequences in high salt and/or alcohol concentrations favor the Z form above the "classical" B helix. It was found that methylation or bromination of cytosine at the C-5 position also has a stabilizing effect on the Z structure (10,11). For example the DNA hexamer $d(m5C-G)_3$ in 100 % ^2H_2O adopts a normal B helix. Addition of $MgCl_2$ and/or methanol causes dramatic changes in the ^1H-NMR spectrum. In a 5 mM $MgCl_2$ 30% $C^2H_3O^2H$/70% ^2H_2O solution the DNA fragment occurs in an equimolar mixture of the B and Z forms, which are in slow exchange on the NMR time scale. By increasing the $MgCl_2$ and/or the methanol concentration, it was found possible to force the hexamer into essentially pure Z form.

X-ray crystallography on single crystals of the DNA hexamer $d(m5C-G)_3$ showed pronounced differences between the Z and B form in the solid state (12). The most important differences can be summarized as follows; B DNA has an S-S-S-S-... sequence of sugar-ring conformers with all the bases adopting the anti conformation (see figure 1), whereas Z DNA has an alternating S-N-S-N-... sequence of sugar-ring conformers with the dC residues S, anti, and the dG residues N, syn.

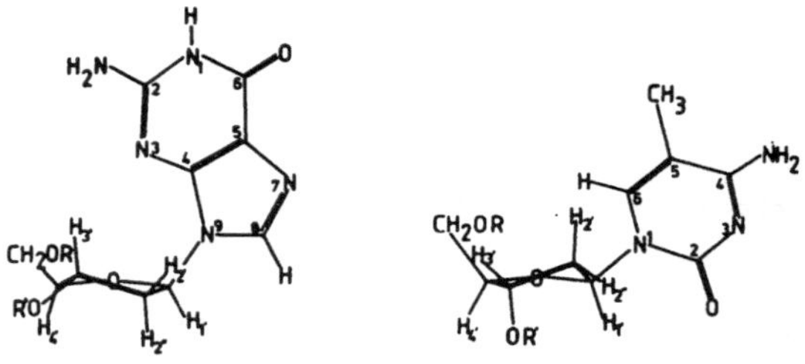

GUANINE [syn]
N-conformer (P=0°)

5-METHYL CYTOSINE [anti]
S-conformer (P=180°)

Figure 1

Thus far, the geometrical details of the Z form in solution have not been studied thoroughly. One of the most powerful methods in the conformational analysis of oligonucleotides in solution is NMR spectroscopy. By means of (2-dimensional) NOE spectroscopy it is possible to discriminate between the three major species of DNA, i.e. A, B and Z DNA (13, this work). 2-Dimensional NOE spectroscopy (NOESY), in combination with correlation spectroscopy (COSY), can at the same time be used for the assignment of the ^{1}H-NMR spectrum. The assignment strategy for the B form has already been described extensively in the literature (14,15). With the aid of COSY all coupled spin systems (i.e. sugar-ring protons belonging to one single residue and the H-6/5-CH$_3$ protons of each cytidine) can be distinguished. For example, via the NOEs between the base protons of residue (n) and the H-1' of residues (n) and (n-1), the sequential assignment can be completed (see figure 2, right-hand side). In Z DNA a completely different NOE pattern is observed. Similar as in the B duplex, the Z form exhibitis internucleotide NOEs between the base protons of residues dm5C(n) and the H-1' protons of the residues dG(n-1). However, there are no such internucleotide NOEs between the base protons of residues dG(n) and dm5C(n-1). There exists only a very strong intranucleotide NOE between the H-8 and the H-1' of the dG residues (figure 2, left-hand side).

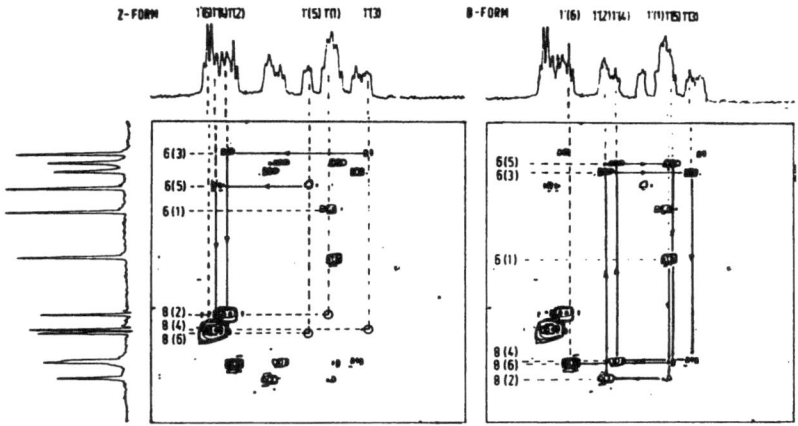

Figure 2 - Part of the NOESY, showing the NOEs between the base protons and the H-1' protons of d(m5C-G)$_3$ in the B$_2$ (right) and Z$_2$ form (left) (10mM DNA, 5 mM MgCl$_2$, 30% C^{2}H$_3$O^{2}H/70% ^{2}H$_2$O, p^{2}H 7.1, 290 K)

This is more clearly shown in figure 3. Such a strong NOE is indicative for a syn position of the guanine base in Z DNA. Notwithstanding the difficulty described above, a continuous mapping of the ^{1}H resonances of the Z form is still possible via an alternative route, thanks to short internucleotide proton-proton distances between the H-8 of dG(n) and the H-4',H-5', H-5" of residues dm5C(n-1) (figure 4). Fortunately, the proton resonances of the H-4', H-5' and H-5" of the dm5C residues in the Z form show, compared to those in the B form, a pronounced upfield shift (up to 1.7 ppm), which makes these proton signals easy to distinguish among the other H-4', H-5' and H-5" protons resonances. The final assi-

gnment of the ^{1}H-NMR spectrum of the equimolar B/Z mixture of the hexamer d(m5C-G)$_3$ is shown in figure 5.

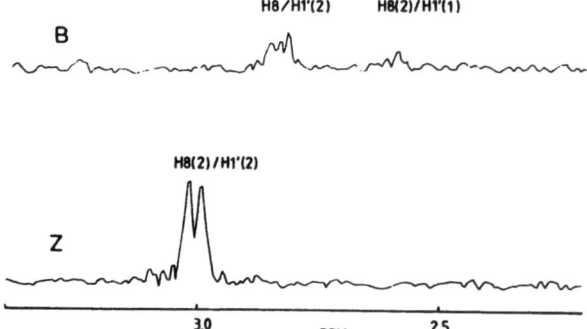

Figure 3 – Cross sections of the NOESY, showing the NOE peaks between
the H-8(2) and the H-1'(2) in the B and Z form (conditions
as in fig.2)

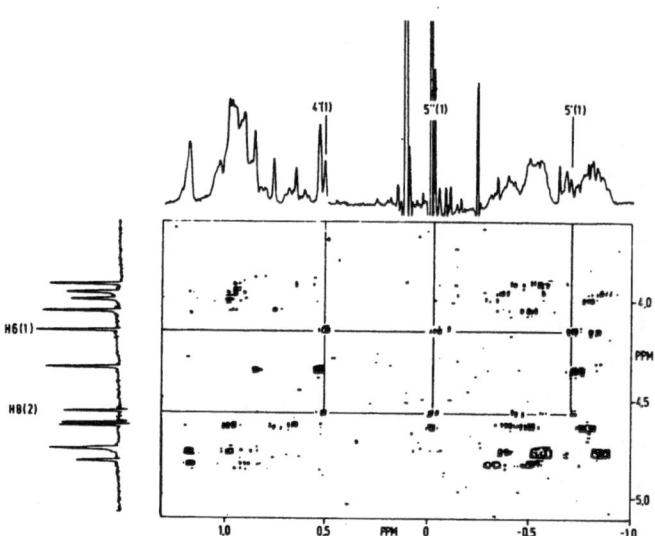

Figure 4 – Part of the NOESY, showing the NOE cross peaks between the
H-8(2) and the H-4'(1), H-5'(1) in the Z form (conditions
as in fig.2)

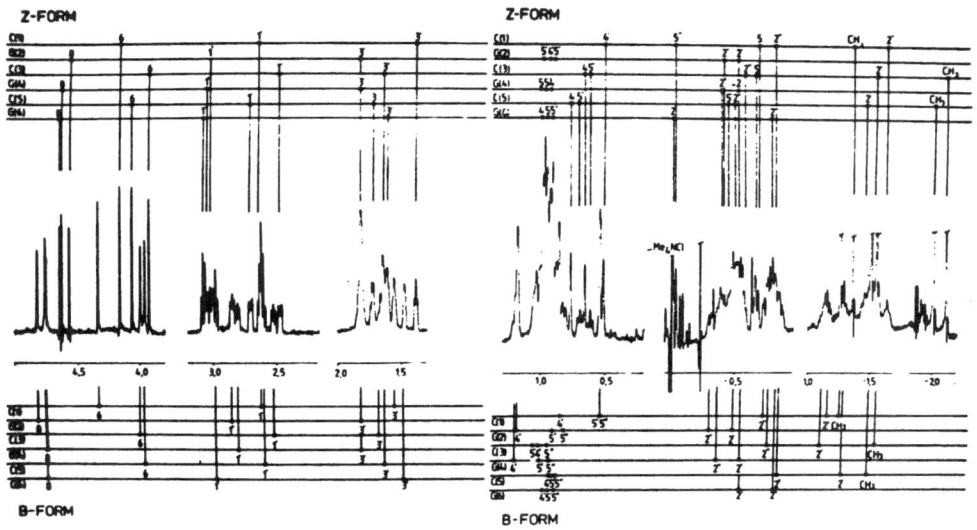

Figure 5 - Final assignment of the B and Z forms of d(m5C-G)$_3$ (conditions as in fig.2)

A second method for the study of geometrical details of oligonucleotides

is a coupling constant analysis (15). By means of a generalized Karplus

type equation, three-bond proton-proton coupling constants can be trans-

lated into the torsional angle around the H-C-C-H bond. With the aid of

the program PSEUROT (16) the sugar-ring conformation can be described in

terms of percentages of N and S conformers. Figure 6 shows the percenta-

ges N and S, based on the J1'2', J1'2" and on the width of the H-2' and

H-2" proton resonances, for each residue of d(m5C-G)$_3$ in the B and Z

form. It can be seen that in the B form all sugar rings prefer the S

conformer, but not exclusively, except the 3' terminal residue which has more conformational freedom. However, in the Z form only the dC residues adopt the S conformer (in the same extent as in the B form), the dG residues (2) and (4) prefer the N conformer (up to 100%). The sugar ring of the terminal dG(6) residue behaves the same as in the B form, approximately 67% S.

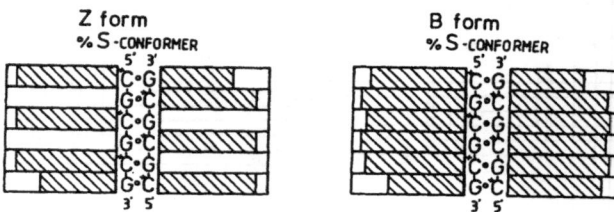

Figure 6 – Percentages of N and S conformers for each residue in the B and Z form at 295 K

It was already shown by the 2D-NOE experiment that in solution the bases of the dG residues in the Z form adopt the syn position and those of the dC residues the anti conformation. Hence it can be concluded that major distinct geometrical features of Z DNA in the crystal are maintained in solution.

REFERENCES

1) F.M.Pohl and T.M.Jovin, J.Mol.Biol. **67**, 375 (1972).

2) A.H.-J.Wang,G.J.Quigley,F.J.Kolpak, J.L.Crawford, J.H.van Boom,G.A.

van der Marel and A.Rich, Nature **282**, 680 (1979).

3) H.R.Drew and R.E.Dickerson, J.Mol.Biol. **152**, 723 (1981).

4) M.Leng,B.Hartmann,B.Malfoy,J.Pilet and M.Ptak, "Nucleic Acids: The Vectors of Life", (B.Pullman and J.Jortner Eds), Reidel Publishing Company, 1983, p.49.

5) J.A.Cavailles,J.-M.Neumann,J.Taboury,B.Langlois d'Estaintot,T.Huynh- -Dinh, J.Igolen and S.Tran-Dinh, J.Biomol.Struct. Dyns. **1**, 1347 (1984).

6) A.H.-J.Wang,G.J.Quigley,F.J.Kolpak, G.A.van der Marel, J.H.van Boom and A.Rich, Science **211**, 171 (1981).

7) J.Klysik,S.M.Stirdivant,J.E.Larson,P.A.Hart and R.D.Wells, Nature **290**, 672 (1981).

8) A.Nordheim and A.Rich, Nature **303**, 674 (1983).

9) J.H.van de Sande and T.M.Jovin, The EMBO J. **1**, 115 (1980).

10) M.Behe and G.Felsenfeld, Proc.Natl.Acad.Sci. USA **78**,1619 (1983).

11) C.-W.Chen,J.S.Cohen and M.Behe, Biochemistry **22**, 2136 (1983).

12) S.Fujii, A.H.-J.Wang, G.A.van der Marel, J.H.van Boom and A.Rich, Nucl.Acids Res. **10**, 7979 (1984).

13) C.A.G.Haasnoot,H.P.Westerink,G.A.van der Marel and J.H.van Boom, J.Biomol.Struct.Dyns. **2**, 3445 (1984).

14) R.M.Scheek,R.Boelens,N.Russo,J.H.van Boom and R.Kaptein, Biochemistry **23**, 1371 (1984).

15) C.Altona, Recl.Trav.Chim.Pays-bas **101**, 413 (1982).

16) F.A.A.M.de Leeuw and C.Altona, J.Comp.Chem. **4**, 428 (1983).

PBB, Vol. 2
Advanced Magnetic Resonance Techniques
in Systems of High Molecular Complexity
© 1986 Birkhäuser Boston, Inc.

NMR STUDY OF A TRIDECANUCLEOTIDE CONTAINING A NO-BASE RESIDUE

S.Pochet,T.Huynh-Dinh,J.M.Neumann,S.Tran-Dinh and J.Igolen

Unité de Chimie Organique, Institut Pasteur

28 rue du Dr. Roux, 75724 Paris Cedex France

Service de Biophysique, C.E.N. Saclay

91191 Gif-sur-Yvette Cedex France

INTRODUCTION

The structure of the tridecadeoxynucleotide d(CGm^{5}CGCGxACATGT) has been investigated in aqueous solution (0.1 M NaCl) by 500 MHz ^{1}H-NMR. In this molecule, the B and Z promoters (d(ACATGT) and d(CGm^{5}CGCG) respectively) are separated by a residue without base d(x) (1-cyano-2-deoxy-β-D-ery-thropentofuranose)(1). One might expect that a single no-base site per-mits the B and Z helices to coexist in this sequence.

RESULTS

Figure 1 shows the 500 MHz spectrum of d(CGm^{5}CGCGxACATGT) base protons at 69°C. Also shown in this figure is the spectral region where the apu-rinic residue H1' proton signal is located and easily identified by its triplet structure. The lack of base deshielding effect explains the low chemical shift value (4.5 ppm) of this proton compared to those of the other H1' proton signals (between 6.4 and 6.0 ppm). Base proton assign-ment was performed i) by comparing at high temperature (90°C) the 13-mer spectrum with those of the corresponding hexamers, d(CGm^{5}CGCG) (2) and d(ACATGT) (3) ii) by recording a 2D-NOESY spectrum in order to study the

intra- and inter-base NOEs; this experiment was performed at 45°C where
the signals of the three methyl group protons (those of the methylated
cytidine and of the two thymidines) are well resolved.

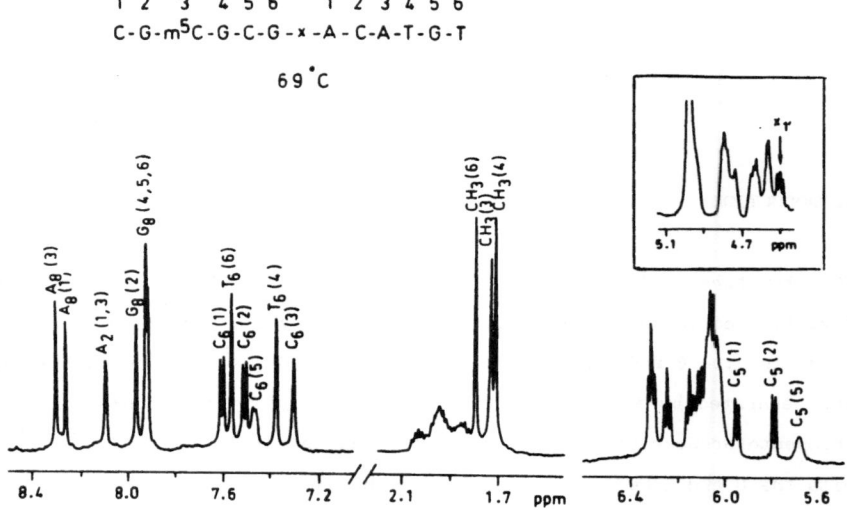

Figure 1 - 500 MHz ^{1}H-NMR spectrum of the 13-mer (2mM, 0.1 M NaCl) at
69°C

In general, the differences in chemical shift between 13-mer and hexamer
protons are slight: the most signficant variations involve the two residues surrounding the apurinic residue (dC, dG on the left; dA, dC on the
right with respect to the 5'-3' direction); these variations are in the
range 0.03 to 0.07 ppm.

On lowering the temperature, the base proton signals of the 13-mer are shifted to higher field with various amplitudes depending on the residue whereas new resonances appear when t° is less than 50°C. These new resonances are poorly resolved in the aromatic region because of their low intensities, the large number of signals already present and their increasing linewidths but a new resonance is clearly observable in the methyl proton region (figure 2). This additional resonance is located at

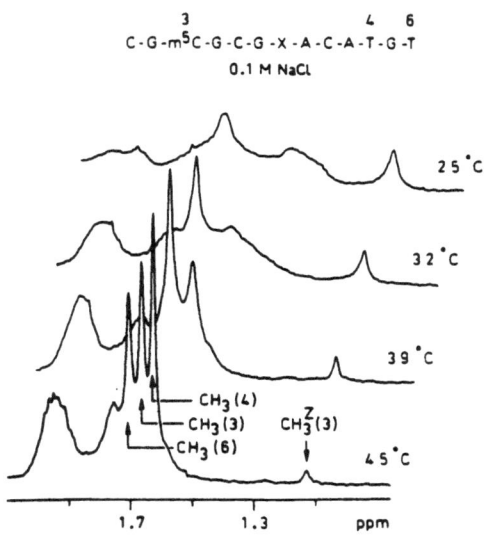

Figure 2 - Temperature effect on the 13-mer methyl protons

higher field with respect to the three CH_3 signals of methylated pyrimidines and its intensity increases on lowering the temperature suggesting the occurrence of a slow exchange process (at the ^{1}H-NMR time scale). It is now well known that 2D-NOESY experiments provide useful information on slowly exchanging species: the longitudinal magnetizations of two signals corresponding to the same proton in two exchanging forms are coupled together and off-diagonal cross peaks connecting these signals can be observed in a 2D-NOESY spectrum (4,5). Figure 3 shows the (1.0-2.2) ppm part of the 13-mer 2D-NOESY spectrum recorded at 45°C.

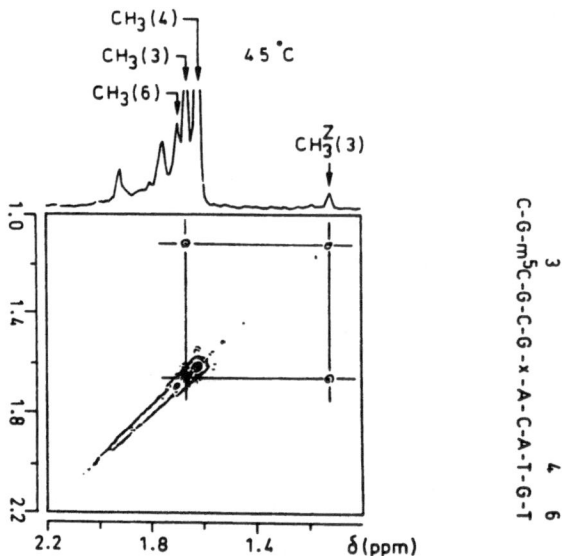

Figure 3 - Contour plot of the 13-mer NOESY spectrum (CH_3 protons) at 45°C

A cross peak connecting the additional resonance to the $d(m^5C)$ CH_3 signal is observed whereas no connectivities involving the CH_3 signals relative to the two thymidines are detected.

The chemical shift variations versus temperature of the major signals of the 13-mer base protons display a sigmoidal form (when the variations are sufficiently significant) and are very similar to the corresponding curves observed for the B helix-coil transition of d(CGm^{5}CGCG)(2) and d(ACATGT) (3) in 0.1 M NaCl solutions. As in the case of high temperatures, at room temperature some chemical shift differences between the 13-mer and the corresponding hexamers are observed for the residues surrounding the apurinic site; the two extreme residues are also affected. Moreover the linewidths of the 13-mer proton signals are much larger than those usually observed for the self-complementary oligonucleotides. These results show that the d(CGm^{5}CGCG) and d(ACATGT) parts of the 13mer undergo a B helix-coil transition as their corresponding self-complementary hexamers; this duplex formation occurs via a base pairing as shown in the following scheme:

$$..T\ CGm^5CGCGxACATGT\ CGm^5CG..$$
$$..AxGCGm^5CGC\ TGTACAxGCGm^5C..$$

However the observed additional resonances demonstrate via 2D-NOESY experiment the occurrence of another structure in slow equilibrium with the B form. The additional d(m^{5}C) CH$_3$ resonance is located at 1.1 ppm (fig.2) i.e. the same value as that observed for the same protons in the Z duplex of d(CGm^{5}CGCG) in high salt concentration (2). This value (0.5-0.6 ppm lower than in the B form) is one of the most significant features of Z proton spectra and has also been observed for the methylated pyrimidine (dm^{5}C,dT) CH$_3$ signals in the following Z duplexes: d(m^{5}CGCG-m^{5}CG)(6), d(C2aminoACGTG), d(m^{5}CGCAm^{5}CGTGCG), d(CGCAm^{5}CGTGm^{5}CG)(7). This shift has been explained theoretically by Giessner-Prettre et al.(8). Since in the 13-mer, the two thymidines of the d(ACATGT) part do not exhibit exchanging CH$_3$ signals and considering that the elementary unit of Z helices is a pyrimidine-purine dinucleotide, the present results show that the d(CGm^{5}CGCG) part of the 13-mer is involved in a B-Z like equilibrium whereas the d(ACATGT) part only adopts the B conformation.

Signal integration shows that the Z-like form represents 30% of the d(CGm^{5}CGCG) duplexes at room temperature. The slope of the curve Log((B)/(Z)) versus reciprocal absolute temperature (where (B) and (Z) are the B and Z proportions determined by signal integration) gives a Z-B transition enthalpy of about 13 Kcal/mole. This value obtained in a 0.1 M NaCl solution is greater than that observed for the Z-B equilibrium of d(CGm^{5}CGCG)(2) and d(m^{5}CGCGm^{5}CG) (6) in 2 M NaCl solutions (7.5 and 8 Kcal/mole respectively).

DISCUSSION

Our results justify the choice of the 13-mer sequence and confirm our prediction: the B and Z helices can coexist within the same duplex via a single apurinic residue. Examination of the base pairing process of the 13-mer (see the above scheme) shows that each of the two duplex units d(ACATGT) and d(CGm^{5}CGCG) are surrounded by a missing residue on one side and by a no-base residue on the other. Such alterations -of biological importance- involving one base pair lead to a flexible junction between B and Z forms in a appropriate base sequence.

The occurrence of a Z form observed at low ionic strength (0.1 M NaCl) necessitates an explanation since in the case of the d(CGm^{5}CGCG) hexamer (2) formation of a Z duplex requires a higher salt concentration (at least 1 M). In fact, the 13-mer base pairing process leads to a pseudo polymerization and the increase of the chain length favors the Z form (poly d(m^{5}CG) adopts a Z form in 0.1 M NaCl solution (9)). However, by contrast to the polynucleotide, the enthalpy variation of the B-Z transition is clearly positive (13 Kcal/mole).

REFERENCES

1) S.Pochet, T.Huynh-Dinh, J.M.Neumann, S.Tran-Dinh, J.A.Taboury , E.Taillandier and J.Igolen, Tetrahedron Lett. 2085 (1985).

2) J.A.Cavaillès, J.M.Neumann, J.A.Taboury, B.Langlois D'Estaintot, T. Huynh-Dinh,J.Igolen and S.Tran-Dinh, J.Biomol.Struct.Dyns **1**, 1347 (1984).

3) S.Tran-Dinh, J.M.Neumann, T.Huynh-Dinh, B.Genissel, J.Igolen and G. Simonnot, Eur.J.Biochem. **124**, 415 (1982)

4) J.Jeener,B.H.Meier,P.Bachmann and R.R.Ernst, J.Chem.Phys. **71**, 4546 (1979).

5) J.Feigon,A.H.-J.Wang,G.A.van der Marel,J.H.van Boom and A.Rich, Nucl.Acids Res. **12**, 1243 (1984)

6) S.Tran-Dinh, J.A.Taboury, J.M.Neumann, T.Huynh-Dinh, B.Genissel, B. Langlois d'Estaintot and J.Igolen, Biochemistry **23**, 1362 (1984).

7) J.A.Cavaillès,J.M.Neumann,T.Huynh-Dinh,B.Langlois d'Estaintot and J.Igolen, Eur.J.Biochem. **147**, 183 (1985).

8) C.Giessner-Prettre,B.Pullman,S.Tran-Dinh,J.M.Neumann,T.Huynh-Dinh and J.Igolen, Nucl.Acids Res. **12**, 3271 (1984).

9) M.Behe and G.Felsenfeld, Proc.Natl.Acad.Sci USA **78**, 1619 (1981).

PBB, Vol. 2
Advanced Magnetic Resonance Techniques
in Systems of High Molecular Complexity
© 1986 Birkhäuser Boston, Inc.

DYNAMIC NUCLEAR MAGNETIC RESONANCE OF CAPRIMIDYL SULFATE

G.Gurato[a], G.Nadali[a] and M.F.Coletta[b]

[a]CENTRO RICERCHE MONTEDIPE

Stabilimento Petrolchimico, P.Marghera, Venezia

[b]Centro di Studio sugli Stati Molecolari Radicalici ed Eccitati

c.o.Dipartimento di Chimica Fisica dell'Università

via Loredan 2, Padova

INTRODUCTION

The capramidyl sulfate (CLS), an intermediate product in the production

of caprolactam (1), was prepared by reacting a caprolactam solution in

CCl_4 with gaseous SO_3 (2,3). The ^1H and ^{13}C NMR spectra in C_6D_5Cl at 80

and 20 MHz were examined in the temperature range from -50° to 130°C to

confirm the following molecular structure for CLS:

$$
\begin{array}{cccc}
3 & 2 & 1 & O \\
 & & & \parallel \\
\diagup CH_2{-}CH_2{-}C{-}O{-}S{-}OH \\
CH_2{-}CH_2{-}CH_2{-}N \quad\quad O \\
4 & 5 & 6 &
\end{array}
\qquad (I)
$$

^1H SPECTRA

Using TMS as a reference, the assignments of the different peaks of the

^1H spectrum for a 0.3 M solution in $CDCl_3$ were carried out (Table I).

The presence of δ_4 peak indicates that the proton of the $-SO_3H$ group ex-

changes with the nitrogen atom and suggests a dynamic equilibrium betwe-

en structures (I) and (II)

$$\underset{4}{CH_2}-\underset{5}{CH_2}-\underset{6}{CH_2}-\overset{+}{N}-H$$

with structure showing:

positions 3 2 1 and 4 5 6:

$$\underset{3}{CH_2}-\underset{2}{CH_2}-\underset{1}{C}-O-\overset{O}{\underset{O}{S}}-O^-$$

(II)

Table I

Chemical shifts ppm	Proton number	Assignments
1 = 1.82	6	$-(CH_2)_3-$ (3,4,5)
2 = 2.75	2	$-(CH_2)-CO$ (2)
3 = 3.55	2	$-(CH_2)-N$ (6)
4 = 9.79		$=\overset{+}{N}H-$
5 = 14.25	1	$-SO_3H$

Numbers 2-6 refer to protons attached to the corresponding carbon nuclei

The analysis of ^1H spectra for CLS in 0.3 M C_6D_5Cl solutions at the various temperatures indicates an equilibrium (eq.(1))among at least three structures [1]at low temperatures (-50°C).

$$[1] \quad R-SO_3H \rightleftharpoons \text{(I)} \rightleftharpoons \text{(II)} \rightleftharpoons \text{(III)}$$

(I)   (II)   (III)

Only structure (III) presents a hindered rotation around the C-O bond due to the intramolecular hydrogen bond. The presence of the three forms

Table II

Chemical shifts and assignments of the lines in the
^{1}H spectra of CLS in C_6D_5Cl at different temperatures

T °C	$-(CH_2)_3$	CH_2-CO	CH_2-N	CH_2-N	$=N^+H$	$-SO_3H$
-50.2	1.02	2.06	3.12	3.68	9.98-10.19	14.21
-39.9	1.02	2.05	3.11	3.68	9.98-10.19	14.21
-29.6	1.03	2.08	3.09	3.67	9.95-10.18	14.21
-19.3	1.04	2.08	3.10	3.67	9.48-10.18	14.21
- 9.0	1.06	2.10	3.05	3.59-3.71	9.90-10.19	14.25
1.3	1.07	2.04-2.10	3.00	3.57-3.68	9.85-10.22	14.28
11.6	1.06	2.00	2.91	3.53-3.63	9.80-10.25	14.35
28.0	1.14	2.11-2.18	3.04	3.58-3.69	9.81-10.07	14.25
39.5	1.12	2.04-2.12	2.91	3.53-3.65	9.94	14.36
50.0	1.16	2.08-2.16	2.97	3.55-3.66	9.89	14.31
60.4	1.20	2.14-2.21	3.03	3.58-3.69	9.83	14.26
70.9	1.22	2.16-2.23	3.03	3.58-3.69	9.83	14.26
81.4	1.24	2.16-2.23	3.01	3.58-3.65	9.81	14.20
91.8	1.24	2.17-2.25	3.02	3.62	9.96	14.23
102.3	1.26	2.18-2.27	3.03	3.63	10.27	13.89
112.7	1.27	2.17-2.27	3.04	3.59	12.49	
123.2	1.29	2.17-2.27	3.12	3.46	11.89	
133.7	1.30	2.18-2.28	3.23		11.90	
144.1	1.31	2.18-2.29	3.25		11.87	

is evidenced by the 14.21 ppm peak of $-SO_3H$ as well as by the presence
of the $=N^+-H$ peaks at 9.98 and 10.19 ppm (Table II). The free rotation

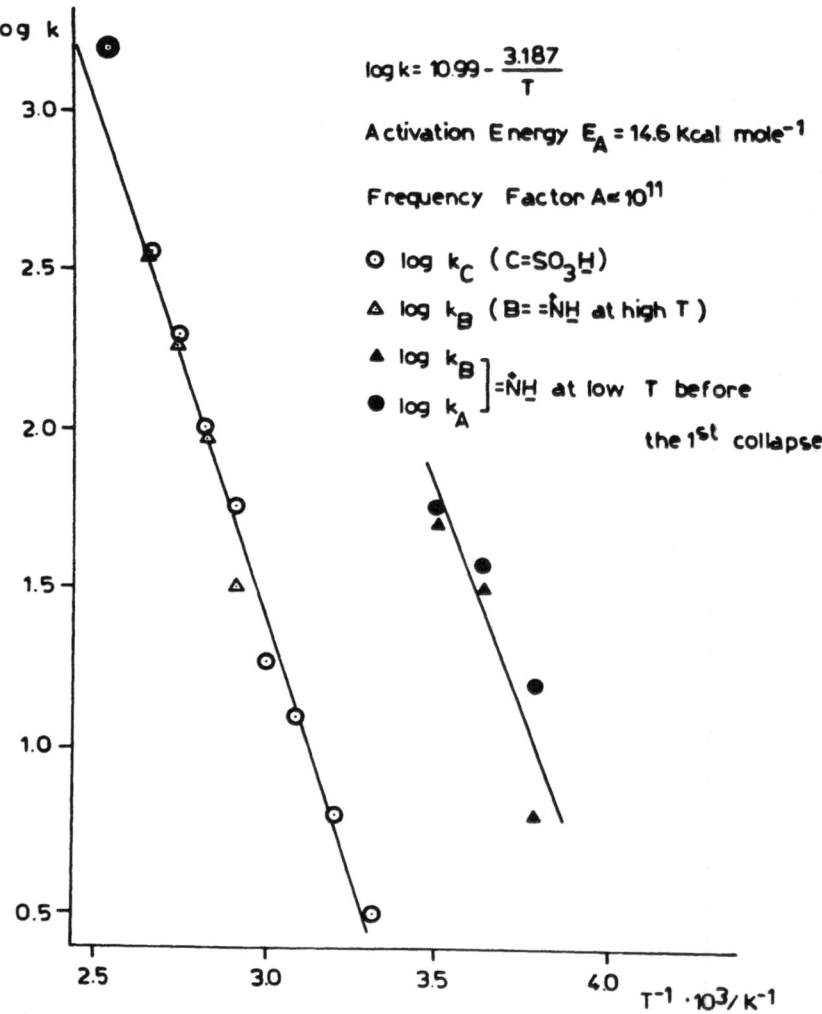

Figure 1

around the C$\overset{\downarrow}{C}$O-S bond increases with temperature, giving rise to structure (II) still in thermodynamic equilibrium with structure (I). A first
collapse temperature can be observed at $\simeq 40°C$ (δ = 9.94). At high temperatures, there occurs a fast exchange of the proton between $-SO_3H$ and
$=N-CH_2$ groups, meant as an intermediate system between (I) and (II). The
collapse temperature is around 123°C; at higher temperature further sha-
rpening of the proton line at $\simeq 11.9$ ppm, goes on. At 144°C the sharpe-
ning is not complete ($\Delta\nu_{\frac{1}{2}}$=74 Hz), but the temperature cannot be raised
further due to the solvent b.p. Owing to some uncertainties in measurement, the collapse can be better followed observing what happens to the
resonances of the CH_2-N group. The collapse of the two c.s. of the CH_2-N
group occurs at 123°C at δ=3.23 ppm (table II). The proton residence times in the three sites were calculated on the basis of the width of the
proton NMR lines. By plotting the log of the rate constants vs.$10^3/T$
(K^{-1}), an activation energy of 15 kcal/mole and a frequency factor of
$\approx 10^{11}$ were obtained (figure 1).

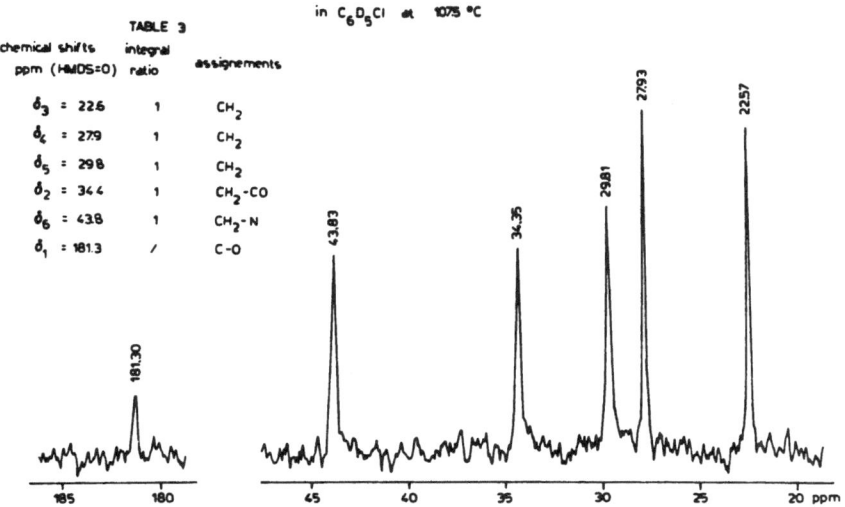

Figure 2 - ^{13}C NMR spectrum of CLS in C6D5Cl at 107.5°C

^{13}C SPECTRA

The ^{13}C spectra of a 0.7 M CLS solution in C_6D_5Cl were recorded for a temperature range from -50° to 108°C. At high temperatures a 6-line spectrum was obtained (figure 2). The line number was 12 at 29°C and 15 at temperatures lower than 29°C. The c.s. and the assignments for CLS resonances in C_6D_5Cl at 108°C were determined. The -50°C spectrum displays three c.s. for the C_1 of the three different structures. The 15 li-

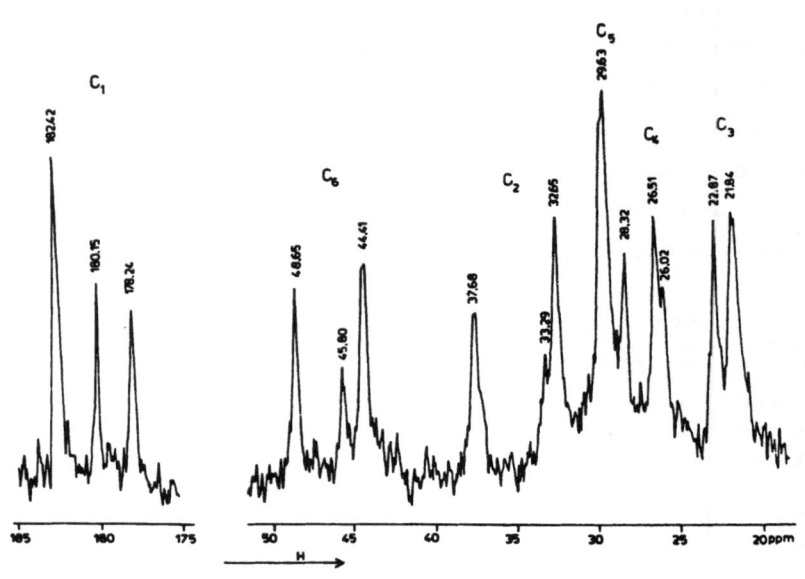

Figure 3

nes and the 3 C-O species (figure 3) agree with the findings of ^1H NMR spectroscopy, that is, the occurrence of equilibrium (1) at low temperature. The 12-lines spectrum at 29°C is characterized by two C-O species, which corresponds to the equilibrium between forms (I) and (II). The 6-line spectrum at 108°C cannot be assigned to structure (I) or to structure (II), but to an intermediate system.

REFERENCES

1)   A.F.Turbak, Ind.Eng.Chem.Prod.Res.Develop. **7**, 189 (1968).

2)   G.Nadali, Tesi di Laurea in Chimica, Università di Padova, 1983/84.

3)   L.Giuffrè,G.Sioli. and E.Losio, Chim.Ind. **50**, 983 (1968).

PBB, Vol. 2
Advanced Magnetic Resonance Techniques
in Systems of High Molecular Complexity
© 1986 Birkhäuser Boston, Inc.

SOLVENT PROTON NUCLEAR MAGNETIC RELAXATION DISPERSION (NMRD)

IN SOLUTIONS OF PARAMAGNETIC MACROMOLECULES

Ivano Bertini, Fabrizio Briganti, and Claudio Luchinat

Department of Chemistry, University of Florence,

Via G. Capponi 7, 50121 Florence, ITALY

1. INTRODUCTION

It has long been realized that the theory describing nuclear relaxation in the presence of unpaired electron(s)-nucleus coupling, while approximately correct for systems where the correlation time for the interaction is determined by molecular tumbling, breaks down in slow rotating systems for a number of reasons (1-8). This inadequacy has severely limited the use of nuclear relaxation data as a source of quantitative structural and dynamic information on paramagnetic macromolecules of biological interest (9,10). One important aspect of this issue would be the understanding of water proton longitudinal relaxation rates as a function of magnetic field (NMRD) in solutions of paramagnetic metalloproteins. With no exceptions, the paramagnetic contribution to such rates is predominantly point-dipolar in origin (4, 11) and therefore potentially informative on the geometry of the water proton-metal ion interaction.

The classic theory for electron-nucleus coupling will be reviewed and all the pitfalls underlined. Then, it will be shown how a theoretical approach recently developed by us to include the proper spin-Hamiltonian terms of the system is able to overcome most of the difficulties and to rationalize most of the experimental data.

The deeper understanding of the influence of the electronic parameters on nuclear relaxation allows us also to propose a phenomenological approach to electron—nucleus coupling in the non-Redfield limit; finally, it will be shown how a knowledge of the factors influencing nuclear relaxation in exchange-coupled systems can lead to the design of modified metalloenzymes where a copper(II) ion can act as an NMR shift reagent.

2. PARAMETERS INVOLVED IN ELECTRON—NUCLEUS COUPLING

The interaction of unpaired electrons with nuclear spins causes an increase of the longitudinal and transverse relaxation rates of the latter. The paramagnetic enhancements of the rate constants are defined as (7,8,12):

$$T_{1p}^{-1} = T_1^{-1} - T_{1(dia)}^{-1} \qquad (1)$$

$$T_{2p}^{-1} = T_2^{-1} - T_{2(dia)}^{-1} \qquad (2)$$

where T_1^{-1} and T_2^{-1} are the nuclear relaxation rates in the system of interest and $T_{1(dia)}^{-1}$ and $T_{2(dia)}^{-1}$ are the nuclear relaxation rates in its diamagnetic analogue. T_{1p}^{-1} and T_{2p}^{-1} are thus the paramagnetic contributions to the relaxation rates. These quantities reflect the average behavior of an ensemble of dynamically equivalent nuclei. If each nucleus in the ensemble experiences the same kind of interaction with the paramagnetic center in a non labile chemical species, then T_{1p}^{-1} and T_{2p}^{-1} are a direct measure of the paramagnetic effects, which are termed T_{1M}^{-1} and T_{2M}^{-1}. On the other hand, if only a small fraction, f_M, of nuclei interacts with the paramagnetic center by forming labile adducts of lifetime τ_M, then the bulk nuclei still experience paramagnetic relaxation enhancements given by:

$$T_{1p}^{-1} = f_M (T_{1M} + \tau_M)^{-1} \tag{3}$$

$$T_{2p}^{-1} = (f_M/\tau_M) \left[T_{2M}^{-2} + T_{2M}^{-1} \tau_M^{-1} + \Delta\omega_M^2 / (T_{2M}^{-1} + \tau_M^{-1})^2 + \Delta\omega_M^2 \right] \tag{4}$$

where $\Delta\omega_M$ is the difference in the chemical shifts experienced by the nucleus in the paramagnetic and in the bulk diamagnetic environments; for short τ_M values ($\tau_M^{-1} \gg T_{1M}^{-1}$, T_{2M}^{-1}, $\Delta\omega_M$) the above equations reduce to:

$$T_{1p}^{-1} = f_M T_{1M}^{-1} \tag{5}$$

$$T_{2p}^{-1} = f_M T_{2M}^{-1} \tag{6}$$

and the paramagnetic effects can still be directly obtained. Extraction of T_{1M}^{-1} and T_{2M}^{-1} from experimental quantities may lead to relevant chemical information on the system.

There are two different contributions to both T_{1M}^{-1} and T_{2M}^{-1}, that are dipolar and contact in origin. The first is particularly suited to provide structural information since it arises from a through-space interaction and thus depends on the reciprocal sixth power of the electron-nucleus distance. If electron delocalization on ligand nuclei is not too strong, T_{1M}^{-1} and T_{2M}^{-1} are mainly dipolar in origin, and the nucleus-electron distance is practically the distance of the nucleus from the paramagnetic center, for example the metal ion. Protons of water molecules coordinated to 3d metal ions have been calculated to undergo an essentially pure metal-centered (11) dipolar (4) interaction. Therefore, from now on we will focus on theoretically analyzing the dipolar interaction; also, the discussion will be restricted to nuclear longitudinal relaxation, because most of the experimental data are longitudinal relaxation rates. Analogous arguments could be however used for transverse relaxation rates.

3. DIPOLAR COUPLING IN A SIMPLE SYSTEM

An equation for T_{1M}^{-1} has been derived more than 30 years ago by calculating the transition probabilities for an $I=1/2$ nuclear spin dipolarly coupled with an $S=1/2$ isolated electron spin. T_{1M}^{-1} is proportional to the products of the squared electronic and nuclear magnetic moments divided by the sixth power of the electron–nucleus distances. This treatment, due to Solomon (13), has been further extended for any value of I and S (8,14,15), in the assumption that the electronic spin system is still adequately described by the following spin Hamiltonian:

$$\hat{\mathcal{H}}_o = g_e \mu_B \underset{\sim}{B}_o \cdot \hat{\underset{\sim}{S}} \tag{7}$$

the resulting equation for T_{1M}^{-1} is:

$$T_{1M}^{-1} = (2/15)(\mu_o/4\pi)^2 \gamma_I^2 g_e^2 \mu_B^2 S(S+1) r^{-6} \left\{ (7\tau_c/1+\omega_S^2 \tau_c^2) + (3\tau_c/1+\omega_I^2 \tau_c^2) \right\} \tag{8}$$

where γ_I is the nuclear magnetogyric ratio, and τ_c is the correlation time for the electron–nucleus interaction.

The correlation time τ_c is a further important parameter in the analysis of electron–nucleus coupling. It can be defined as the time constant for the change of the magnetic field induced by the electron at the nucleus. Indeed, it is the modulation of the electronic magnetic field at the nucleus, brought about by all the time-dependent phenomena operative in the system, that induces transitions between nuclear spin levels. τ_c^{-1} is thus defined as the sum of the rate constants for all such processes:

$$\tau_c^{-1} = \tau_M^{-1} + \tau_r^{-1} + \tau_s^{-1} + \dots \qquad (9)$$

In equation 9 are shown the three rate constants which are most likely to be operative, i.e., the dissociation rate constant for the electron-nucleus complex, already defined in the preceding section, the rate constant for the rotation of the complex, τ_r^{-1}, and the electronic relaxation rate τ_s^{-1}. It is clear from equation 9 that, when the latter dominates the modulation process, nuclear relaxation measurements can also give information on the electronic relaxation processes.

4. NUCLEAR MAGNETIC RELAXATION DISPERSION (NMRD)

Even in the simple theoretical frame described in the preceding section, the equation for T_{1M}^{-1} contains two parameters, namely r and τ_c. Either of them can in principle be obtained independently, but in most of the cases of practical importance both are generally unknown. A straightforward way to obtain their values consists in measuring T_{1M}^{-1} at various magnetic fields and fit the data to equation 8 using the appropriate ω_I and ω_s values and r and τ_c as unknown parameters. Of course the experimentally accessible quantity is T_1^{-1}, which is related to T_{1M}^{-1} through equations 1 and 3. Therefore in systems undergoing chemical exchange, τ_M, and sometimes f_M, have to be considered as additional parameters. Furthermore, in dealing with solvent nuclei interacting with complex systems, the possibility should be considered of having more than one site of interaction with the paramagnetic center, each with its own occupancy f_M, and with its own τ_M, r, and possibly τ_c values. In the most general case the experimental T_{1p}^{-1} would be given by:

$$T_{1p}^{-1} = \sum f_M(i)(T_{1M}(i) + \tau_M(i))^{-1} \qquad (10)$$

where

$$T_{1M}(i)^{-1} = (2/15)(\mu_0/4\pi)^2 \gamma_I^2 g_e^2 \mu_B^2 S(S+1) r(i)^{-6}.$$

$$\cdot \left| (7\tau_c(i)/1 + \omega_S^2 \tau_c(i)^2) + (3\tau_c(i)/1 + \omega_I^2 \tau_c(i)^2) \right| \qquad (11)$$

The field dependence of T_1^{-1} is termed nuclear magnetic relaxation dispersion (NMRD) (4). It is clear from the functional form of equation 8 that the measurements need to extend over a large field range in order to be meaningful. Also, the field regions where drops in T_{1M}^{-1} occur (dispersions) depend on the τ_c of the system, as exemplified in Fig. 1.

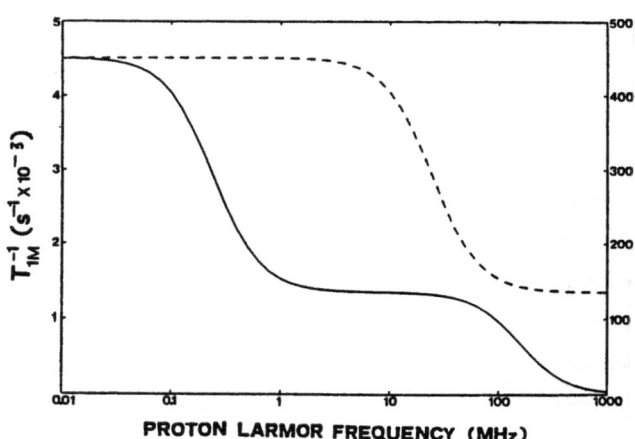

PROTON LARMOR FREQUENCY (MHz)

Figure 1 – NMRD profiles for an S=1/2, I=1/2 (proton) system calculated with equation 8 for r=285 pm, and τ_c=1x10^{-9} s (——)(right scale), or τ_c=1x10^{-11} s (-----)(left scale).

Therefore, the wider the magnetic field range accessible, the better the chances to observe a complete dispersion of T_1^{-1}.

Commercial NMR instruments based on electromagnets can be modified to allow the magnetic field to be varied. If the system has one predominant NMR signal, as in the case of solvent protons, the decrease in resolution with decreasing magnetic field is not a problem, but the limiting factor is sensitivity. In practice, solvent proton relaxation measurements using commercial electromagnets can be performed from 24 mT (1 MHz) up to 2.4 T (100 MHz) (1, 16), and further extended up to 14 T (600 MHz) by using several commercial instruments based on superconducting magnets. On the other hand, home-made instruments are available to extend the range down to 240 μT (0.01 MHz) (1, 16-18), or even to the earth's magnetic field (19). These instruments solve the

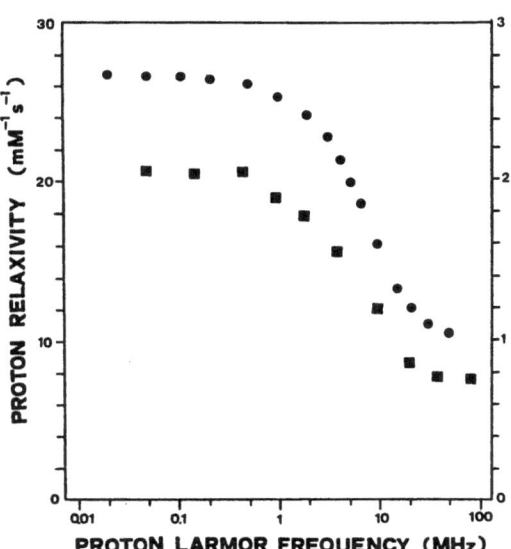

Figure 2 – Water proton NMRD data of solutions of $Cu(ClO_4)_2$ $6H_2O$ (■) (right scale) (20), and $GdCl_3$ $6H_2O$ (●) (left scale) (21) at 25 °C. The data are expressed in relaxivity, i.e. in T_{1p}^{-1} per millimolar metal ion concentration. The dispersions correspond to $\omega_s \tau_c = 1$, with $\tau_c = \tau_r = 3 \times 10^{-11}$ s for Cu(II) and 4.5×10^{-11} s for Gd(III).

problem of sensitivity by first inducing a suitable nuclear spin population difference with a relatively strong magnetic field, and then letting the spin system to relax to a new equilibrium situation at the low magnetic field of interest. With the aid of such instruments a wealth of experimental data has been obtained during the last two decades on solvent (mainly water) proton relaxation in solutions of paramagnetic metal complexes and metalloproteins (1). However, only some data on hexaaqua metal complexes (see, for example, Fig. 2), and virtually no data on metalloproteins, could be rationalized using the simple theoretical frame of the preceding paragraph. This finding has strongly limited the potential and the popularity of NMRD, while relaxation measurements at single fields have been still widely used in a qualitative, and sometimes arbitrary, fashion.

5. A GENERAL APPROACH TO DIPOLAR COUPLING

Fig. 3 shows the typical ranges of τ_M, τ_r, and τ_s values for paramagnetic metal complexes having one or more coordinated water molecules (1).

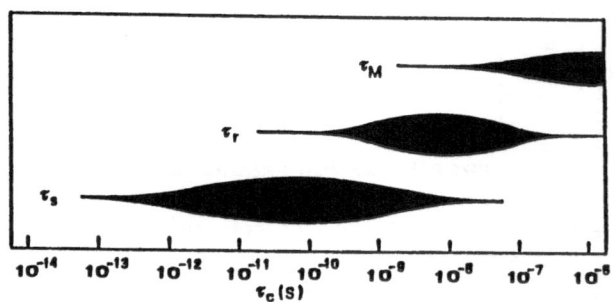

Figure 3 - Common ranges of values for τ_s, τ_r, and τ_M. τ_r is referred to biological molecules (1).

The chemical exchange time may vary from 10^{-8} s for very labile water molecules, like apical water in some tetragonal copper(II) complexes (22), to 10^3 s for inert metal complexes like those of chromium(III) (23). Rotational times may range from 10^{-11} s for hexaaqua metal complexes to 10^{-6} s for proteins of m. w. 1,000,000; bejond this limit the physical properties of the system deviate more and more from those of a regular solution. Electronic relaxation times range from about 10^{-8} s for some oxovanadium(IV) complexes (24) to 10^{-13} s for lanthanides (25, 26) and some six-coordinated cobalt(II) complexes (27, 28). The few experimental cases that can be analyzed in terms of the Solomon equation are all characterized by a very short rotational correlation time compared to τ_M and τ_s (Fig. 2). This finding is a very good starting point to consider the origin of the dramatic deviations from the simple Solomon behavior observed in all the other cases. When τ_r is long, as in the case of macromolecules, either τ_M or τ_s may become the relevant correlation time. Since τ_M cannot be much shorter than 10^{-8} s it can dominate τ_c only when the electronic relaxation times are very long. Very few cases of this type have ever been encountered. However, before considering the many cases of dominance of τ_s, it is worth mentioning the situation of solvent molecules not coordinated to the metal ions, but interacting with them through dipolar coupling when they happen to be in their proximity upon diffusion in solution. This relaxation mechanism is termed outer sphere (6-8). In this case, if the electronic relaxation time is long, the correlation time is the diffusional time τ_D, defined as:

$$\tau_D = 2d^2/(D_N + D_M) \tag{12}$$

where d is the distance of closest approach, and D_N and D_M are the diffusion coefficients of the molecules containing the nucleus under investigation and the metal ion, respectively.

The diffusion process modulates the dipolar coupling energy by essentially varying r, which is thus no longer a constant in the equation for T_{1p}. A widely used equation is of the form (29):

$$T_{1p}^{-1} = N_M (\mu_o/4\pi)^2 (16\pi/225) \, \gamma_I^2 g_e^2 \mu_B^2 S(S+1) d^{-1} (D_N + D_M)^{-1}.$$

$$\cdot \left[7f(\omega_S, \tau_D) + 3f(\omega_I, \tau_D) \right] \tag{13}$$

where N_M is the number of metal ions per cubic meter of solution,

$$f(\omega, \tau_D) = (15/2)I(u) \qquad\qquad u = |\omega \tau_D|^{\frac{1}{2}}$$

$$I(u) = u^{-5} \left\{ u^2 - 2 + e^{-u} \left[(u^2 - 2)\sin(u) + (u^2 + 4u + 2)\cos(u) \right] \right\}$$

Note that in this case a T_{1M} cannot be defined, because there is not a defined paramagnetic site but all the molecules in solution simultaneously feel the paramagnetic centers, though to different extents. Somewhat different equations that take into account random jumps models for rotation have also been proposed (30-32). An outer sphere contribution to T_{1p}^{-1} is always present even when there are coordinated water molecules, although its relative weight is often small. Since the functional form of equation 13, or its analogues, is different from that of equation 8, namely the dispersions are smoother (Fig. 4), sizeable outer sphere contributions may alter the overall NMRD profiles considerably. This has been proposed to be the case for methemoglobin (Fig. 5) (33).

When τ_s is the dominant correlation time there are several possible reasons for a system to deviate from the Solomon behavior. Some of these reasons were considered since the early times of NMRD measurements (5,9), some other have been only recently analyzed in detail. The most obvious possibility to take into consideration is that τ_s may depend on

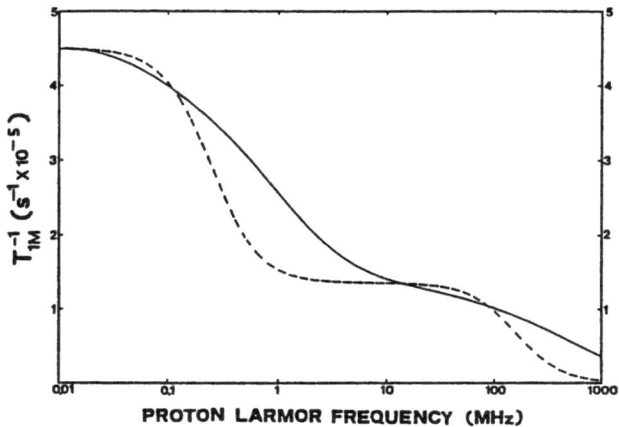

Figure 4 – Comparison between NMRD profiles calculated with equation 8 (-----), and equation 13 (——), using $\tau_c = \tau_D = 1\times10^{-9}$ s.

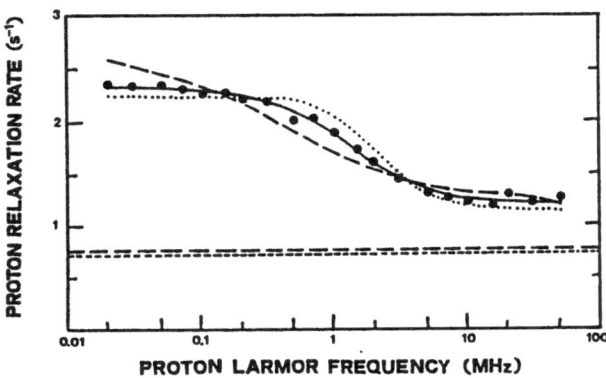

Figure 5 – Comparison among the best fit (——) of the paramagnetic contribution to the relaxation rates of methemoglobin, 1.67 mM protein in 0.2 M phosphate buffer at 6 °C and pH 6.55, and two other fits corresponding to the limits of no diffusion (only τ_s is important) (·····), and no paramagnetic relaxation (only τ_D is important) (-----) (33).

the applied magnetic field. Indeed, at least some electron relaxation mechanisms arise from coupling of the electron spins with fluctuating magnetic fields in solutions, and can be shown to depend on the electron Larmor frequency, ω_S. In the particular case of electron relaxation determined by fluctuations of the quadratic zero field splitting induced by collisions with solvent molecules the following equation has been proposed (34,35):

$$\tau_s^{-1} = (1/5\tau_{s0})\{(1/1+\omega_S^2\tau_v^2)+(4/1+4\omega_S^2\tau_v^2)\} \qquad (14)$$

where $1/(5\tau_{s0})$ is proportional to the quadratic zero field splitting and τ_v is the correlation time for the collision process.

The field dependence of τ_s is strongly reflected in the NMRD profiles, giving often rise to an increase of T_1^{-1} in the intermediate field regions where $\omega_S\tau_c > 1 > \omega_I\tau_c$. Examples of this behavior will be shown later. Although some of the NMRD data on macromolecular systems could be at least qualitatively rationalized by invoking a field dependence of τ_s (4,36), many others show more or less marked "anomalies" which still cannot be accounted for. It should be recalled however, that none of these systems meets the assumption made in deriving equation 8. In fact, in real coordination compounds the electron-spin system cannot be adequately described by the spin Hamiltonian 7. The low symmetry of the complexes and spin-orbit coupling effects cause the g factor to deviate from g_e and be anisotropic; for $S > 1/2$ zero field splitting may be present; moreover, if the metal nucleus is magnetically active, its hyperfine coupling with the electron spin system should be taken into account. Finally, when a second paramagnetic metal ion is close to the paramagnetic center of interest, exchange coupling should also be included in the overall spin Hamiltonian, which then becomes:

$$\hat{\mathcal{H}}_o = \mu_B \underset{\sim}{B}_o \cdot \underset{\approx}{g} \cdot \hat{\underset{\sim}{S}}_1 + \hat{\underset{\sim}{S}}_1 \cdot \underset{\approx}{D} \cdot \hat{\underset{\sim}{S}}_1 + \hat{\underset{\sim}{I}} \cdot \underset{\approx}{A} \cdot \hat{\underset{\sim}{S}}_1 + \hat{\underset{\sim}{S}}_1 \cdot \underset{\approx}{J} \cdot \hat{\underset{\sim}{S}}_2 \qquad (15)$$

Therefore, a new equation should be developed in the place of equation 8 using the correct description of the electronic spin system (37). According to the Kubo and Tomita formalism (14), the nuclear relaxation rate enhancement, T_{1M}^{-1}, due to dipolar coupling, can be expressed in general terms as:

$$T_{1M}^{-1} = \hbar^{-2} \int_0^\infty <[\hat{I}_z, \hat{\mathcal{H}}_{dip}(t)][\hat{\mathcal{H}}_{dip}(0), \hat{I}_z]> e^{i\gamma\omega_I t} \Big/ \langle \hat{I}_z^2 \rangle \quad (\gamma = -1, 0, +1) \quad (16)$$

where

$$\hat{\mathcal{H}}_{dip}(t) = e^{(i/\hbar)\hat{\mathcal{H}}_o t} \hat{\mathcal{H}}_{dip}(0) e^{-(i/\hbar)\hat{\mathcal{H}}_o t} \qquad (17)$$

here $\hat{\mathcal{H}}_{dip}$ represents the dipolar interaction Hamiltonian between the nuclear and the electronic spins and can be expressed as:

$$\hat{\mathcal{H}}_{dip}(0) = \hat{A}^{-\gamma} \hat{F}^{\gamma} \qquad (18)$$

where $\hat{A}^{-\gamma}$ contains nuclear spin operators and $\hat{F}^{\gamma}$ contains electron spin operators (38).

Using the Wigner-Eckart theorem electronic and nuclear spin operators can be separated, to give:

$$<[\hat{I}_z, \hat{\mathcal{H}}_{dip}(t)][\hat{\mathcal{H}}_{dip}(0), \hat{I}_z]> = (-)^\gamma \gamma^2 (1/3) I(I+1) g^{\gamma, -\gamma}(t) \qquad (19)$$

where

$$g^{\gamma, -\gamma}(t) = <\hat{F}^{\gamma}(t) \hat{F}^{\gamma}(0)> \qquad (20)$$

so that the problem reduces to the diagonalization of $\hat{\mathcal{H}}_0$. It can be

shown that the additional terms in $\hat{\mathcal{H}}_0$ will produce a significant deviation from the Solomon behavior only when their associated energy is
larger than $\hbar\tau_c^{-1}$; furthermore, with the exception of exchange coupling,
the effect can be observed only when their energy is larger than the
electron Zeeman energy, i.e. at low magnetic fields.

Several general considerations can be made: first, g anisotropy is not a
major source of perturbation, as long as $\Delta g < g_e$ (37,39,40). This is
almost always true, if one considers that large experimental departures
from g_e and strong anisotropies are only observed when interpreting with
fictitious spin Hamiltonians systems having zero field splitting, and do
not reflect the actual g values of the S manifold. Typical A values for
3d metal ions are of the order of 10^{-2} cm^{-1}, i.e. 2×10^9 s^{-1}. Therefore,
only those systems having $\tau_c > 5 \times 10^{-10}$ s will experience deviations from
the Solomon behavior due to hyperfine coupling. This happens, for
instance, for copper(II) and oxovanadium(IV) proteins, for which satisfactory fitting of NMRD data have been lastly achieved with the use of
equation 16 and the diagonalization of the proper spin Hamiltonian $\hat{\mathcal{H}}_0$
(40-42). Fig. 6 shows examples of such highly satisfactory fittings; for
the first time reliable τ_s values and structural information (r values)
could be obtained for this class of metalloproteins (Table I). It is
also fully understood why hexaaqua copper(II) can be fitted with the
standard equation (20), because in this case the short τ_c overcomes the
effect of hyperfine coupling. The smooth transition from the Solomon to
the non-Solomon behavior is nicely demonstrated by ethyleneglycol
solutions of copper(II) complexes at various temperatures (Fig. 7)
(43,44). Due to the high viscosity of ethyleneglycol the complexes
rotate slower than in water at room temperature, and τ_r can be increased
so much by lowering the temperature that hyperfine coupling starts being
effective, until the correlation time is dominated by τ_s and the system
behaves like a metalloprotein (43,44). It should be mentioned that when

$\hbar\tau_r^{-1} < A$ the electron spin system behaves like a frozen solution; there-

---

Table I

Best fit parameters for the water proton NMRD data on Cu(II)-
Transferrin (TFN)(41), VO(IV)-TFN (41), Cu(II)-Bovine
Carbonic Anhydrase II (BCA II)(42), Cu(II)-BCA II+N$_3^-$ (42),
Cu(II)-BCA II+HCO$_3^-$ (42), and Cu(II)-Superoxide Dismutase (SOD) (40).

sample	TEMP (°C)	G$\stackrel{a}{=}$ (pm^{-6})	τ_c (ns)	ϑ (degrees)
Cu^{2+}-TFN	8	5.48x10^{-16}	7.52	45.5
	25	7.36x10^{-16}	5.70	50.0
	38	7.47x10^{-16}	5.40	51.7
VO^{2+}-TFN	8	7.95x10^{-16}	22.4	26.7
	25	7.81x10^{-16}	19.2	29.5
	38	4.74x10^{-16}	20.5	23.6
Cu^{2+}-BCA II	25	3.9 x10^{-15}	1.9	53.0
Cu^{2+}-BCA II + N$_3^-$	25	4.7 x10^{-15}	2.6	50.0
Cu^{2+}-BCA II + HCO$_3^-$	25	1.2 x10^{-15}	2.1	52.0
Cu$_2$Zn$_2$-SOD	25	1.7 x10^{-15}	1.8	20.0

$\underline{a}$G= Σ_i (n$_i$/r$_i^6$), where n$_i$ is the number of (equivalent) protons interacting with the metal ion at distance r$_i$. For a single water molecule in the usual coordination geometry, r=2.8 Å and G=4.15x10^{-15} pm^{-6}.

---

fore T$_{1M}^{-1}$ calculated through equation 16 has to be averaged over all the orientations of the z axis of the complex with respect to the magnetic field (40). Furthermore, since the system is anisotropic, the angle ϑ between the metal-nucleus vector and the molecular z axis appears as a further parameter in the final equation for T_{1M}^{-1}.

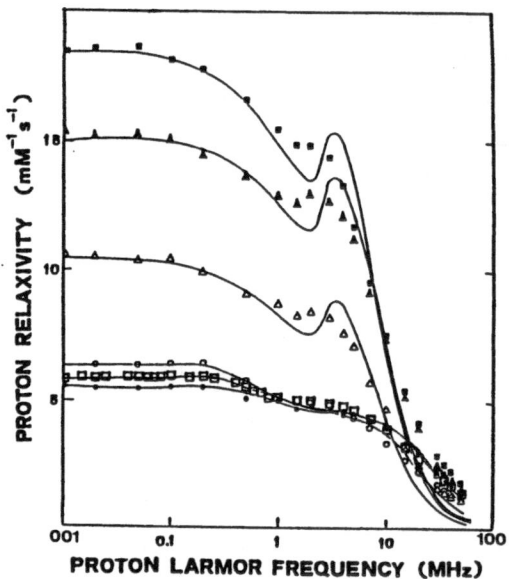

Figure 6 – Water proton NMRD profiles of solutions of Cu^{2+}-Transferrin at 8 (○), 25 (□), and 38 °C (●), and VO^{2+}-Transferrin at 8 (■), 25 (▲), and 38 °C (△). The solid lines are best fit curves obtained using anisotropic A values obtained from EPR data (40–42).

Of course analytical solutions for T_{1M}^{-1} could only be obtained if $\hat{\mathscr{H}}_0$ can be diagonalized analytically. Otherwise, a numerical fitting is performed.

As an example of zero field splitting being the dominant term in $\hat{\mathscr{H}}_0$, fig. 8 shows the NMRD data for a cobalt(II) substituted metalloprotein (45,46). High spin cobalt(II) has very short electronic relaxation times, ranging from 10^{-11} to 10^{-13} s depending on the coordination number and geometry. τ_s thus dominates the correlation time for the system.

Under these conditions, the hyperfine coupling with the I=7/2 metal nu-

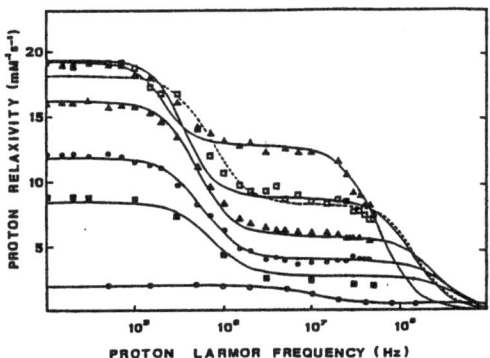

Figure 7 - Solvent proton NMRD profiles of ethyleneglycol solutions of Cu(ClO$_4$)$_2$ 6H$_2$O at -9 (▲), 5 (□), 15 (△), 25 (●), and 39 °C (■) as compared to those of water solutions at 25°C (O). The solid lines are best fit curves obtained using an isotropic A value (or the Solomon equation for the water solution); the dashed line is the best fit curve of the data at 25°C using anisotropic A values (43,44).

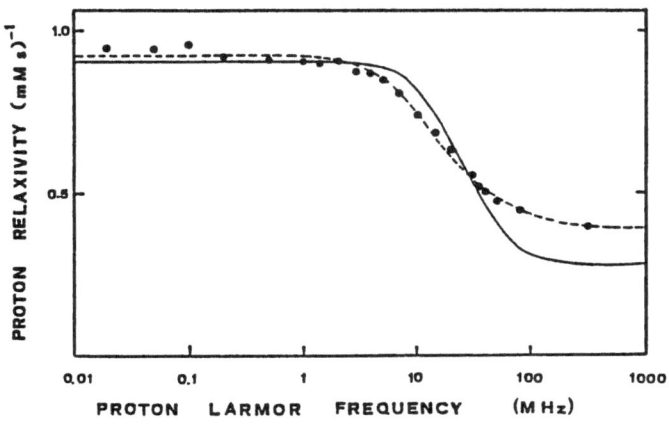

Figure 8 - Water proton NMRD profiles of Co^{2+}-Human Carbonic Anhydrase I at 25 °C and pH 9.9 (45). The best fit curves calculated with the Solomon equation (——), and equation 16 (-----) are also reported (46).

clear spin is ineffective, since $A < \hbar \tau_s^{-1}$. On the other hand, zero field splitting values in cobalt(II) complexes are often larger than a few
wavenumbers, and therefore the relation $D > \hbar \tau_s^{-1}$ is likely to hold. The
NMRD data in fig. 8 are fitted using the Zeeman plus a large $\hat{\underset{\sim}{S}} \underset{\approx}{D} \hat{\underset{\sim}{S}}$ term
in $\hat{\mathcal{H}}_0$ (46). For comparison purposes, the best fit using the Solomon
equation (solid line), is also shown. Again, it is clear that the consideration of the proper spin Hamiltonian fully rationalizes the experimental data. It is worth noting that the best fit parameters differ
sizeably from those obtained using the Solomon equation (Table II), both
in the value of τ_s and in the structural G parameter. The Solomon value
for the latter would indicate the presence of two water molecules in the
coordination sphere, while both X-ray and spectroscopic data point to
only one solvent-accessible coordination position (47).

Table II

Best fit parameters for the water proton NMRD data on the
high pH form of Co(II)-carbonic anhydrase, (reported in Fig. 8),
according to the Solomon equation and equation 16 (46)

PARAMETER	SOLOMON EQUATION	EQUATION 16
G (pm^{-6})	9.0×10^{-15} ($\pm 8.4\%$)	5.0×10^{-15} ($\pm 8.1\%$)
τ_c (s)	9.5×10^{-12} ($\pm 9.1\%$)	3.3×10^{-11} ($\pm 7.6\%$)
	----	$72°$ ($\pm 2.4\%$)
$\sum (\Delta T_{1p}^{-1})^2$ (s^{-2})	5.6×10^{-1}	6.1×10^{-2}

The copper(II), cobalt(II), and probably oxovanadium(IV) proteins analyzed so far have τ_s as the dominant correlation time; nevertheless, their
NMRD data could be satisfactorily analyzed in the above theoretical
frame without introducing a field dependence of τ_s. On the other hand,

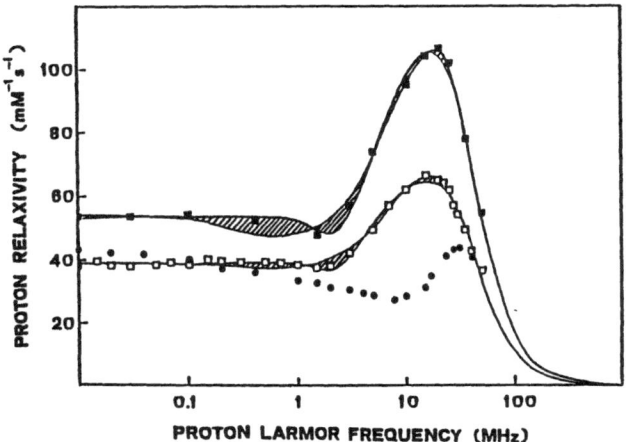

Figure 9 - Water proton NMRD profiles of Mn^{2+}-Concanavalin A at 25 ($\square$), and 5 °C ($\blacksquare$)(4). The solid curves are best fit curves calculated for D=0 (upper curve at 0.7 MHz), and D=0.04 cm^{-1} (lower curve at 0.7 MHz)(49). NMRD data for Mn^{2+}-Carboxy Peptidase (CPA) at 25 °C ($\bullet$) and pH 6.7 are also reported (48).

such phenomenon is clearly evident in manganese(II) containing proteins, that are thus much more difficult to analyze: a typical set of NMRD data for this class of proteins is shown in Fig. 9 (48,49). A proper analysis in this case should take into account a large number of parameters (4, 48-50). From the functional form of the field dependence of τ_s^{-1} proposed for manganese(II) (equation 14) it appears that the increase in τ_s with magnetic field would be endless. This does not occur in real systems, however, because other less efficient electron mechanisms will always prevent τ_s^{-1} to decrease to zero. In any case, if τ_s becomes too long, either τ_r or τ_M would become the dominant correlation times of the system. It has been shown that a complete fitting, even in the Solomon frame, would require f, τ_M, r, τ_r, τ_{s0}, and τ_v as adjustable parameters, only five of which can be treated independently (4,50). Moderately satisfactory fits have been obtained in this way for some manganese(II) proteins, while for some others the best fitting parameters indicated a too large number of water molecules interacting with the manganese(II) ion (4). In the latter case it has been shown that introduction of a static zero field splitting term and of the metal nucleus-electrons hyperfine coupling in $\hat{\mathcal{H}}_0$ substantially reduces the number of interacting water molecules, in much better agreement with the X-ray data, while the goodness of the fitting is about the same as that obtained in the Solomon frame (Fig. 9 and Table III) (49).

Table III

Best fit parameters (49) for the water proton NMRD
data on Mn(II)-concanavalin A (Con A)(4,50), (reported in Fig. 9)

D (cm^{-1})	n^{a}	τ_M (s)	τ_v (s)	τ_{s0} (s)	τ_r (s)	ϑ (deg)	$\Sigma\Delta^2$ (mMs)$^{-2}$ ^{b}
			25°C				
0	13.1	9.9×10^{-7}	6.9×10^{-11}	1.4×10^{-10}	3.3×10^{-8}	--	55.1
0.02	9.0	6.3×10^{-7}	4.2×10^{-11}	3.4×10^{-10}	2.2×10^{-8}	54	57.4
0.04	8.3	5.7×10^{-7}	3.2×10^{-11}	4.7×10^{-10}	2.2×10^{-8}	67	64.6
			5°C				
0	12.0	1.5×10^{-6}	7.5×10^{-11}	1.4×10^{-10}	4.6×10^{-8}	--	34.9
0.02	7.5	9.1×10^{-7}	5.7×10^{-11}	2.7×10^{-10}	2.4×10^{-8}	15	24.2
0.04	5.3	5.9×10^{-7}	5.0×10^{-11}	3.5×10^{-10}	1.6×10^{-8}	0	41.9

[a] Number of protons at 2.8 Å from the manganese(II) ion.

[b] Sum of the squared deviations between experimental and calculated data.

These results indicate that, while once again the proper spin Hamiltonian has to be used in order to obtain physically meaningful structural parameters, the analysis of NMRD data of manganese(II) proteins relies on a too large number of parameters to be confidently performed without the aid of independent information on the system. This conclusion is somewhat surprising because manganese(II) has been the most popular paramagnetic probe to obtain information on ligands and substrates binding in metalloproteins. Although at the high magnetic fields usually employed in the latter experiments zero field splitting and hyperfine coupling effects are probably negligible (51), manganese(II) still does not seem the most suitable metal ion for this kind of experiments.

A comment is appropriate on the meaning of the τ_s parameter for a very fast relaxing system. The cobalt(II) system discussed above is four coordinated and has $\tau_s = 10^{-11}$ s; five or six-coordinated cobalt(II) systems behave as having even shorter electronic relaxation times (52), as do lanthanide ions, with the exception of gadolinium(III). For these and other fast relaxing ions the problem arises of whether electron relaxation can be described as a pure exponential process with time constant τ_s (2,53-55). If electron relaxation arises from the coupling of the electron magnetic moment with fluctuating magnetic fields in solution, the electron relaxation rate constant τ_s^{-1} can be defined by equations of the type of equation 14 only when $\tau_v \ll \tau_{s0}$, i.e. when the fluctuation rate is much larger than the interaction energy (2). This is known as the Redfield limit. Within this limit, the electron-nucleus coupling can be analyzed as shown in the foregoing, irrespectively of the actual electron relaxation mechanism involved. The only quantity of interest from the nuclear spin point of view is the electronic relaxation time. Outside the Redfield limit the treatment of nuclear relaxation properties requires the knowledge of the electron-lattice coupling mechanism. A theoretical treatment is available where the modulation of the zero field splitting by rotation is the dominant electron relaxation mechanism (55-59). Theoretical NMRD profiles have been calculated in this way for several S>1/2 systems (Fig. 10).

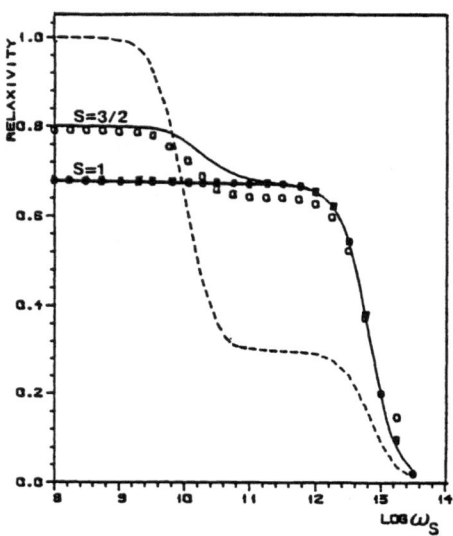

Figure 10 - Comparison between NMRD curves calculated using the theoretical treatment in references (56) and (59) (■ and □), and equation 16 (——)for S=1 and S=3/2,and ϑ=0. The curves in the non-Redfield limit are calculated for τ_r =2/3 (S=1) or 4/5 (S=3/2) of the τ_s values used for the curves in the Redfield limit. The curve corresponding to the Solomon equation with the correlation time equal to τ_s is also shown (------) (60).

Although a τ_s for the systems cannot be defined, the curves looked surprisingly similar to NMRD profiles calculated for the corresponding S>1/2 systems in the Redfield limit and in the presence of a static zero field splitting term in $\hat{\mathcal{H}}_0$ (37,46). A detailed comparison of the two kinds of calculations (60) has shown that the curves can be made superimposable, for ϑ=0, when the τ_r used in the non-Redfield limit is equal to 2/3, 4/5, or 79/105 of the τ_s used in the Redfield limit for S=1, 3/2, 5/2, respectively (60–62). This striking result implies that a non-Redfield system still behaves as a Redfield system from the nuclear spin point of view, and that a τ_s value can be always operatively defined (60). A further, more fundamental implication would be that even a non-Redfield system does have a well defined electronic relaxation time, linked to the time constant for the fluctuation of the electron–lattice interaction energy by a simple proportionality constant (60–62). The latter proposal would however need direct experimental and/or theoretical verification.

6. EXCHANGE COUPLED SYSTEMS

Nuclear relaxation can be dramatically altered when the paramagnetic metal ion interacting with the nucleus under investigation is on its turn magnetically coupled with another paramagnetic metal ion. The origin of this alteration of NMR parameters is twofold. The first reason lies in the influence that the $\hat{S}_1 \ J \ \hat{S}_2$ term plays in the $\hat{\mathcal{H}}_0$ Hamiltonian (equation 15). Calculations have been performed, for the case of two S=1/2 metal ions coupled by an isotropic exchange coupling constant J, of the relaxation parameters of a nucleus which is only coupled with one of the two metals (63). These calculations show that the overall effect of the coupling would be to reduce by a factor of two the total relaxation rate at every magnetic field in the limit of strong J. Similar considerations have led to compute reduction coefficients for every other S_1-S_2 pairs: these coefficients are listed in Table IV (64). According to these coefficients the coupling between two paramagnetic metal ions always brings a reduction in the nuclear relaxation caused by the presence of the paramagnetic entities, and the reduction is greater the smaller the electron spin with which the nucleus is interacting.

The second effect caused by exchange coupling phenomena on nuclear relaxation is even more dramatic. This effect comes from the alteration in the electron relaxation rates brought about by the exchange coupling itself. In particular, when the two metal ions which are exchange coupled have electronic relaxation times differing by orders of magnitude, as it may be the case for a Cu(II)-high spin Co(II) pair or Mn(II)-low

---

Table IV

Reduction coefficients (X_1 and X_2) calculated for M_1-M_2 exchange-coupled pairs according to ref. (64)

S_1 \ S_2	5/2	2	3/2	1	1/2
1/2	17/18 · 19/54	23/25 · 9/25	7/8 · 3/8	7/9 · 11/27	1/2 · 1/2
1	$\frac{1049}{1225}$ · $\frac{13}{35}$	$\frac{43}{54}$ · $\frac{7}{18}$	$\frac{467}{675}$ · $\frac{19}{45}$	1/2 · 1/2	
3/2	$\frac{107}{144}$ · $\frac{173}{432}$	$\frac{113}{175}$ · $\frac{379}{875}$	1/2 · 1/2		
2	$\frac{102031}{165375}$ · $\frac{6257}{14175}$	1/2 · 1/2			
5/2	1/2 · 1/2				

(In each cell the upper-right value is X_1 and the lower-left value is X_2; s_2 labels the columns and s_1 labels the rows.)

---

spin Fe(III) pair, and so on, the fast relaxing metal ion can be considered as a source of further electron relaxation for the slow relaxing ion.

An equation has been proposed for the limit of weak coupling (63), according to which the electronic relaxation rate of the slow relaxing ion in the pair is increased by the presence of the other metal ion:

$$\tau_{ss}^{-1}(J) = \tau_{ss}^{-1}(0) + (J^2/\hbar^2)\tau_{ss}(0) \qquad (21)$$

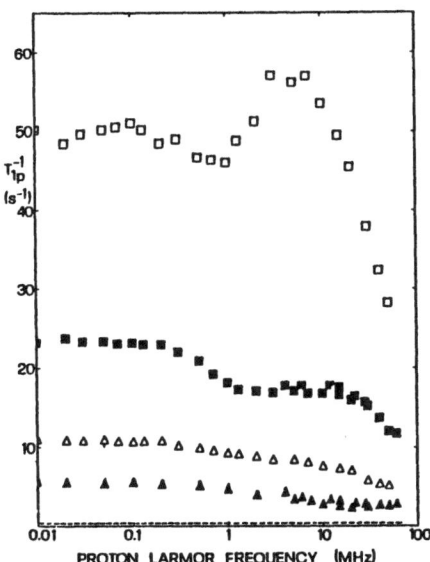

Figure 11 - Solvent proton NMRD profiles for ethyleneglycol solutions of $[Cu(en)_2\text{-}Fe(CN)_6]^-$ 10 mM at 0 ($\triangle$) and 25 °C ($\blacktriangle$). To appreciate the effect of magnetic coupling, the profiles for the ethyleneglycol solutions containing the analogous adduct of $Cu(en)_2^{2+}$ with the diamagnetic complex $Co(CN)_6^{3-}$ at 0 ($\square$) and 25 °C ($\blacksquare$) are also reported (63). The dashed line represents the solvent proton NMRD profile of the $Fe(CN)_6^{3-}$ in ethyleneglycol solution in the same concentration at 25 °C.

where $\tau_{ss}^{-1}(J)$ is the electronic relaxation rate of the coupled, slowly relaxing ion, and $\tau_{ss}^{-1}(0)$ is the value of τ_{ss}^{-1} at J=0 (isolated ion). In the limit of strong coupling ($J \gg \hbar\tau_{sf}^{-1}$, where τ_{sf} is the electron relaxation time for the fast relaxing ion) it may be predicted that the electronic relaxation rate of the other ion will tend to the value of the fast relaxing one (63). Therefore nuclear relaxation for nuclei interacting mainly with the slow relaxing ion is strongly decreased because of the coupling with the fast relaxing metal. Fig. 11 shows the dramatic change that is observed on the solvent protons interacting with Cu(II) in a small inorganic complex when the latter ion is on its turn interacting with hexacyanoferrate(III); this effect is of course much more relevant when the system under investigation is under slow rotation conditions. Otherwise, for a small complex, the correlation time would be the rotational correlation time and therefore the decrease in electron relaxation rate is of no consequence as far as nuclear relaxation is concerned. Such effect has been evidenciated by performing the above experiment in a viscous solvent like ethyleneglycol.

These findings have important bearing for biologically relevant macromolecules. In nature there are several metalloproteins containing more than one paramagnetic metal ion, either as native or as metal substituted proteins. The presence, or the induction, of exchange coupling between these ions may be exploited to obtain sharp NMR signals from nuclei nearby a metal like Cu(II) which by itself would cause such a large broadening of NMR signals to make them undetectable. As an example, fig. 12 shows the proton NMR spectra of Co(II)-substituted superoxide dismutase (SOD)(65). The native protein contains a Zn(II) and a Cu(II) ions bridged by a histidinate residue (66); the Cu(II) ion is in the active site of the molecule. Co(II) can be substituted for Zn with virtually no alteration of the physico-chemical and enzymatic properties

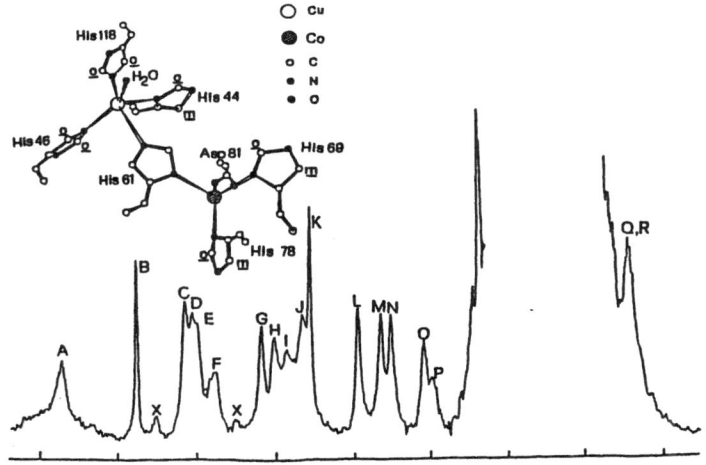

Figure 12 - Proton NMR spectrum (300 MHz) at 30 °C of Cu_2Co_2SOD in 10 mM acetate buffer at pH 5.5 in H_2O. The dashed signals have been assigned to ring protons of histidins coordinated to copper(II) (His 44, 46, 118).

of the molecule. However, the substitution with Co(II) makes the proton nuclei of the histidine residues coordinated to Cu(II) accessible to NMR investigation. The exchange coupling constant for this system has been recently measured and found to be 12 cm^{-1} (67), i.e. strong enough to sizeably shorten the electron relaxation rate of Cu(II) and make it very close to that of Co(II) in the native zinc site.

7. REFERENCES

1) I.Bertini and C.Luchinat, "NMR of Paramagnetic Species in Biological Systems", Benjamin/Cummings, Boston (1986).

2) J.Kowalewski,L.Nordenskiöld,N.Benetis and P.-O.Westlund, Progr.NMR Spectrosc. **17**, 141 (1985).

3) D.R.Burton,S.Forsen,G.Karlström and R.A.Dwek, Progr.NMR Spectrosc. **13**, 1 (1980).

4) S.H.Koenig and R.D.Brown, in "ESR and NMR of Paramagnetic Species in Biological and Related Systems", (I.Bertini and R.S.Drago Eds.), p.89, Reidel, Dordrecht (1980).

5) D.R.Burton, in "ESR and NMR of Paramagnetic Species in Biological and Related Systems", (I.Bertini and R.S.Drago Eds.), Reidel, Dordrecht (1980).

6) R.A.Dwek, "Nuclear Magnetic Resonance in Biochemistry", Claredon Press, Oxford (1973).

7) H.G.Hertz, in "Water.A Comprehensive Treatise", Vol. 3, Chapter 7, (F.Franks Ed.), Plenum, New York (1973).

8) A.Abragam, "The Principles of Nuclear Magnetism", Clarendon Press, Oxford (1961).

9) S.H.Koenig and R.D.Brown, Ann.N.Y.Acad.Sci. **222**, 752 (1973).

10) S.H.Koenig and R.D.Brown, in "The Coordination Chemistry of Metalloenzymes", (I.Bertini,R.S.Drago and C.Luchinat Eds.), p.19, Reidel, Dordrecht (1983).

11) L.Nordenskiöld,A.Laaksonen and J.Kowalewski, J.Amer.Chem.Soc. **104**, 379 (1982).

12) G.N.La Mar,W.DeW.Horrocks Jr. and R.H.Holm, "NMR of Paramagnetic Molecules", Academic Press, New York/London (1973).

13) I.Solomon, Phys.Rev. **99**, 559 (1955).

14) R.Kubo and K.Tomita, J.Phys.Soc.Jpn. **9**, 888 (1964).

15) S.H.Koenig, J.Magn.Reson. **31**, 1 (1978).

16) A.S.Mildvan and R.K.Gupta, Methods Enzymol. **49**, 322 (1978).

17) A.G.Anderson and A.G.Redfield, Phys.Rev. **116**, 583 (1959).

18) K.Hallenga and S.H.Koenig, Biochemistry **15**, 4255 (1976).

19) G.J.Bene, Phys.Rep. **58(4)**, 213 (1980).

20) R.Hausser and F.Noack, Z.Physik. **182**, 93 (1964).

21) S.H.Koenig and M.Epstein, J.Chem.Phys. **63**, 2279 (1975).

22) T.J.Swift and R.E.Connick, J.Chem.Phys. **37**, 307 (1962); Err. **41**, 2553 (1964).

23) M.V.Olson,Y.Kanazawa and H.Taube, J.Chem.Phys. **51**, 289 (1969).

24) I.Bertini,G.Canti,C.Luchinat and A.Scozzafava, Inorg.Chim.Acta **36**, 9 (1979).

25) B.M.Alsaadi,J.C.Rossotti and R.J.P.Williams, J.Chem.Soc. Dalton Trans. 2147 (1980).

26) P.D.Burns and G.N.La Mar, J.Magn.Reson. **46**, 61 (1982).

27) I.Bertini,G.Lanini and C.Luchinat, Inorg.Chim.Acta **80**, 123 (1983).

28) I.Bertini,C.Luchinat,L.Messori and A.Scozzafava, Eur.J.Biochem. **141**, 375 (1984).

29) O.S.Hubbard, Proc.Roy.Soc. **A291**, 537 (1966).

30) L.P.Hwang and J.H.Freed, J.Chem.Phys. **63**, 4017 (1975).

31) J.H.Freed, J.Chem.Phys. **68**, 4034 (1978).

32) C.A.Sholl, J.Phys.C **14**, 447 (1981).

33) S.H.Koenig,R.D.Brown and T.R.Lindstrom, Biophys.J. **34**, 397 (1981).

34) N.Bloembergen and L.O.Morgan, J.Chem.Phys. **34**, 842 (1961).

35) M.Rubinstein,A.Baram and Z.Luz, Mol.Phys. **20**, 67 (1971).

36) R.D.Brown,C.F.Brewer and S.H.Koenig, Biochemistry **16**, 3883 (1977).

37) I.Bertini,C.Luchinat,M.Mancini and G.Spina), in "Magneto-Structural Correlations in Exchange-Coupled Systems", (R.D.Willett,D.Gatteschi and O.Kahn Eds.), p.421, Reidel, Dordrecht (1985).

38) U.Lindner, Ann.Phys.Lpz. **16**, 619 (1965).

39) H.Sternlicht, J.Chem.Phys. **42**, 2250 (1965).

40) I.Bertini,F.Briganti,C.Luchinat,M.Mancini and G.Spina, J.Magn. Reson. **63**, 41 (1985).

41) I.Bertini,F.Briganti,C.Luchinat and S.H.Koenig, Biochemistry in press.

42) I.Bertini and C.Luchinat, in "Biochemical and Inorganic Aspects of Copper Coordination Chemistry", (K.D.Karlin and J.Zubieta Eds.), Adenine Press, in press.

43) L.Banci,I.Bertini and C.Luchinat, Inorg.Chim.Acta **100**, 173 (1985).

44) L.Banci,I.Bertini and C.Luchinat, Chem.Phys.Letters **118**, 345 (1985)

45) S.H.Koenig,R.D.Brown,I.Bertini and C.Luchinat, Biophys.J. **41**, 139 (1983).

46) I.Bertini,C.Luchinat,M.Mancini and G.Spina, J.Magn.Reson. **59**, 213 (1984).

47) I.Bertini and C.Luchinat, Acc.Chem.Res. **16**, 272 (1983).

48) S.H.Koenig,R.D.Brown and J.Studebaker, Cold Spring Harbor Symp. Quant.Biol. **36**, 561 (1971).

49) L.Banci,I.Bertini,F.Briganti and C.Luchinat, J.Magn.Reson. in press

50) S.H.Koenig and R.D.Brown, J.Magn.Reson. **61**, 426 (1985).

51) T.Kushnir and G.Navon, J.Magn.Reson. **56**, 373 (1984).

52) I.Bertini and C.Luchinat, in "Advances in Inorganic Biochemistry", (G.L.Eichhorn and L.G.Marzilli Eds.), Vol. 6, in press.

53) D.T.Pegg and D.M.Doddrell, Aust.J.Chem. **29**, 1869 (1976).

54) D.T.Pegg and D.M.Doddrell, Aust.J.Chem. **29**, 1885 (1976).

55) J.Kowalewski,A.Laaksonen,L.Nordenskiöld and M.Blomberg, J.Chem. Phys. **74**, 2927 (1981).

56) N.Benetis, J.Kowalewski, L.Nordenskiöld, H.Wennerström and P.-O. Westlund, Mol.Phys. **48**, 329 (1983).

57) N.Benetis, J.Kowalewski, L.Nordenskiöld, H.Wennerström and P.-O. Westlund, Mol.Phys. **50**, 515 (1983).

58) N.Benetis, J.Kowalewski, L.Nordenskiöld, H.Wennerström and P.-O. Westlund, J.Magn.Reson. **58**, 261 (1984).

59) P.-O.Westlund, H.Wennerström, L.Nordenskiöld, J.Kowalewski and N. Benetis, J.Magn.Reson. **59**, 91 (1984).

60) I.Bertini,C.Luchinat and J.Kowalewski, J.Magn.Reson. **62**, 235 (1985)

61) I.Bertini,C.Luchinat and L.Messori, in "Metal Ions in Biological Systems", Vol. 21, (H.Sigel Ed.), M.Dekker Inc., New York and Basel, in press.

62) L.Banci,I.Bertini and C.Luchinat, Magn.Reson.Rev. in press.

63) I.Bertini,G.Lanini,C.Luchinat,M.Mancini and G.Spina, J.Magn.Reson. **63**, 56 (1985).

64) C.Owens,R.S.Drago,I.Bertini,C.Luchinat and L.Banci, submitted.

65) I.Bertini, G.Lanini, C.Luchinat, L.Messori, R.Monnanni and A. Scozzafava, J.Amer.Chem.Soc. **107**, 4391 (1985).

66) J.A.Tainer,E.D.Getzoff,K.M.Beem,J.S.Richardson and D.C.Richardson, J.Mol.Biol. **160**, 181 (1982).

67) I.Morgenstern-Badarau,D.Cocco,A.Desideri,G.Rotilio,J.Jordanov and N.Duprè, submitted.

PBB, Vol. 2
Advanced Magnetic Resonance Techniques
in Systems of High Molecular Complexity
© 1986 Birkhäuser Boston, Inc.

NUCLEAR SPIN RELAXATION IN PARAMAGNETIC NICKEL (II) COMPLEXES

J. Kowalewski, N. Benetis, T. Larsson, P.-O. Westlund.

Division of Physical Chemistry, Arrhenius Laboratory

University of Stockholm, S-106 91 Stockholm,Sweden

and

H. Wennerström

Division of Physical Chemistry I, Chemical Center

P.O.Box 124, S-221 00 Lund, Sweden

INTRODUCTION

The main features of nuclear spin relaxation in paramagnetic systems in solution have been described in the late fifties and early sixties by Solomon, Bloembergen and Morgan (1-4). Briefly, two spins (I = nuclear spin, S = electron spin) in a paramagnetic complex are allowed to interact through hyperfine interaction (dipolar, DD, and scalar, SC). The reorientation of the complex modulates the DD part of the interaction and results in relaxation. Due to strong coupling to the thermal bath (trough zero-field splitting, ZFS, for S > 1/2) the electron spin is itself characterized by rapid relaxation. The relaxation of the electron spin modulates the SC part of the hyperfine interaction and, if it is fast enough, can compete with reorientation as the modulation process for the DD interaction. In addition, chemical exchange can under certain circumstances compete with both relaxation and reorientation as the random modulating process. The set of equations describing the nuclear spin-lattice and spin-spin relaxation rates in a paramagnetic complex are known in the literature as modified Solomon-Bloembergen (MSB) equations (5, 6).

Several problems arise when the MSB equations are applied to interpret the relaxation of nuclear spins in nickel (II) complexes.Two of these problems are related to the high zero-field splitting (ZFS) generally believed to occur in nickel (II) complexes. The high ZFS affects the electron spin energy levels making the simple Zeeman-split energy level diagram assumed by Solomon (1) invalid. This situation can be handled by introducing corrections to the MSB equations (7-9). Further, the large ZFS can make the Redfield-type description of the electron spin relaxation, implicit in the MSB approach, invalid. This complication, referred to as the slow-motion regime for the electron spin, is more difficult to circumvent; although it has been recognized in connection with the ESR lineshapes for some time (10), its implication for the NMR relaxation problem has been first treated in detail by Friedman et al. (11). Another problem, the cross-correlation between the DD and ZFS interactions, arises if the two interactions are modulated by the same type of motion (12). Finally, the deviations from the point-dipole nature of the electron spin can complicate the interpretation of the strength of the DD interaction between the nuclear and the electron spin (13). According to Kowalewski et al. (14-16), this problem should not be critical for the case of proton relaxation.

THEORY

A formalism allowing to treat in a consistent manner the three first problems mentioned above (energy levels, slow motion, cross-correlation) has recently been developed in our laboratory (17-19). Briefly, the electron spin is considered to be a part of the lattice. This means that the lattice becomes a complicated system, containing the classical and the quantized degrees of freedom. Such systems can conveniently be handled using the Liouville superoperator formalism (10). The nuclear spin is coupled to the lattice by the hyperfine hamiltonian:

$$\mathcal{H}_{IL} = \sum (-1)^n I_n^{(1)} T_{-n}^{(1)} \tag{1}$$

For the SC part of the interaction, Eq.(1) is equivalent to the conventional formulation. For the DD part, the operators T are rank one irreducible tensor operators, obtained by coupling the electron spin operators (rank one) and the rank two spherical harmonics.

The nuclear spin relaxation rates can be treated by a Redfield-type approach and can be expressed in terms of complex spectral densities:

$$T_{1I}^{-1} = -2\mathrm{Re}\ K_{1,-1}(\omega_I) \tag{2}$$

and

$$T_{2I}^{-1} = \mathrm{Re}\left[K_{0,0}(0) - K_{1,-1}(\omega_I)\right] \tag{3}$$

which are Fourier-Laplace transforms of the correlation functions:

$$G_{n,-n}(\tau) = (-1)^n \mathrm{Tr}_L \left\{ T_n^\dagger \exp(-iL_L\tau) T_n \sigma^T \right\} / \mathrm{Tr}_L(1) \tag{4}$$

L_L is the Liouville superoperator for the lattice, to be specified below, and σ^T is the thermal equilibrium density operator for the lattice. The evaluation of the spectral densities is performed by introducing a complete and orthonormal basis in the Liouville space and taking the matrix inverse:

$$K_{n,-n}(\omega) = (-1)^n c_n *(iL_L + i\omega 1)^{-1} c_n /Tr_L(1) \qquad (5)$$

In the practical implementation, the basis set has to be truncated, of course. The theory predicts that the total spectral densities in Eqs. (2), (3) and (5) are sums of three terms: the pure DD and SC contribution and an interference (DD-SC) term.

THE DESCRIPTION OF THE LATTICE

The lattice Liouville superoperator consists in general of three terms:

$$L_L = L_{SO} + L_R + L_{RS} \qquad (6)$$

L_{SO} is the superoperator generated by the electron spin Zeeman hamiltonian. L_R represents the classical degrees of freedom (the thermal bath). In the simplest model, it is the Markov operator representing the rotational diffusion (17-19). The term L_{RS} represents the interaction between the electron spin system and the reorientational degrees of freedom, the ZFS interaction. For the case of permanent, cylindrically symmetric ZFS interaction with the principal axis coincident with the DD axis, the ZFS hamiltonian is

$$\mathcal{H}_{ZFS} = f_0 \sum_\mu D_{-\mu' 0}^2 (\Omega_{ML}) A_\mu^{2,L} \qquad (7)$$

where f_0 is the ZFS parameter (the notation Δ is also used), $D(\Omega_{ML})$ are Wigner rotation matrices, the Ω_{ML} is the set of eulerian angles specifying the transformation from laboratory (L) to molecular (M) coordinate system and A are second order electron spin operators. For this simple case, the basis set in the Liouville space is formed as a

direct product of symmetric top eigenfunctions and suitably symmetrized
spin vectors.

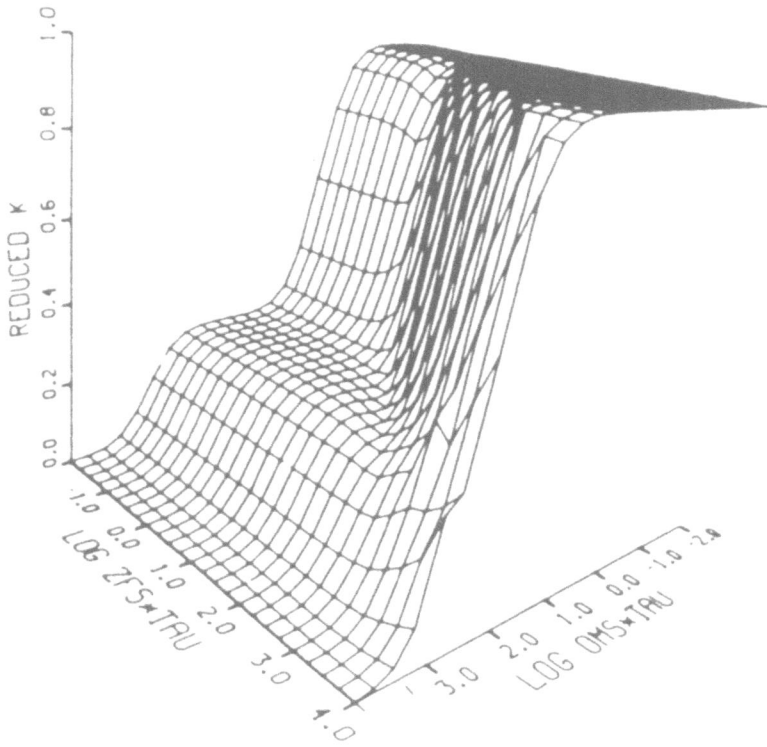

Figure 1 – Plot of the completely reduced spectral density $K_{1,-1}^{DD}$ for
the isotropic rotation-permanent ZFS model and coincident
ZFS and DD principal axes (from ref.18)

The results of the calculations for the DD term using this model are
shown in Fig. 1, (taken from ref. (18)) where the completely reduced
spectral density $K_{1,-1}^{DD}$ (the spectral density appearing in Eqs.(2) and
(3) divided by the DD interaction strength constant squared and the
rotational correlation time) is displayed as a function of the dimensionless variables $\omega_S \tau_R$ and $f_0 \tau_R$. It should be noted that the low ZFS

limit of $K^{DD}_{1,-1}$ is identical to the Solomon equation (1) and that the behaviour of the function in the slow motion regime for the electron spin $(f_0 \tau_R > 1)$ is very different from the predictions of the MSB approach.

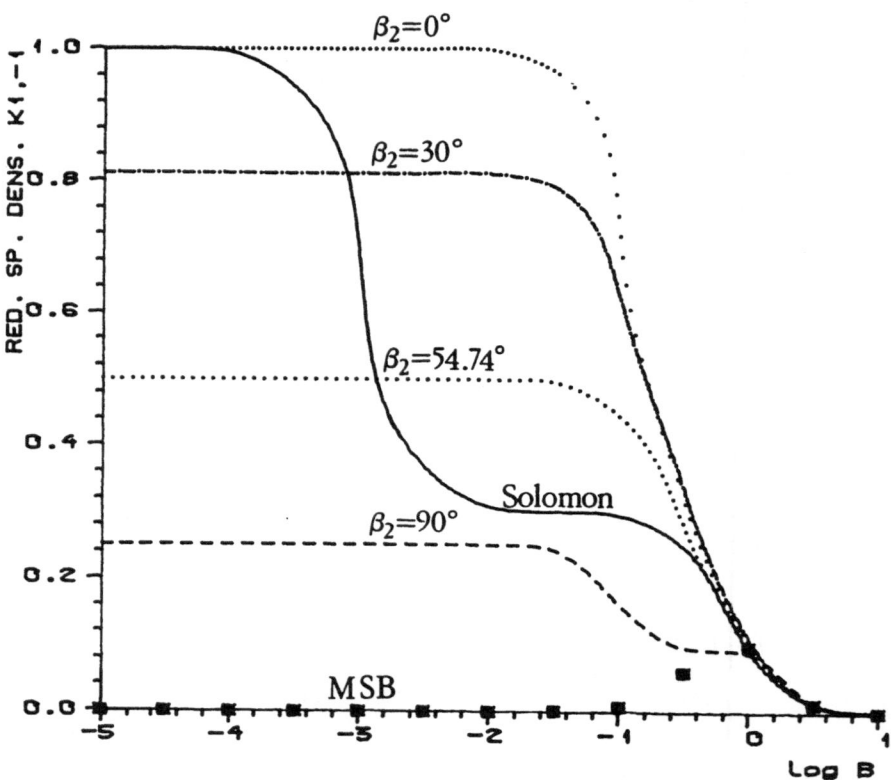

Figure 2 – The completely reduced spectral density $K^{DD}_{1,-1}$ at high ZFS as a function of the magnetic field (in Tesla) for the iso-tropic rotation–permanent ZFS model. The angle β_2 is the angle between the principal axes of the ZFS and DD tensors. The Solomon and MSB results are also shown. The rotational correlation time is 5 ns (from ref.20)

More recently, the model for systems of low symmetry (thus possessing a permanent ZFS) has been generalized to allow non-coincidence of the DD and ZFS tensors (20). This case has earlier been studied by Lindner (8) and Bertini et al. (9). In fact, these theories and the present approach predict the same trend in the dependence of the nuclear spin-lattice relaxation rate on the angle β_2 between the ZFS and DD principal axes, i.e. a decrease of the rate as the angle increases. This can be seen in Fig. 2, taken from reference (20).The main difference compared to the earlier theories (8,9) is that the angle dependence of the present theory is stronger. This is related to the fact that the Lindner and Bertini et al. theories simplify the treatment of the electron spin relaxation and thus do not take into account the cross-correlation effects. The map of $K_{1,-1}^{DD}$, analogous to Fig. 1, for $\beta_2 = 90°$ is shown in Fig. 3.

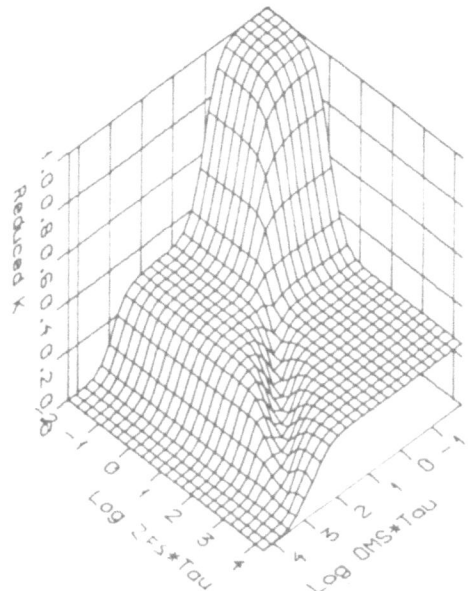

Figure 3 — Plot of $K_{1,-1}^{DD}$ for the isotropic rotation-permanent ZFS model for $\beta_2'=90°$ (from ref.20)

The effects of anisotropic reorientation on nuclear relaxation rates have recently been reviewed (21). These effects can also be dealt with in the present approach to the nuclear spin relaxation in paramagnetic systems. The liouvillian for the thermal bath can easily be modified to cover the symmetric top rotational diffusion (with the principal diffusion axis coincident with the principal ZFS axis) (20), but the remainder of the formalism is the same. The dynamic parameters now include two correlation times, $\tau_\parallel$ and $\tau_\perp$, for the reorientation parallel and perpendicular to the principal diffusion axis. An example of the results, obtained for $\tau_\parallel / \tau_\perp = 10$ and $\beta_2 = 90°$ is displayed in Fig.4. As a general comment one should mention that, in the case of anisotropic rotational diffusion, different nuclear relaxation rates in the same complex can have different temperature dependence (20).

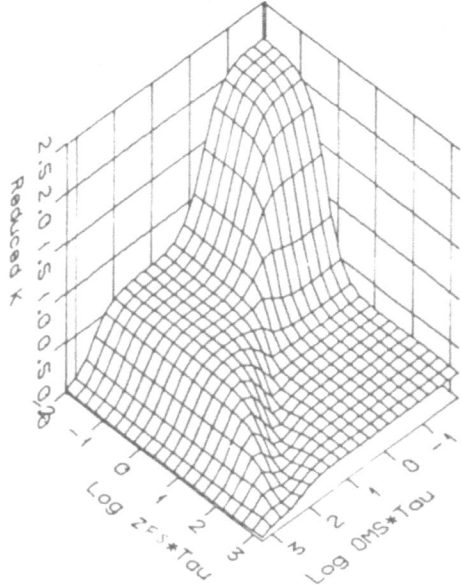

Figure 4 - Plot of $K_{1,-1}^{DD}$ for the anisotropic rotation-permanent ZFS model for $\beta_2 = 90°$ and $\tau_\parallel / \tau_\perp = 10$ (from ref.20)

The class of systems that has early attracted the attention of numerous authors (6) are complexes of high (octahedral or tetrahedral) symmetry which thus cannot possess any permanent ZFS, such as aqueous metal ions; Bloembergen and Morgan (4) have suggested that the electron spin relaxation in these systems is caused by the distortion of the high symmetry of the complexes resulting in a transient ZFS, characterized by a correlation time τ_V shorter than the rotational correlation time. For the aqueous Ni (II) ion, they found the relaxation of the water protons to be rather inefficient, indicating very rapid electron relaxation. They also warned that this might be indicative of the slow-motion regime. More recently, the case of the aqueous Ni(II) ion has been treated by Friedman and coworkers (11) and at our laboratory (22). In this communication, we concentrate on one particular dynamic model for the distortion, the pseudorotation model (22, 23). In this model, the ZFS parameter is assumed to have a constant magnitude and a random orientation in the molecular frame. The random process is characterized by a distortion correlation time τ_D, analogous to the τ_V mentioned above. In the correlation function of eq.(4), the exponential in the rotational correlation time can be factored out and the remainder treated in a way very similar to the case of isotropic rotation (22). In the limit of low ZFS, the predictions of the model become identical to those of Solomon (1) or, if the reorientation is sufficiently slow, to the MSB equations (5, 6). As the dynamics in the system approaches the slow motion regime, the present theory deviates from the MSB equations. Among other features, one should mention that the dependence of the nuclear spin-lattice relaxation rate on the magnitude of the ZFS is weaker than in the MSB approach. In the range of $f_0\tau_D$ close to unity, the low field limit of the relaxation rate predicted by the present model is higher than that of the MSB approach. The field dependence of the nuclear spin-lattice relaxation rate obtained by the two methods is however quite similar (cf. Fig. 5).It can also be seen in the figure that the present pseudorotation model for the low field coincides with the results of Friedman et al. The figure also contains some preliminary results obtained using a dynamic model in which the distortion dynamics is described by a Smoluchowski-type equation (22).

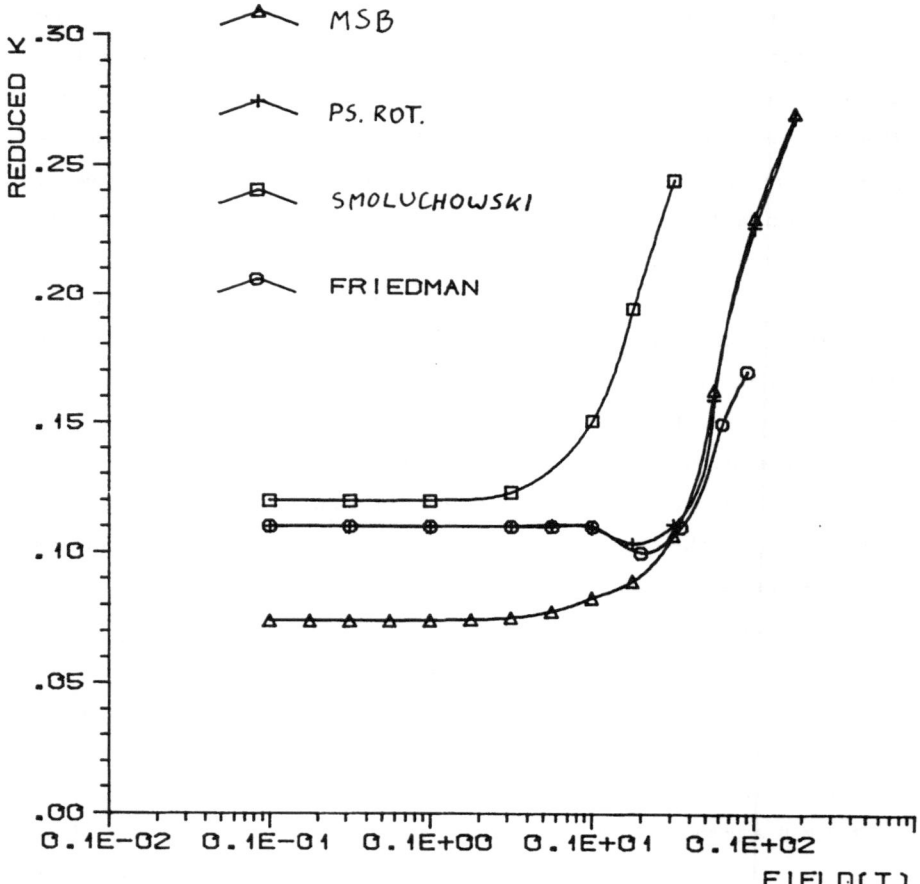

Figure 5 – The predicted field dependence of the proton spin-lattice relaxation rate in aqueous Ni(II) solution using different models. The parameters are from Friedman et al.(11)

COMPARISON WITH EXPERIMENTAL DATA

Benetis et al. (24) have studied ^{15}N and ^{1}H relaxation in the NH$_2$ group in aniline in the presence of bis (2,2,6,6-tetramethyl-heptanedionato) nickel (II) (Ni (dpm)$_2$) at two magnetic fields and over a wide temperature range and fitted the data to the permanent ZFS isotropic reorientation model described above. It was possible to obtain an excellent fit (cf. Fig. 6), but the resulting parameters (ZFS of about 70 cm^{-1} and τ_R at room temperature of few picoseconds) have been subject to criticism (25). Moreover, subsequent additional ^{15}N experiments at still

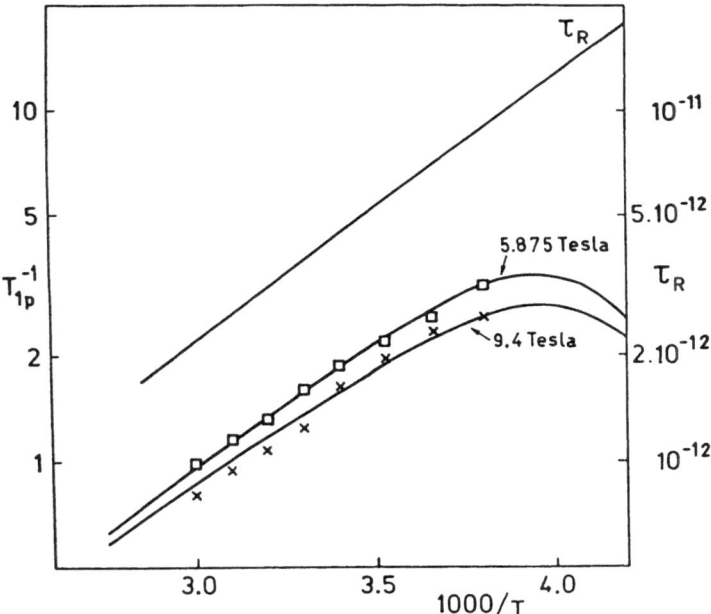

Figure 6 – The NH$_2$ proton spin-lattice relaxation in aniline in the presence of Ni(dpm)$_2$ as a function of temperature at two magnetic fields (from ref.24)

higher field did not agree with the published best-fit parameters. Possibly, the models of reference (20) could turn out to be better suited for the case at hand; in order to apply these models with their additional parameters, one would however need a more extensive set of variable-field data.

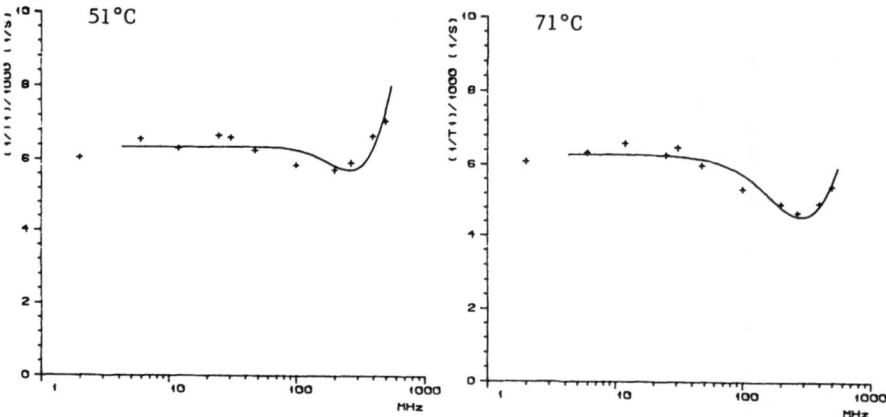

Figure 7 - The proton spin-lattice relaxation in aqueous Ni(II) solution as a function of the magnetic field at two temperatures. The relaxation rate is normalized to 1 mM Ni(II) concentration

The second experimental example is a "classic": the proton spin-lattice relaxation in the aqueous nickel (II) ion (4, 11). New experimental data for this system at low pH and low to moderate magnetic field have recently been reported by Hertz and Holz (26); in addition, high field data (2.35 to 11.7 Tesla) have been obtained at our laboratory during last months. The high temperature data (51 and 71°C), where the exchange contributions can be neglected (26) were fitted using the pseudorotation model discussed above and the MSB equations. The constancy of the relaxation rate at the low field, a dip at the intermediate field and the rise at high field are reproduced correctly (cf. Fig. 7). However, the relevance of the resulting best-fit parameters can also here be questioned.

Table I

Best-fit parameters for the proton spin-lattice relaxation rates in aqueous Ni(II) solutions at pH=0.1 obtained at two temperatures and using the MSB and the present pseudorotation models.The rotational correlation times were fixed at 14.9 ps at 51°C and 10.4 ps at 71° and the effective g value was 2.25

	r(NiH), Å	τ_D,ps	ZFSx10^{-12}, rad/s
51°C			
====			
MSB	2.17+0.04	0.45+0.03	1.63+0.13
Ps.rot.	2.24+0.04	0.57+0.04	1.55+0.14
71°C			
====			
MSB	2.27+0.03	0.38+0.02	1.50+0.10
Ps. rot.	2.32+0.04	0.42+0.02	1.51+0.10

Notably, the interspin distance obtained using the slow motion theory turns out to be very short (cf. Table I) and the fact that the MSB approach gives even shorter distance is poor consolation.It should be noted that the low-field proton relaxation rate for the aqueous nickel (II) ion can be explained by reasonable sets of parameters (for example, the parameters similar to those of Friedman et al. (11)). However in order to describe the field dependence of the relaxation rate within the pseudorotation model, the fitted parameters migrate to questionable values. A noteworthy feature of the data in Fig. 7 is that the low-field limit of the nuclear spin-lattice relaxation rate is practically temperature independent (this can also be seen in the rest of the Hertz and Holz data set (26)). Since the relaxation rate at this limit depends, on the one hand, on the DD interactions strength (presumably temperature independent) and, on the other, on the product $\Delta^2 \tau_D$ (should physically be interpreted as rms average of the "transient" ZFS), the experimental data indicate a mutual compensation of the temperature effect on τ_D and Δ.

ACKNOWLEDGEMENTS

This work has been supported by the Swedish Natural Science Research Council. Friendly assistance of Dr Kjell Ankner, Dr. Tomas Klason and Dr. Karin Leontein who helped to measure proton relaxation for the aqueous nickel ion on their spectrometers is gratefully acknowledged.

REFERENCES

1) I.Solomon, Phys.Rev., **99**, 559 (1955).

2) I.Solomon and N.Bloembergen, J.Chem.Phys., **25**, 261 (1956).

3) N.Bloembergen, J.Chem.Phys., **27**, 572 (1957).

4) N.Bloembergen and L.O.Morgan, J.Chem.Phys., **34**, 842 (1961).

5) D.R.Burton,S.Forsén,G.Karlström and R.A.Dwek, Progr.NMR Spectrosc., **13**, 1 (1980).

6) J.Kowalewski,L.Nordenskiöld,N.Benetis and P.-O.Westlund, Progr.NMR Spectrosc., **17**, 141 (1985).

7) S.Koenig,R.D.Brown and J.Studebaker, Cold Spring Harbor Symp. Quant. Biol., **36**, 551 (1971).

8) U.Lindner, Ann.Phys. (Leipzig), **16**, 319 (1965).

9) I.Bertini,C.Luchinat,M.Mancini and G.Spina, in "Magneto-structural correlations in exchange coupled systems" (D. Gatteschi, O. Kahn, and R.D. Willett, eds.), D. Reidel, Dordrecht, 1985.

10) L.T.Muus and P.W.Atkins, eds., "Electron Spin Relaxation in Liquids", Plenum Press, New York and London, 1972.

11) H.L.Friedman,M.Holz, and H.G.Hertz, J.Chem.Phys., **70**, 3369 (1979).

12) N.Benetis,J.Kowalewski,L.Nordenskiöld,H.Wennerström and P.-O. Westlund, Mol.Phys., **50**, 515 (1983).

13) H.P.W.Gottlieb,M.Barfield and D.M.Doddrell, J.Chem.Phys., **67**, 3785 (1977).

14) J.Kowalewski,A.Laaksonen,L.Nordenskiöld and M. Blomberg, J.Chem. Phys., **74**, 2927 (1981).

15) L.Nordenskiöld,A.Laaksonen and J.Kowalewski, J.Amer.Chem.Soc., **104**, 379 (1982).

16) J.Kowalewski,A.Laaksonen,L.Nordenskiöld and V.R.Saunders, J.Magn. Reson., **53**, 346 (1983).

17) N.Benetis,J.Kowalewski,L.Nordenskiöld,H.Wennerström and P.-O. Westlund, Mol.Phys., **48**, 329 (1983).

18) N.Benetis,J.Kowalewski, L. Nordenskiöld, H. Wennerström, and P.-O. Westlund, J. Magn. Reson., **58**, 261 (1984).

19) P.-O.Westlund,H.Wennerström,L.Nordenskiöld,J.Kowalewski and N. Benetis, J.Magn.Reson., **59**, 91 (1984).

20) N.Benetis and J.Kowalewski, J.Magn.Reson., in press.

21) H.G.Hertz, Progr. NMR Spectrosc., **16**, 115 (1983).

22) P.-O.Westlund, Ph.D. Thesis, University of Stockholm, 1985; P.-O Westlund, N.Benetis and H.Wennerström, to be published.

23) M.Rubinstein,A.Baram and Z.Luz, Mol.Phys., **20**, 67 (1971).

24) N.Benetis,J.Kowalewski,L.Nordenskiöld and U.Edlund, J.Magn.Reson., **58**, 282 (1984).

25) I.Bertini, private communication.

26) H.G.Hertz and M.Holz, J.Magn.Reson., in press.

PBB, Vol. 2
Advanced Magnetic Resonance Techniques
in Systems of High Molecular Complexity
© 1986 Birkhäuser Boston, Inc.

MULTIPULSE DYNAMIC NUCLEAR MAGNETIC RESONANCE

THEORY AND APPLICATIONS TO LIQUID CRYSTAL POLYMERS

K.Müller,P.Meier and G.Kothe

Institut für Physikalische Chemie, Universität Stuttgart

Pfaffenwaldring 55, D-7000 Stuttgart 80, Germany

INTRODUCTION

Pulsed nuclear magnetic resonance has been established as a valuable tool to study molecular order and dynamics in complex chemical systems (1-3). So far, however, most investigations have been concerned with conventional relaxation rates and single quantum spectra ignoring phenomena, which arise in multipulse sequences. Moreover, the methods of analysis are often limited to the fast-motion or high-field regime. In this study we present a more comprehensive approach, which is also applicable in the slow-motional and low-field region (4-6). In addition, by using multipulse sequences, arbitrary relaxation rates and lineshapes of single and multiple quantum transitions (7,8) are considered.

Multipulse dynamic NMR is a time domain technique. The spin system is subject to a sequence of non-selective rf pulses and the response after the last pulse is used to characterize the molecular dynamics of the system. According to the sequences, different NMR responses are obtained. Moreover, significant signal changes occur when the pulse separations are varied. Apparently, variation of typical NMR parameters such as pulse sequence or pulse separation provides a large number of independent experiments.

However, analysis of these experiments in terms of molecular order and dynamics is often hampered by the well-known difficulties in handling the various couplings of a multispin system. Thus, in order to simplify the analysis, nuclear spin labels with isolated magnetic interactions, are introduced. For the studies, presented in this paper, only deuteron spin labels have been employed. The technique, however, is easily extended to other nuclei, appropriate for dynamic NMR investigations.

In the first section the thoretical method is developed. Then typical examples are given to illustrate the applicability of the model and to determine the limiting conditions under which the simpler approaches used previously are valid. In the main section multipulse dynamic NMR experiments of specifically deuterated liquid crystal polymers are presented. Computer simulations provide the orientational distributions and conformations of the polymer chains and the correlation times of the various motions. They are related to molecular order and dynamics of the liquid crystal polymers. The discussion clearly demonstrates the advantages of multipulse dynamic NMR in characterizing complex chemical and biological systems.

THEORY

Basis of our dynamic NMR model is the density matrix formalism. In order to describe the time evolution of the density matrix during some arbitrary pulse sequence, we divide the sequence into regions where a pulse is present and regions where there is no pulse. Figure 1 schematically depicts various states of the density matrix of an I=1 spin system in multipulse dynamic NMR. The squares symbolize different matrix elements, hatched according to their population. Diagonal elements represent populations of the individual spin levels, while off-diagonal elements represent coherences belonging to transitions between them.

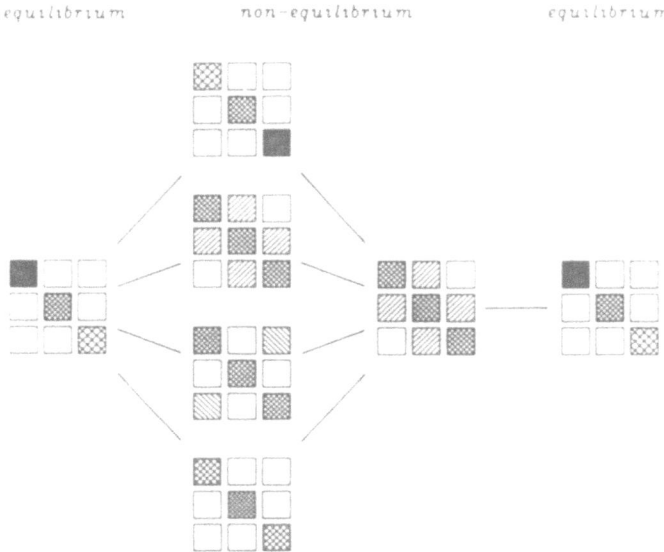

Figure 1 – Density matrix of an I=1 spin system, characterizing various states in multipulse dynamic NMR The squares symbolize
different matrix elements, hatched according to their population. Equilibrium state: Boltzmann population. Non-equilibrium states: Zeeman polarization (top), single quantum
excitation (upper centre), double quantum excitation (lower
centre), and quadrupole polarization (bottom)

Before any rf pulse is applied , the spin system is at thermal equilibrium, according to a Boltzmann population of the spin states. Application of the first pulse creates a defined non-equilibrium state, such
as the Zeeman polarization (top of Fig. 1) or single quantum excitation
(upper center of Fig. 1). After the pulse the density matrix evolves in
time under the influence of the magnetic interactions of the spin system. Then a second pulse is applied, preparing a new initial condition,
followed by a second evolution period and so on (see e.g. double quantum
excitation at lower center of Fig. 1 and quadrupole polarization at bottom of Fig. 1). The last pulse (reading pulse) finally populates those

density matrix elements, which can be detected in a conventional NMR experiment. After a time long compared to the relaxation times of the spin system, the density matrix has returned to its equilibrium state.

Generally, the observable NMR signal after the n-th (reading) pulse is given by

$$L(t, \tau_1, \tau_2, \ldots, \tau_{n-1}) = \mathrm{Tr} \left[\underset{\sim}{\varrho}(t, \tau_1, \tau_2, \ldots, \tau_{n-1}) \cdot \underset{\sim}{I}_+ \right] \cdot$$
$$\cdot \exp\left[-i\omega(t - \tau_1 - \tau_2 - \ldots - \tau_{n-1})\right] \qquad (1)$$

Where $\underset{\sim}{\varrho}(t, \tau_1, \tau_2, \ldots, \imath_{n-1})$ is the time-dependent density matrix, $\underset{\sim}{I}_+$ is the nuclear spin raising operator, ω is the angular frequency of the radiations field and $\tau_1, \tau_2, \ldots, \tau_{n-1}$ are the various pulse separation times. Fourier transformation of $L(t, \tau_1, \tau_2, \ldots, \tau_{n-1})$ starting from $t = \tau_1 + \tau_2 + \ldots + \tau_{n-1}$ (FID) or $t = \tau_1 + \tau_2 + \ldots + \tau_{n-1} + \tau_i$ (spin echo) yields single quantum frequency spectra, which sensitively depend on the actual pulse sequence. From the decay of the FID or echo amplitude as a function of a chosen τ_i various relaxation times T_{ij} can be evaluated. Finally, by Fourier transforming $L(t, \tau_1, \tau_2, \ldots, \tau_{n-1})$ for a fixed $t > \tau_1 + \tau_2 + \ldots + \tau_{n-1}$ with respect to a particular τ_i multiple quantum spectra are obtained.

Analysis of these experiments in terms of molecular order and dynamics requires a comprehensive model. We have developed such a model (4-6), based on the stochastic Liouville approach (9-12). The action of the different pulses on the density matrix $\underset{\sim}{\varrho}(t)$ is considered by unitary transformations, employing Wigner rotation matrices $\underset{\sim}{D}(\varphi, \vartheta, -\varphi)$:

$$\underset{\sim}{\varrho}^+(t) = \underset{\sim}{D}(\varphi, \vartheta, -\varphi) \cdot \underset{\sim}{\varrho}^-(t) \cdot \underset{\sim}{D^*}^{-1}(\varphi, \vartheta, -\varphi) \qquad (2)$$

Between the pulses the density matrix is assumed to obey the stochastic Liouville equation (9-12)

$$\partial/\partial t \; \varrho_{AB}(t) = (i/\hbar)\left[\varrho_{AB}(t),H_{AB}\right] + \partial/\partial t(\varrho_{AB}(t))_{inter} + \\ +\partial/\partial t(\varrho_{AB}(t))_{intra} \tag{3}$$

which we solve using a finite grid point method. The spin Hamiltonian H_{AB} employed accounts for Zeeman and quadrupolar (dipolar) interactions, including non-secular contributions.

Molecular motion in our model is considered by discrete Master equations such as

$$\partial/\partial t(\varrho_{AB}(t))_{inter} = \sum_{A'}\left[k_{A'ABB}\cdot(\varrho_{A'B}(t)-\varrho_{A'B}^{(eq)}) - \right. \\ \left. -k_{AA'BB}\cdot(\varrho_{AB}(t)-\varrho_{AB}^{(eq)})\right] \tag{4}$$

where the matrices $k_{AA'BB}$ are multiples of the unit matrix, characterizing the rate at which spins at site AB move into site A'B. The values of the transition rates depend upon the model used to describe the motion. For the intermolecular motion a diffusive process is assumed (rotation through a sequence of infinitesimally small angular steps). In that case the rotational correlation time τ_R is related to the rate constants by (4-6)

$$\tau_R = N_A^2/\left[3\pi^2(k_{AA+1BB}+k_{AA-1BB})\right] \tag{5}$$

Where N_A is the total number of sites. Note that anisotropic rotation requires two different rotational correlation times $\tau_{R\perp}$ and $\tau_{R\|}$. $\tau_{R\perp}$ is the correlation time for reorientation of the symmetry axis of the molecular diffusion tensor, while $\tau_{R\|}$ refers to rotation about it. For the intramolecular motion a jump process is assumed. Thus, isomerization occurs trhough jumps between different conformations with an avarage lifetime τ_J.

The equilibrium distribution of the spin system is described by an orientational distribution function, depending on internal and external coordinates. The internal part considers different conformations and the external part (4-6)

$$f(\varphi,\vartheta,\psi) = N_1 \ \exp A(\cos\vartheta \ \cos\xi - \sin\vartheta \ \cos\psi \ \sin\xi \ )^2$$
$$\cos\xi = \cos\delta \ \cos\varrho - \sin\delta \ \cos\varepsilon \ \sin\varrho \qquad\qquad (6)$$
$$f(\delta,\varepsilon) = N_2 \ \exp(B\cos^2\delta \ )$$

different orientations. Here φ, ϑ, ψ, δ, ε, ϱ are Euler angles, relating various molecular and laboratory systems (4-6). The coefficient A characterizes the orientation with respect to a local director (microorder), while the parameter B specifies the orientation of the director axes in a laboratory frame (macroorder). Micro- and macroorder parameters S_{zz} and $S_{z''z''}$ are related to the coefficients A and B by mean value integrals:

$$S_{zz} = \tfrac{1}{2} N_1 \int_0^\pi (3\cos^2\beta-1)\exp(A\cos^2\beta)\sin\beta d\beta$$
$$\qquad\qquad (7)$$
$$S_{z''z''} = \tfrac{1}{2} N_2 \int_0^\pi (3\cos^2\delta-1)\exp(B\cos^2\delta)\sin\delta d\delta$$

Typical multipulse sequences, employed in dynamic NMR of I=1 spin systems, are shown in Figure 2. The quadrupole echo sequence (13) (top of Fig.2) consists of two $\pi/2$ pulses, which are separated by a time τ_1 and have a $\pi/2$ relative phase shift

$$(\pi/2)_x - \tau_1 - (\pi/2)_y$$

At another time τ_1 later, a refocussing of the transverse magnetization occurs. The amplitude of the echo decreases with increasing τ_1 because of irreversible losses of phase coherence due to fluctuating quadrupole interactions. Measurement of this amplitude as a function of τ_1 provides the spin-spin relaxation time T_{2E} (14). By Fourier transforming the echo signal, a quadrupole echo spectrum is obtained (15). Since T_{2E} is most sensitive to motions with correlation times

$$\tau_R \cong (e^2 qQ/h)^{-1}, \tag{8}$$

where $e^2 qQ/h$ is the quadrupole coupling constant, quadrupole echo sequences offer a means to study molecular dynamics in the range $10^{-8} s < \tau_R < 10^{-4} s$.

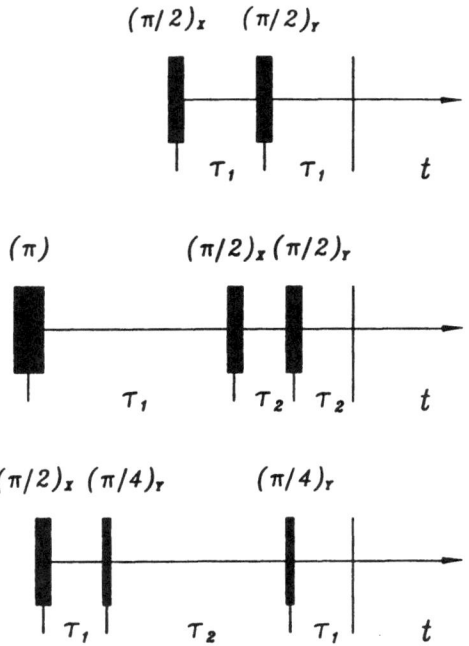

Figure 2 – Schematic representation of various multipulse sequences, employed in dynamic NMR of I=1 spin systems: Quadrupole echo sequence (top), inversion recovery sequence (center) and Jeener-Broekaert sequence (bottom)

Faster motions are accessible by employing the inversion recovery sequence (center of Fig.2)

$$\pi - \tau_1 - (\pi/2)_x - \tau_2 - (\pi/2)_y$$

in a high magnetic field. The first pulse inverts the equilibrium population of the density matrix, creating Zeeman order. The quadrupole echo sequence, applied at time τ_1 after the π pulse, transforms Zeeman order back into observable single quantum coherence. From measurements of the echo amplitude as a function of τ_1, the spin lattice relaxation times T_{1Z} can be obtained. They are particularly sensitive to motions with correlation times

$$\tau_R \cong \omega_o^{-1}, \tag{9}$$

where ω_o is the Larmor frequency. Thus, by employing high magnetic fields (B > 5 T) fast molecular dynamics in the range $10^{-10} s < \tau_R < 10^{-8} s$ can be studied.

In contrast, spin alignment permits the study of extremely slow molecular motions (16). It is created through the Jeener-Broekaert sequence (17)

$$(\pi/2)_x - \tau_1 - (\pi/4)_y - \tau_2 - (\pi/4)_y$$

depicted at the bottom of Fig.2. The first pulse generates transverse magnetization. During the evolution period τ_1 the spin system evolves in the presence of the quadrupole interactions. Spin alignment is created with the second pulse. During the long waiting period τ_2, spin alignment is subject to spin lattice relaxation with time constant T_{1Q} being comparable but not identical with the spin lattice relaxation time T_{1Z}. The third pulse transforms spin alignment back into transverse coherence. This evolves again in the presence of the quadrupolar interactions, leading to an alignment echo in phase with the reading pulse after a refocussing time τ_1. The decay of the alignment echo as a function of τ_2 generally deviates from a single exponential. Analysis of this relaxation curve yields information about type and time scale of extremely slow motions with correlation times as long as $\tau_R \cong 100$ s limited only by the condition (16)

$$\tau_R \leq T_{1Q} \tag{10}$$

Thus, by combining analysis of quadrupole echo, inversion recovery and spin alignment studies, it is possible to follow dynamic processes over 12 orders of magnitude of correlation times. Evidently, multipulse dynamic NMR represents a powerful method for studying molecular order and dynamics of complex chemical and biological systems. In the following the theory will be applied in the analysis of angular and pulse dependent NMR studies of partially deuterated liquid crystal polymers.

EXPERIMENTS AND METHODS

The family of liquid crystal polymers for this study has the general structure shown in Figure 3 (18). The Roman numerals, I-V, refer to five different polymers, deuterated at different sites in the repeating unit, as indicated in the formula. The polyesters I-III exhibit a glass temperature (T_g) at 303 K, a melting point (T_m) at 433 K and a clearing temperature (T_{ni}) at 553 K, forming a stable nematic melt over the latter temperature range (DSC and polarization microscopy). The nematic range of the polymers IV and V is smaller, extending from 429 K to 543 K. All clearing temperatures slightly depend on the average molecular weight $\overline{M}_n$

of the samples, varying between $5000 \leq \overline{M}_n \leq 30000$ (vapor pressure osmo-metry). Note that the two groups of polymers differ only in the alipha-tic spacer, containing ten and nine methylene segments, respectively.

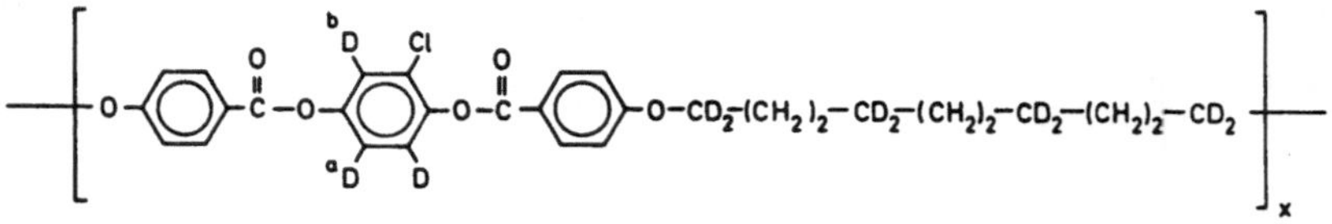

Figure 3 - Molecular structures of the liquid crystal polyesters stu-died. The Roman numerals refer to five different polye-sters, specifically deuterated at the sites indicated in the formula

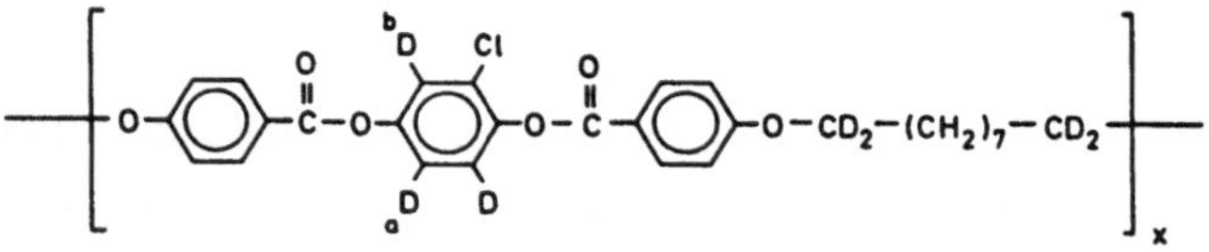

Deuteron spin labels were either attached to the central phenyl ring of the mesogenic unit or to various positions in the aliphatic spacer, as described elsewhere (19). Macroscopic alignment of the samples was achi-eved with an electric (E = 48 kV/cm) or a magnetic field (B = 7.0 T) 20 K above T_m and by solid-state extrusion 60 K below T_m.

The ^{2}H NMR experiments were carried out at 46.1 MHz, using quadrupole echo, inversion recovery and Jeener-Broekaert sequences. The width for a

$\pi/2$ pulse was 3.5 μs, employing a home-built probe (10 mm coil), equipped with a goniometer. If necessary, lineshapes were corrected for distortion, due to finite pulse width. All experiments were recorded using quadrature detection with a digitizing rate of 2 MHz and appropriate phase cycling schemes. The number of scans varied between 500 and 5000.

A Fortran program package was employed to analyze the ^{2}H NMR experiments. The programs DEUROTJUMP simulate multipulse dynamic NMR experiments of I=1 spin systems, undergoing inter- and intramolecular motion in an anisotropic medium. Numerical diagonalization of the complex symmetric matrices was achieved employing either the Rutishauser (20) or the Lanczos algorithm (21). In case of large matrix dimensions (N $\geq$ 1000) the Lanczos algorithm was found to yield reliable numerical results with considerable reduction in computing time and computer storage requirements.

Analysis of the ^{2}H NMR experiments requires knowledge of the orientation of the various molecular tensors in the polymers I-V. Angular dependent studies of macroscopically aligned samples indicate that each repeating unit can be characterized by a single order tensor, axially symmetric about its major axis. These findings correspond to the overall shape of the repeating unit, which is also expected to exhibit axially symmetric rotational diffusion about this axis. The orientation of the order (diffusion) tensor relative to the magnetic tensor of a particular ^{2}H spin label was determined from angular dependent lineshapes. Because of internal mobility there are several discrete orientations (conformations) for each deuteron site.

RESULTS AND DISCUSSION

Macroscopically aligned samples of the liquid crystal polymers I-V were studied over a wide temperature range, using multipulse dynamic NMR techniques. The observed ^{2}H NMR lineshapes and relaxation curves, varying drastically with magnetic field orientation, were simulated, employing the NMR model outlined above. An iterative fit of several angular and pulse dependent experiments for any given temperature provided reliable values for the simulation parameters, i.e. the micro- and macro-order parameters, the rotational correlation times, and the lifetimes and populations of particular conformations. Generally good agreement between experiment and simulation was found.

In Figure 4 the correlation times for the various motions of the liquid crystal polymers are plotted as a function of 1/T. They refer to reorientation of the chain axis (full circles), rotation about the chain axis (full triangles) and trans-gauche isomerization (full and open squares) of the first spacer segment, respectively. Dashed lines indicate different phase transitions (DSC). Inspection of the logarithmic plot reveals a large dynamic range, extending over five orders of magnitude.

The correlation times of Fig.4 reflect the complex molecular dynamics of thermotropic polymers in the liquid crystal, solid and glassy state. Note, that this detailed information could only be obtained employing multipulse dynamic NMR techniques. For any given temperature at least two different relaxation experiments were carried out. Moreover, variation of pulse separation and magnetic field orientation provided additional experiments for a proper dynamic characterization of the systems.

In the anisotropic melt the correlation times for chain rotation and chain fluctuation are of the order of 10^{-8} s, while trans-gauche isomerization occurs even faster. Apparently, these rapid motions are responsible for the unusual rheological behavior of liquid crystal polymers. At the melting point the situation is more complicated. Two components

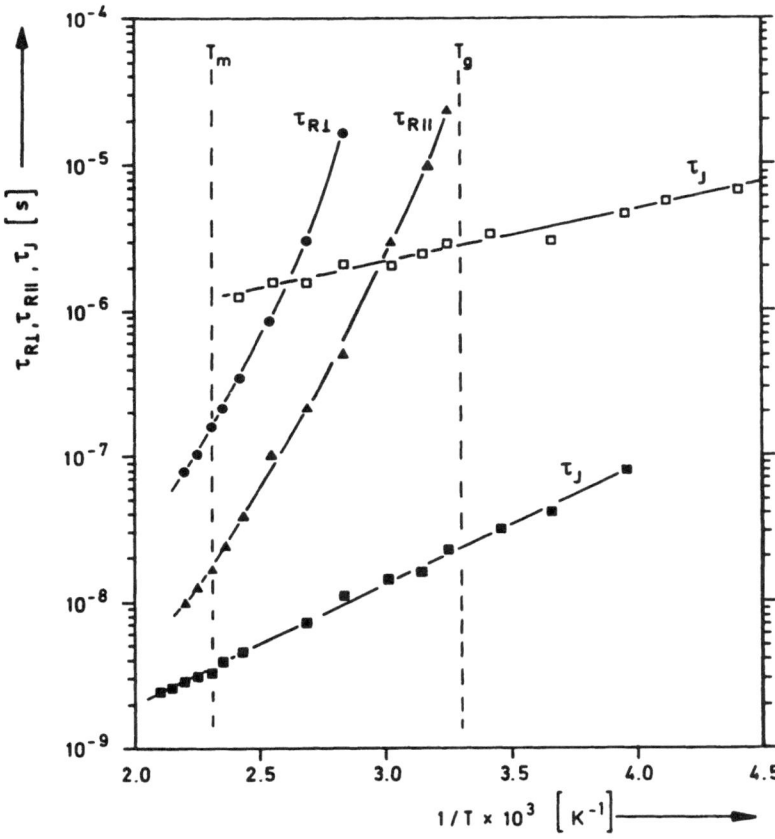

Figure 4 – Arrhenius plot of various correlation times, characterizing the molecular dynamics of the polymers I–V. Circles refer to chain fluctuation ($\tau_{R\perp}$), triangles denote chain rotation ($\tau_{R\parallel}$), and squares denote trans-gauche isomerization (τ_J) of the first spacer segment. Dashed lines indicate different phase transitions of the polymers (T_m=melting point, T_g=glass transition temperature). The liquid crystalline and crystalline components, observed below the melting point, are distinguished by full and open symbols.

are now observed, which we assign to a liquid crystalline (full symbols) and a crystalline phase (open symbols). Decomposition of various relaxation curves into two components yields a crystallinity of (60±5)%, practically independent of temperature. Note the drastic motional decrease of the crystalline component at the melting point. Only slow trans-gauche isomerization can be detected anymore.

In contrast, the dynamics of the liquid crystalline component continues into the biphasic region. One sees, however, that the Arrhenius plot for chain rotation and chain fluctuation is not linear, the apparent activation energy increasing with decreasing temperature. Thus, all intermolecular motions gradually freeze and at temperatures $T < T_g$ intramolecular motions are the dominant processes. In fact, we have been able to detect trans-gauche isomerization even at $T = 130$ K with a correlation time of $\tau_J \cong 10^{-4}$ s.

The molecular order of semiflexible thermotropic polymers comprises the conformational order of the spacer and the orientational order of the mesogenic unit. Conformational order in these systems is conveniently described in terms of trans populations n_t, giving the probability of finding a particular spacer segment in the trans state. In figure 5 these trans populations for polymers I-III are plotted as a function of the reduced temperature $T^*=T/T_{ni}$. They refer to the δ-methylene groups and are indicated by full (liquid crystalline component) and open (crystalline component) triangles.

At the isotropic-nematic transition the trans population is $n_t=0.46$. Decreasing the temperature significantly increases the conformational order to a high limiting value of $n_t=0.76$, exhibited throughout the spacer (5,6). Evidently, highly extended conformers prevail in the nematic phase of these polymers. These findings, corroborated by ^{2}H NMR studies

of other thermotropic polyesters (22,23), are the most prominent feature
distinguishing liquid crystal main chain polymers from their monomeric
analogues. It appears that a number of unique properties, exhibited by
main chain polymers can be attributed to the conformational order,
restricted to highly extended configurations (23).

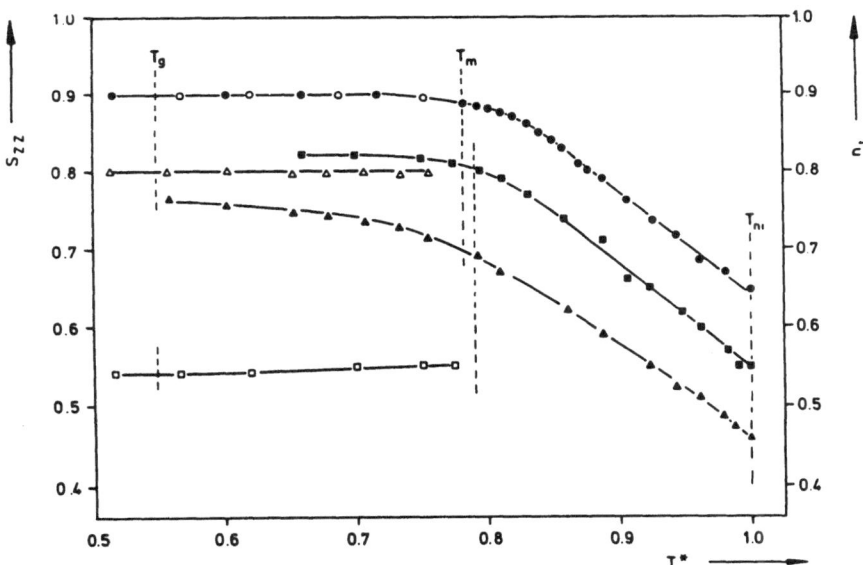

Figure 5 — Temperature dependence of the microorder parameter S_{zz} and
trans population n_t of the liquid crystal polymers studied.
Circles denote S_{zz} of polymers I-III (ten spacer segments),
while squares correspond to S_{zz} of the polymers IV and V
(nine spacer segments). Triangles indicate the trans
population n_t of the δ-methylene group of polymer III. The
liquid crystalline and crystalline components, observed
below the melting point, are distinguished by full and open
symbols. Dashed lines indicate different phase transitions.
Reduced temperature $T^*=T/T_{ni}$.

We now discuss the orientational order of the polymers in terms of the familiar microorder parameter S_{ZZ}, characterizing the average orientation of the repeating units with respect to the director. In Fig.5 this order parameters are plotted as a function of the reduced temperature $T^*=T/T_{ni}$. Circles refer to the polymers I-III (ten spacer segments), while squares correspond to the polymers IV and V (nine spacer segments). As before, liquid crystalline and crystalline components are distinguished by full and open symbols.

In the isotropic melt $S_{ZZ}=0$, indicating a random oriantation of the chains. At the isotropic-nematic transition ($T^*=1$) the order parameters jump to finite values and then increase with decreasing temperature to limiting values, which depend on the number of spacer segments. In case of ten methylene groups, liquid crystalline and crystalline components show the same limiting order parameter of $S_{ZZ}=0.9$, a value considerably larger than those observed in low molecular weight nematogens. In contrast, polymers with nine spacer segments exhibit significant lower order parameters in the nematic phase and a further decrease in the crystalline state. This pronounced odd-even effect, also observed in other thermotropic polyesters (24,25), presents a challenging theoretical problem. Statistical mechanical treatments of this phenomenon have recently appeared (26). They are in substantial agreement with the present results.

The degree of macroscopic alignment $S_{z''z''}$ of the liquid crystal polymers depends on the orientation method. Because the dielectric anisotropy of the polyester is negative, only a two-dimensional distribution of director axes is achieved using a high electric field. However, a uniform alignment of the domains is obtained by a magnetic field of 7.0 T. A detailed analysis of angular dependent ^{2}H NMR experiments yielded a macroorder parameter of $S_{z''z''}=1.0$. Likewise, solid state extrusion of the liquid crystal polyester produces fibers with $S_{z''z''}=0.9$. High modulus and strength may result from this highly oriented chain configuration (27).

Generally, the macroorder of the liquid crystal polymers is frozen in at the melting point and glass transition, respectively. No external forces are required to maintain the director distribution in the sample. Therefore, liquid crystal polymers can be used as storage material. The information, which is inserted in the nematic state by an electric field can be stored permanently in the solid state of the material. This exceptional property of liquid crystal polymers has recently been applied in a new laser-addressed thermo-optic storage device (28). Further applications in non-linear optics are currently being developed.

ACKNOWLEDGEMENTS

It is a pleasure to thank Dr.B.Hisgen (University of Mainz), Dr.A. Schneller (University of Massachusetts) and Dr.C.Eisenbach (University of Freiburg) for advice and help in preparing the specifically deuterated polymers. The authors are also grateful to Dr.E.Ohmes (University of Stuttgart) for assistance in the numerical computations. Financial support of this work by the Deutsche Forschungsgemeinschaft and Fonds der Chemischen Industrie is gratefully acknowledged.

REFERENCES

1) J.Jeener,B.H.Meier,P.Bachmann and R.R.Ernst, J.Chem.Phys. **71**, 4546 (1979)

2) J.W.Doane, in "Magnetic Resonance of Phase Transitions", (F.J. Owens, C.P.Poole,Jr. and H.A.Farah, eds.), Academic Press, New

York, 1979, p.171.

3) J.W.Emsley,Ed., "Nuclear Magnetic Resonance of Liquid Crystals", NATO Advanced Study Institute, D.Reidel Publ., Dordrecht, 1985.

4) P.Meier,E.Ohmes,G.Kothe,A.Blume,J.Weidner and H.-J.Eibl, J.Phys. Chem. **87**, 4904 (1983).

5) K.Müller,B.Hisgen,H.Ringsdorf,R.W.Lenz and G.Kothe, Mol.Cryst.Liq. Cryst. **113**, 167 (1984).

6) K.Müller,P.Meier and G.Kothe, Progr.NMR Spectrosc.(1985), in press.

7) W.S.Warren,D.P.Weitekamp and A.Pines, J.Magn.Reson. **40**,581 (1980).

8) G.Bodenhausen, Progr.NMR Spectrosc. **14**, 137 (1980).

9) R.Kubo, "Stochastic Processes in Chemical Physics", Advances in Chemical Physics, (K.Shuler ed.), Wiley, New York, 1969, Vol.16, p.101.

10) J.R.Norris and S.I.Weissman, J.Phys.Chem. **73**, 3119 (1969).

11) J.H.Freed,G.V.Bruno and C.F.Polnaszek, J.Phys.Chem. **75**, 3385 (1971).

12) G.Kothe, Mol.Phys., **33** 147 (1977).

13) J.C.Powles and J.H.Strange, Proc.Phys.Soc. **82**, 6 (1963).

14) D.E.Woessner,B.S.Snowden and G.H.Meeyer, J.Chem.Phys. **51**, 2968 (1969).

15) J.H.Davis,K.R.Jeffrey,M.Bloom,M.F.Valic and T.P.Higgs, Chem.Phys. Letters **42**, 390 (1976).

16) H.W.Spiess, J.Chem.Phys. **72**, 6755 (1980).

17) J.Jeener and P.Broekaert, Phys.Rev. **157**, 232 (1967).

18) J.-I.Jin,S.Antoun,C.Ober and R.W.Lenz, Br.Polym.J. 132 (1980).

19) K.Müller,C.Eisenbach,B.Hisgen,H.Ringsdorf,A.Schneller,R.W.Lenz and G.Kothe, to be published.

20) R.G.Gordon and T.Messenger, in "Electron Spin Relaxation in Liquids", (L.T.Muus and P.W.Atkins eds.), Plenum Press, New York, 1972, p.341.

21) G.Moro and J.H.Freed, J.Chem.Phys. **74**, 3757 (1981).

22) E.T.Samulski,M.M.Gauthier,R.B.Blumstein and A.Blumstein, Macromolecules **17**, 479 (1984).

23) D.Y.Yoon,S.Bruckner,W.Volksen,J.C.Scott and A.C.Griffin, Faraday
Discuss.Chem.Soc. (1985) in press.

24) A.C.Griffin and S.J.Havens, J.Polym.Sci.Polym.Phys.Ed. **19**, 951
(1981).

25) A.Blumstein and O.Thomas, Macromolecules **15**, 1264 (1982).

26) A.Abe, Macromolecules **17**, 2280 (1984).

27) D.A.Simoff and R.S.Porter, Mol.Cryst.Liq.Cryst. **110**, 1 (1984).

28) V.P.Shibaev, S.G.Kostromin, N.A.Platé, S.A.Ivanov, V.Y.Vetrov and
I.A.Yakovlev, Polym.Commun. **24**, 364 (1983).

PBB, Vol. 2
Advanced Magnetic Resonance Techniques
in Systems of High Molecular Complexity
© 1986 Birkhäuser Boston, Inc.

DEUTERIUM SPIN RELAXATION AND MOLECULAR MOTION IN LIQUID CRYSTALS

R.L.Vold,R.R.Vold,J.F.Martin*,B.C.Nishida** and L.S.Selwyn***

Department of Chemistry, University of California, San Diego

La Jolla, California 92093

* Department of Radiology, School of Medicine

University of California, San Diego

La Jolla, California 92093

** IBM Instruments, Inc., Orchard Park

Danbury, CT 06810

*** National Research Council of Canada, Division of Chemistry

Ottawa, Canada K1A OR6

INTRODUCTION

Nuclear spin relaxation is a particularly useful tool for investigating
molecular motion in liquid crystals because there are many motional processes with correlation times in the vicinity of nuclear Larmor frequencies. A partial list of such processes includes molecular reorientation,
quasicoherent director fluctuations (1), fluctuations of local order
(2), translation-rotation coupling (3,4), and chemical exchange between
"sites" of different degrees of local order (5). It is generally far
from trivial to decide what combination of processes is the dominant
source of spin relaxation, and measurements over as wide a range of
Larmor frequencies as possible are needed. In addition, different motional processes may have characteristically different temperature depen-

dence, although it must be noted that quantitative analysis of the temperature dependence is often hindered by lack of sufficient information
about hydrodynamic parameters which appear in director fluctuation
theories.

Deuterium spin relaxation offers considerable advantages as a probe of
molecular motion in liquid crystals. The dominant (quadrupolar) mechanism is sensitive only to molecular reorientation so that translational
motion will not contribute to relaxation except indirectly through translation-rotation coupling. This is simpler than the case of proton dipolar relaxation, where separation of inter- from intramolecular contributions is difficult. A second advantage of deuterium relaxation is that
the range of conveniently accessible Larmor frequencies (ca. 4 to 76
MHz) happens to cover the "cutoff region" (6) for director fluctuations
so that this particular process can be studied in detail. Perhaps the
most important advantage of deuterium relaxation is the simple set of
relations (7,8) between observable relaxation rates and spectral
densities of motion. In a monodomain liquid crystal, such that the deuterium spectrum consists of a well resolved doublet, the two spectral
densities $J_1(\omega_o)$ and $J_2(2\omega_o)$ are easy to determine separately: any of
several methods (9-12) can be used to establish an initially asymmetric
population difference, and subsequent decay of the difference magnetization yields a value for $J_1(\omega_o)$ while recovery of the sum magnetization
to thermal equilibrium depends as well on $J_2(2\omega_o)$.

It is important to measure individual spectral densities because they
probe components of fluctuating fields with different symmetry. The symmetry of fluctuating fields which contribute to J_1 is that of d_{xz} and
d_{yz} "orbitals" while those which influence J_2 have d_{xy} and $d_{x^2-y^2}$ symmetry. This distinction is of no consequence in isotropic media, but in
liquid crystal solutions the loss of spherical symmetry breaks the

"degeneracy" of J_1 and J_2. Any process which involves small angle fluctuations with respect to a space fixed axis (e.g. director fluctuations)
will then contribute very differently to the two spectral densities.

Spectral densities derived from ^2H spin relaxation data depend on the
value chosen for the deuterium quadrupole coupling constant, whose square appears as a proportionality factor in relevant formulae (8). As noted elsewhere in this monograph (13), the proper value of quadrupole
coupling for use in relaxation formulae is not easy to determine, even
when rigid lattice (solid state) values are known. This implies that
correlation times (or more generally, spectral density parameters) determined from ^2H spin relaxation must be treated with caution when compared to the results of other techniques. Of course, the spectral density ratio J_1/J_2 does not suffer from this drawback.

Most deuterium relaxation data for liquid crystals has been analyzed in
terms of a combination of molecular reorientation and director fluctuations (2,14), with perphas a small contribution from slow relaxation of
local structure (15,16). The contribution from molecular reorientation
must include explicit effects of the anisotropic restoring potential. In
this paper we present new data for deuterated p-methoxybenzylidene-p-n-
butyl aniline (MBBA) and for p,p'-diethynylbenzene-d$_2$ (DEB) in octylcyanobiphenyl (8CB), which appear to require that the usual approximations be reassessed. For example, standard formulae for director fluctuations are all based on the assumption of infinitesimal angular fluctuations, which leads to the prediction that J_2 is completely insensitive to this process. Another approximation, which may prove to be less
reliable for ordered fluids than for isotropic media, is the use of symmetric rotor correlation functions to describe a molecule such as DEB
which can have a fully asymmetric rotational diffusion tensor.

EXPERIMENTAL METHODS

Most of the spectral density data reported here were obtained using the Jeener-Broekaert pulse sequence (17) to monitor both Zeeman and quadrupolar order (10). For DEB solutions in which the large quadrupolar splitting precluded use of nonselective pulses, modulated pulses (18) were used both for nonselective R_1 measurements and, with appropriate attention to properly offset carrier frequency, for selective R_1 measurements. Precise temperature control was achieved using homebuilt digital temperature controllers (19). Full experimental details will be presented at a later date.

The large quadrupolar splitting encountered in liquid crystal solutions offers the opportunity to investigate non-trivial effects of imperfect pulses. The simplest "imperfection" is inadequate pulse power. For a three level system, a weak pulse does not simply rotate the magnetization about an effective field with a component in the z direction. Instead, the magnetization is rotated in a higher dimensional space (20) and after a nominal 90° pulse, there is transverse quadrupolar order (antiphase magnetization) and double quantum coherence as well as the desired transverse magnetization. Phase cycling techniques (21) suppress double quantum artifacts, but the simultaneous creation of transverse magnetization $\langle I_x \rangle$ and transverse quadrupolar order $\langle Q_x \rangle$ leads to unavoidable phase distortions (22). Fortunately, for the case of a simple quadrupolar doublet these distortions can be removed by an appropriate combination of constant and linear phase corrections.

A second, less obvious effect of imperfect pulses is associated with the finite rise and fall times of the pulse. It has long been known to practitioners of solid state multipulse spectroscopy (23) that the phase distortion introduced by asymmetric rise/fall characteristics of a pulse is equivalent in lowest order to an effective "resonance offset". The situation is slightly different for deuterium, since the quadrupole coupling may be comparable to the rf field strength. In this case, numerical calculations (22) reveal that asymmetric phase transients produce an effective quadrupolar "offset" which decreases with increasing pulse strength. The result is that the effective nutation angle is different for the two components of the quadrupolar doublet, leading to a spectrum in which the two components have unequal intensity. These and other pulse artifacts have been described in a recent paper by Henrichs et al. (24).

For purposes of relaxation measurements, the existence of asymmetric doublet intensities implies an apparent non-zero equilibrium value of quadrupolar order. It is necessary to include such a term when fitting relaxation data to derive spectral density parameters, but except for a (small) reduction in the dynamic range of the experiment this artifact introduces no errors.

RESULTS AND DISCUSSION

Figures 1 and 2 show spectral density data for MBBA deuterated in the linkage position. The deuterium spectrum of this molecule in the isotropic phase consists of a single line. Individual spectral densities for this phase were therefore obtained indirectly by analysing the lineshape of directly bound ^{13}C (25). Isotropic phase measurements were carried out at ^{13}C Larmor frequencies of 15.1, 25.2, and 50.3 MHz, corresponding to J_1 for the deuteron at 9.2, 15.4, and 30.7 MHz respectively, and to J_2 at twice these frequencies. It was not possible to determine separate values of J_1 and J_2 from the 15.1 MHz ^{13}C lineshapes because the line was almost fully collapsed. As expected (and as required by considerations of symmetry), J_1 and J_2 evaluated at the same frequency in the isotropic phase were found to be identical within experimental error. The deuteron spin lattice relaxation rate (proportional to $(J_1(\omega_o)+4J_2(2\omega_o)$ was measured and found to agree with values calculated from data in figures 1 and 2. All the spectral densities in these figures were obtained using an assumed value of 162 kHz for the deuterium quadrupole coupling constant.

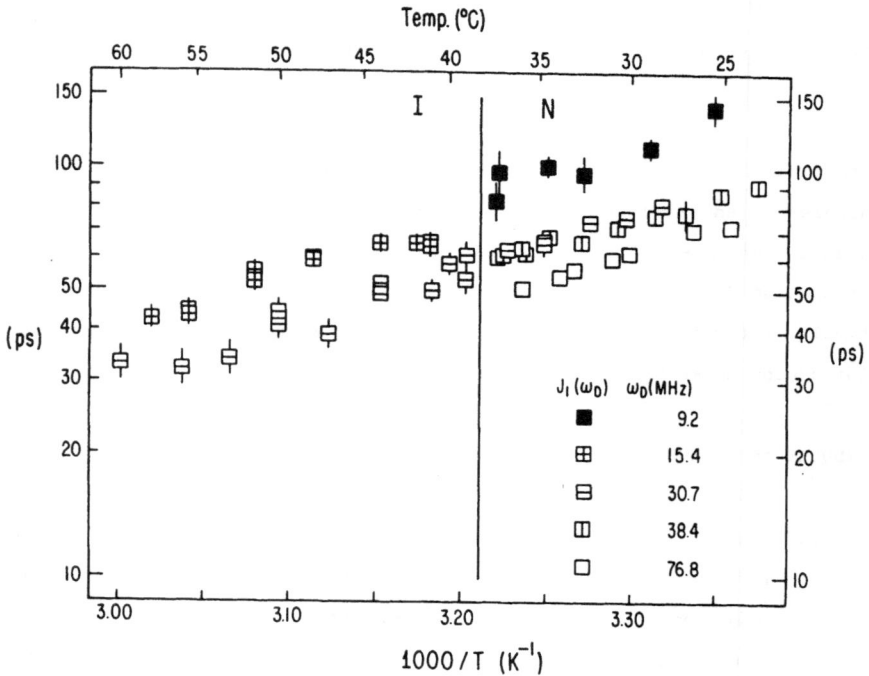

Figure 1 - Spectral density $J_1(\omega_o)$ for MBBA selectively deuterated at the linkage position. Data for the isotropic phase were obtained indirectly from ^{13}C lineshape analysis, and data for the nematic phase were obtained using the Jeener-Broekaert pulse sequence. Note that J_1 is frequency dependent on both sides of the I→N phase transition. In the isotropic phase (only!) this frequency dependence and that of J_2 shown in fig.2 can be accounted for using the Woessner-Huntress (26, 27) symmetric rotor formalism

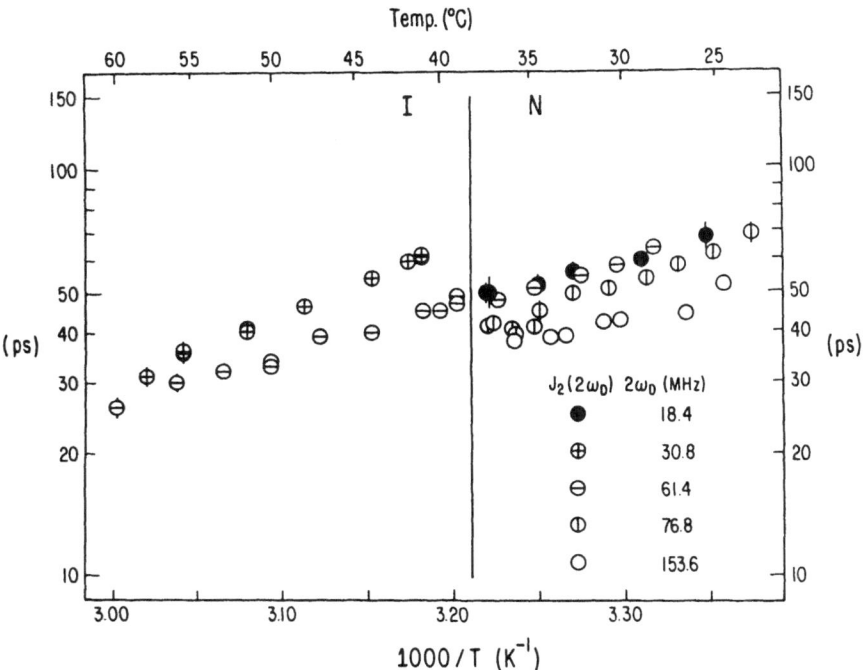

Figure 2 - Spectral density $J_2(2\omega_o)$ for MBBA selectively deuterated at the linkage position. The discontinuity in J_2 observed at the phase transition, combined with the large nematic phase J_1 values shown in fig.1, cannot be accounted for by the simple model of a symmetric rotor moving in a cylindrically symmetric ordering potential

Although MBBA is a flexible molecule with no particular elements of sym-metry, the symmetric rotor formalism of Woessner (26) and Huntress (27) was used to interpret the isotropic phase relaxation data. According to this model, the spectral density $(J_1 = J_2)$ is given at any frequency ω by the relation.

$$J(\omega) = 0.05 \left[(3\cos^2\vartheta - 1)^2 \, \tau_\perp / 1 + \omega^2 \tau_\perp^2 + 12\sin^2\vartheta \cos^2\vartheta \, \tau_b / 1 + \omega^2 \tau_b^2 \right.$$

$$\left. + 3\sin^4\vartheta \tau_c^2 / 1 + \omega^2 \tau_c^2 \right] \tag{1a}$$

with

$$1/\tau_b = 5D_\perp + D_\parallel \tag{1b}$$

$$1/\tau_c = 2D_\perp + 4D_\parallel \tag{1c}$$

where ϑ is the angle between the CD bond and the symmetry axis (principal z-axis of the rotational diffusion tensor), and $\tau_\parallel = 1/6D_\parallel$ and $\tau_\perp = 1/6D_\perp$ are the two principal correlation times. Attempts to fit the relaxation data to a three parameter expression based on Eq.(1) failed; the calculated values of $(dJ/d\vartheta)$ are very small. Therefore, the relaxation data were fit by optimizing values of $\tau_\perp$ and $\tau_\parallel$ for different choices of ϑ. Values of ϑ leading to unphysical results (such as negative correlation times or a wildly non-Arrhenius temperature dependence) were rejected. It was possible to fit the data throughout the isotropic phase (ca. 39 to 60°C) using a single value of ϑ; all values between 60 and 66 degrees gave equally good fits. This result is reasonable in view of the known geometry (28) of MBBA. At all temperatures $\tau_\perp$ was found to be much larger than $\tau_\parallel$, which is to be expected because of the elongated molecular shape. Further details will be reported elsewhere (29). For present purposes it suffices to note that the symmetric rotor model sucessfully accounts for all features of the isotropic phase relaxation data. In particular, there is not need to invoke the existence of slowly relaxing local structures to account for ^{2}H spin lattice relaxation, even very close to the I→N phase transition (5).

A quite different picture emerges from studies of solute relaxation in MBBA (30) and other (31) nematogenic fluids. Figure 3 shows relaxation rates of chloroform-d in methoxymethoxybenzylidene-p-n-butylaniline (MMBBA), an analog of MBBA which exhibits no nematic phase before freezing at ca. -10°C. The solid and dashed lines were calculated by fitting

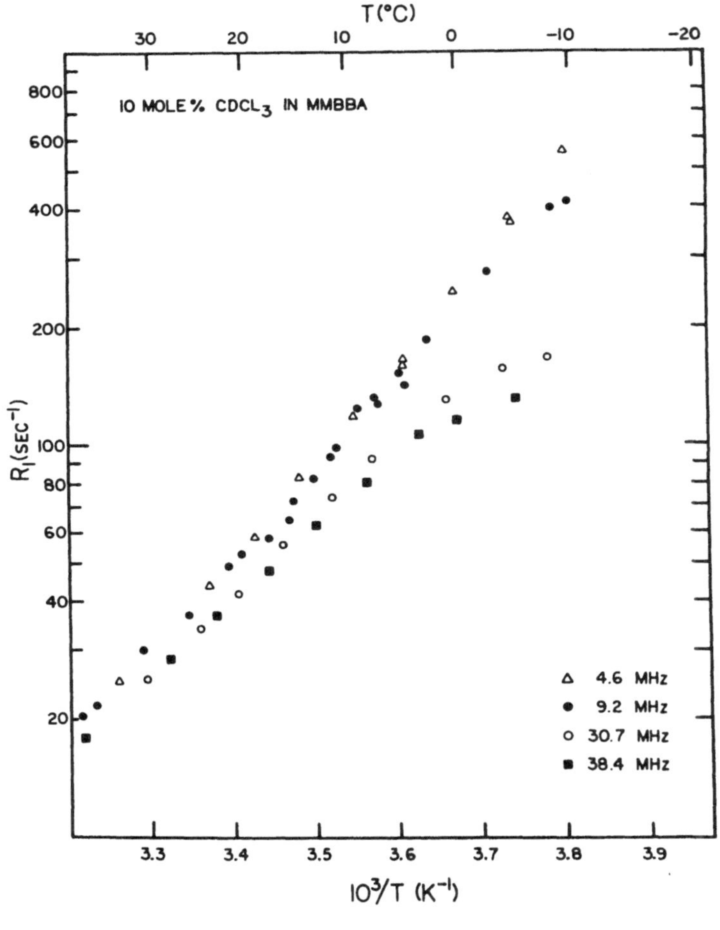

Figure 3 – Deuterium spin lattice relaxation rates of chloroform-d in methoxymethoxybenzylidene-p-n-butylaniline. Points represent experimental data at the indicated Larmor frequencies, and the lines are calculated according to the model of slowly relaxing local structures (2). The solid lines represent a three parameter fit of data at each temperature to Eq.(2), and the dashed extrapolations define a range of temperatures for which local structure relaxation contributes noticeably to the relaxation even though the frequency dependence is not very pronounced

the relaxation data to a spectral density formula

$$J(\omega) = (1-S_1^2)\tau_R/1+\omega^2\tau_R^2 + S_1^2\tau_x/1+\omega^2\tau_x^2 \qquad (2)$$

Here τ_R is the rotational correlation time for the chloroform C_3 axis, S_1 describes the degree of local order, and τ_x is a non-specific relaxation time for the local structure (2). This formula has been used previously to account for relaxation of chloroform in MBBA (30), CD_2Cl_2 in Merck Phase 5 (32), and diethynylbenzene in p-butoxybenzylidene-n-octylaniline (33) and octylcyanobiphenyl (34). Details of the fitting procedure and results for additional solute/solvent combinations will be reported elsewhere (31).

It is interesting to note that the term in Eq.(2) involving $(1-S_1^2)$ appears both in the "model free" formalism of Lipari and Szabo (35) and in the "slowly relaxing local structure" model of Freed (2). In essence, both formalisms partition the observed relaxation into a fast component (associated here with solute reorientation) and a slow component (associated with motion of the surrounding solvent). As noted by Lipari and Szabo, such a formula can be expected to fit relaxation data over a rather wide range of frequencies irrespective of the precise details of the motion. It follows that a good fit of Eq.(2) to experimental data does not by itself lend credence to the notion of slowly relaxing local structures. However, for the case of solutes in nematogenic fluids one can make further progress by comparing the parameters determined from Eq.(2) with independent information such as directly measured correlation times for solvent motion or the order parameter of the solute in a underlying nematic phase. For chloroform in MMBBA, we obtain (31) values for the local order parameter which are similar to those observed directly for chloroform in nematic MBBA at the same temperature, and an activation energy 6.5±0.5 kcal/mole for τ_x, which is not too different from the activation energy (9.4 kcal/mole) of the capillary viscosity of MMBBA.

The dashed lines in fig.3 indicate a range of temperatures in which the observed relaxation rates show a frequency dependence which is too weak for direct analysis, but for which the second term of Eq.(2) nevertheless contributes a significant fraction of the relaxation. The implication is that relaxation data for small solutes in viscous solvents, even when observed to be independent of Larmor frequency, may be difficult to ascribe exclusively to "fast" molecular reorientation.

At this point we return to figures 1 and 2 and attempt to interpret the spectral density data for the nematic phase. For this purpose three models have been considered: strong collision (36), diffusion in a cone (37,38), and rotational diffusion in presence of a restoring pseudo-potential (39). The strong collision model was discarded at an early stage because it involves the untenable assumption that a single correlation time characterizes both J_1 and J_2. (It is possible, though not desirable, to introduce two different correlation times here on an ad hoc basis). As noted elsewhere (38), rotational diffusion in a cone with a specified half angle (chosen to match the observed order parameter S_{zz} for the long molecular axis) is remarkably successful in reproducing the major features of deuterium spin relaxation in ordered phases. The use of a more realistic restoring potential in place of the infinite square well used in the cone model might then be expected to yield quantitatively accurate values for the spectral density parameters.

Solving the rotational diffusion equation including an anisotropic restoring potential is a nontrivial exercise, especially when the motion is slow enough, as in the present case, to render extreme narrowing assumptions untenable. For the case of a symmetric rotor in an axially symmetric restoring potential, each spectral density $J_M(\omega)$ can be expressed (2,39) as a sum of three terms $J_{MK}(\omega)$ where K refers to rotation about a molecule fixed principal axis and M to rotation about a space fixed axis. These calculations may be summarized by the formula.

$$J_{MK}(\omega) = C_{MK} \sum_i \alpha_{MK}^{(i)} \beta_{MK}^{(i)} \tau_K / 1 + (\beta_{MK}^{(i)} \tau_K)^2 \omega^2 \qquad (3)$$

Here the initial amplitude C_{MK}, the number of Lorenztians, their relative weights $\alpha_{MK}^{(i)}$, and the effective correlation times $\beta_{MK} \tau_K$ are functions of the restoring potential. τ_K in Eq.(3) is the appropriate correlation time in absence of any restoring potential; i.e.β_{MK} goes to unity as the restoring potential goes to zero.

First attempts to fit the data in figures 1 and 2 using Eq.(3) were made by extrapolating the correlation times determined for the isotropic phase, in the expectation that the local environment experienced by an MBBA molecule is not very different on either side of the phase transition. It was not possible to account in this fashion for the rather large difference between J_1 and J_2. These calculations were made using a Maier-Saupe restoring potential

$$V(\beta) = \lambda(3\cos^2\beta - 1) \qquad (4)$$

with λ chosen to reproduce the observed quadrupolar splitting (β is the angle between the molecule fixed z axis and the director). Bernassau et al. (40) have pointed out that somewhat larger values of λ might be more appropriate, but we are unable to find any value of λ which fits the data.

Examination of figures 1 and 2 reveals that in the nematic phase, J_1 is larger than J_2 (especially at low frequencies). Director fluctuations certainly contribute far more strongly to J_1 than to J_2, but this cannot account for our MBBA data: the elastic constants (41), viscosities (42) and self-diffusion coefficients (43) of MBBA have all been measured, so that accurate calculations of director fluctuations are possible. For the linkage deuteron, very small contributions to J_1 (no more than ca. 4 ps at 9.2 MHz) are obtained. This happens because the effective order parameter for the CD bond is very small.

It is known (44) that the ordering of the linkage CH bond in MBBA must be described in terms of two order parameters, S_{zz} and $S_{xx}-S_{yy}$. This implies that $V(\beta)$ in Eq.(4) is incomplete: a second term must be added which depends on the Euler angle γ describing the orientation of the principal x axis of the order tensor in a molecule fixed frame. (Terms involving the Euler angle alpha are absent because the nematic phase is itself assumed to be uniaxial). Thus a more realistic solution of the rotational diffusion equation is obtained (15) by including a potential of the form

$$V(\beta,\gamma) = C_1 D_{00}^{(2)}(\Omega) + [C_2 D_{0,2}^{(2)}(\Omega) + D_{0,-2}^{(2)}(\Omega)] \tag{5}$$

where Ω stands for the Euler angles ($\alpha,\beta,$ and γ) which relate a director fixed axis system to one fixed in the molecule. It is easy to show from Eq.(5) that the second term introduces a restoring torque $iM_z V(\beta,)$ about the molecule fixed z-axis, given by

$$T_z = \left(\frac{3}{8}\right)^{\frac{1}{2}} C_2 \sin^2\beta \sin 2\gamma \tag{6}$$

Thus rotational motion about the molecular z-axis is hindered in a fashion which does not happen in the isotropic phase, where $V(\beta,\gamma)=0$. This is reflected in formulae for matrix elements of the symmetrized diffusion operator Γ: terms proportional to $(D_{xx}+D_{yy})C_1$ must be supplemented by terms proportional to $D_{zz}C_2$. Here D_{ii} is the rotational diffusion constant for motion about the i-th principal molecular axis. Even though C_2 for MBBA is expected to be no larger than ca. 10% of C_1, D_{zz} is expected to be sufficiently large compared with the other principal diffusion coefficients that terms involving its product with C_2 are far from negligible. It follows that the effective correlation times appearing in Eq.(6) bear no simple relation to values extrapolated from data for the isotropic phase.

Effects of an "asymmetric" restoring potential should not be confused with complications introduced by an asymmetric diffusion tensor, $D_{xx} \neq D_{yy}$. In absence of either sort of asymmetry, the diffusion operator is diagonal in both indices M and K appearing in Eq.(3), and both sorts of asymmetry couple matrix elements (M,K) to (M,K$\pm$1) and (M,K$\pm$2). However, the fact that for liquid crystals D_{zz} is so much larger than $D_{xx}-D_{yy}$ implies that asymmetric ordering is likely to have a large influence on the form of the spectral densities even when the anisotropy of the diffusion tensor is quite small. Preliminary calculations indicate that solving the rotational diffusion equation subject to the potential given by Eq.(6) can readily account for the observed spectral densities of MBBA, and further calculations are in progress to establish error limits.

Figure 4 shows spectral density data for p,p'-diethynylbenzene-d_2 in octylcyanobiphenyl. Data for the isotropic and nematic phases can be interpreted in essentially the same manner as done previously for rigid solutes in other mesogenic solvents (18,33,45). In the isotropic phase the dominant motional processes are molecular reorientation and (for low Larmor frequencies) slowly relaxing local structures. Upon entering the nematic phase, the local structures are transformed into long range order and the resulting restoring potential, which increases in strength

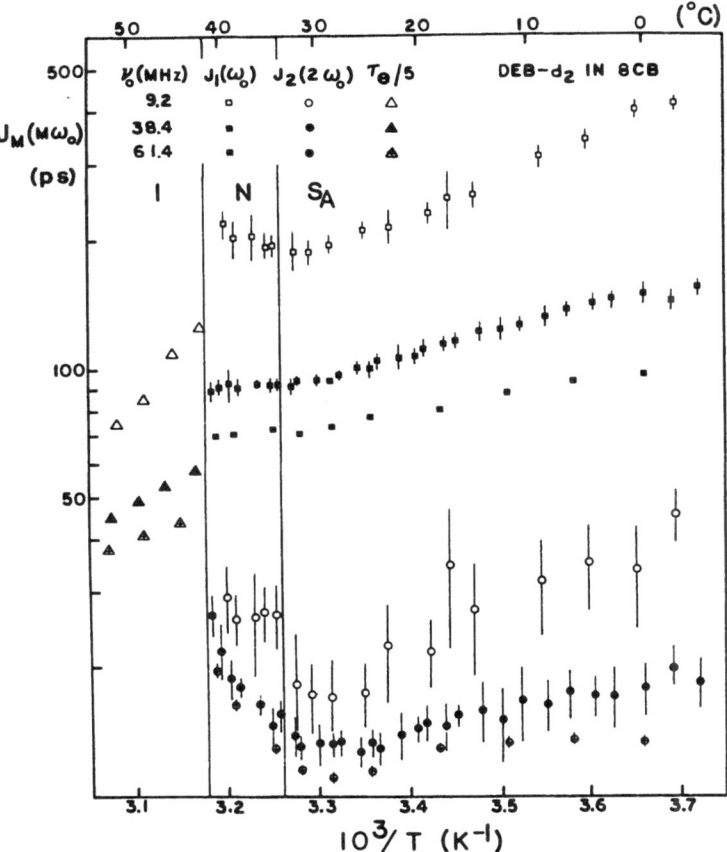

Figure 4 – Spectral density data for diethynylbenzene in octylcyanobiphenyl. Points in the isotropic phase correspond to $0.2\tau_e$,
an effective correlation time derived simply from the measured R_1 values by dividing by an appropriate interaction
strength (34). Spectral densities for the ordered phases
were obtained by a combination of nonselective and selective inversion-recovery experiments. Isotropic and nematic
phase spectral densities can be accounted for in terms of
fast molecular reorientation (both phases), slowly relaxing
local structures (isotropic phase only) and small-angle director fluctuations (nematic and probably also smectic phase). The frequency dependence of J_2 cannot be adequately
reproduced by any of the standard motional models

as temperature decreases, is responsible for the decreasing values of J_2. In essence, the rotational correlation time is becoming longer at lower temperatures but the angular extent of the motion is getting smaller and smaller. These competing influences on J_2 lead to a shallow minimum for this parameter at a temperature somewhat below the $N \rightarrow S_A$ phase transition.

J_1 in fig.4 certainly reflects contributions from director fluctuations, and using reasonable estimates for the relevant hydrodynamic parameters it is possible (34) to account quantitatively for the nematic phase spectral densities. The situation in the smectic phase is less clear, in part because there is less extensive data for elastic constants and viscosities. The most novel feature of fig.4 is the frequency dependence found for J_2. It is not possible to account for this frequency dependence using a symmetric rotor model with a symmetric restoring potential; for any choice of λ in Eq.(4) correlation times which are long enough to produce noticeable frequency dependence also predict spectral densities much larger than those observed. Possible sources of the observed frequency dependence include a) effects of motional asymmetry as discussed above for MBBA, b) a breakdown of the small-angle approximation used in calculations of director fluctuations, and c) coupling between solute orientation and translational diffusion through the smectic layers (4). None of these possibilities can be ruled out at present.

CONCLUSIONS

Qualitative features of deuterium spin lattice relaxation in liquid crystals may be adequately explained by fast molecular reorientation through a range of angles restricted by the restoring pseudopotential, combined with slower motion of the director over a much more restricted angular range. Relaxation of solute spins is similar to that of solvent spins, differing primarily in that rotational correlation times for solutes are typically shorter and less anisotropic than those of solvents. When the ordering is strong, as is normally the case for solvents but only occasionally for solutes, it appears necessary to include effects of asymmetric ordering in modified spectral density functions.

Director fluctuations usually contribute more strongly to relaxation of solute deuterons than to those of the solvent, in part because solvent CD bonds are typically oriented at an unfavorable angle with respect to the long molecular axis, and in part because the shorter rotational correlation times for solutes tend to make rotational contributions less overwhelming. Director fluctuation contributions are best studied via measurements at Larmor frequencies between ca. 4.0 and 30 MHz. Measurements at lower frequencies suffer from low signal/noise ratio and director fluctuation contributions "cut off" at higher frequencies. Although the dominant contributions of director fluctuations occur in J_1, there may be observable higher order contributions to J_2.

ACKNOWLEDGEMENTS

This work was supported by the National Science Foundation (Grant CHE-8122097), the Exxon Educational Foundation, and the Petroleum Research Fund of the American Chemical Society (Grant ACS PRF13167-AC5). We are also grateful to the National Sciences and Engineering Research Council of Canada for a Postgraduate Research Scholarship to L.S. Selwyn. We are also grateful for access to high field intrumentation at NSF-sponsored Regional NMR Centers at the California Institute of Technology and the University of South Carolina.

REFERENCES

1) P.Pincus, Solid State Commun. **7**,415 (1969).

2) J.H.Freed, J.Chem.Phys. **66**, 4183 (1977).

3) R.Blinc, M.Luzar, M.Vilfan and M.Burgar, J.Chem.Phys. **63**, 3445 (1975).

4) G.Moro and P.L.Nordio, J.Phys.Chem., **89**, 997, (1985).

5) J.F.Martin,R.R.Vold and R.L.Vold, J.Chem.Phys. **80**, 2237 (1984).

6) J.W.Doane,C.E.Tarr and M.A.Nickerson, Phys.Rev.Lett. **33**, 620 (1974)

7) H.Bildsoe, J.P.Jacobsen and K.Schaumburg, J.Magn.Reson. **23**, 137 (1976).

8) R.R.Vold and R.L.Vold, J.Chem.Phys. **66**, 4018 (1977).

9) R.L.Vold and R.R.Vold, Progr.NMR Spectrosc. **12**, 79 (1978).

10) R.L.Vold,W.H.Dickerson and R.R.Vold, J.Magn.Reson. **43**, 213 (1981).

11) T.C.Wong and K.R.Jeffrey, Mol.Phys. **46**, 1 (1982).

12) P.A. Beckmann,J.W.Emsley,G.R.Luckhurst and D.L.Turner, Mol.Phys. **50**, 699 (1983).

13) R.R.Vold, in press.

14) P.Ukleja,J.Pirs and J.W.Doane, Phys.Rev. **A14**, 414 (1976).

15) C.F.Polnaszek and J.H.Freed, J.Phys.Chem. **79**, 2283 (1975).

16) H.A.Lopes Cardozo, J.Bulthuis and C.McLean, J.Magn.Reson. **33**, 27 (1977).

17) J.Jeener and P.Broekaert, Phys.Rev. **157**, 232 (1967).

18) W.H.Dickerson,R.R.Vold and R.L.Vold, J.Phys.Chem. **87**, 166 (1983).

19) R.L.Vold and R.R.Vold, J.Magn.Reson. **55**, 78 (1983).

20) M.Mehring,E.K.Wolff and M.E.Stoll, J.Magn.Reson. **37**, 475 (1980).

21) R.R.Vold and G.Bodenhausen, J.Magn.Reson. **39**, 363 (1980).

22) R.L.Vold and R.R.Vold, Poster presented at 26th ENC, Asilomar, Ca. (1985).

23) R.W.Vaughan,D.D.Ellman,L.M.Stacey,W.K.Rhim and J.W.Lee, Rev.Sci. Instr. **43**, 1356 (1972).

24) P.M.Henrichs,J.M.Hewitt and M.Linder, J.Magn.Reson., **60**, 280 (1984)

25) J.F.Martin,R.L.Vold and R.R.Vold, J.Magn.Reson. **51**, 164 (1983).

26) D.E.Woessner, J.Chem.Phys. **36**, 1 (1962).

27) W.T.Huntress,Jr., Adv.Magn.Reson. **4**, 1 (1970).

28) H.Burgi and J.Dunitz, Chem.Comm. 472 (1969).

29) J.F.Martin,R.L.Vold and R.R.Vold, to be published.

30) R.R.Vold,P.H.Kobrin and R.L.Vold, J.Chem.Phys. **69**, 3430 (1978).

31) B.C.Nishida,R.L.Vold and R.R.Vold, to be published.

32) R.Poupko,R.L.Vold and R.R.Vold, J.Phys.Chem. **84**, 3444 (1980).

33) L.S.Selwyn,R.R.Vold and R.L.Vold, Mol.Phys., in press.

34) L.S.Selwyn, Ph.D. Dissertation, University of California, San Diego, 1984.

35) G.Lipari and A.Szabo, J.Amer.Chem.Soc. **104**, 4546 (1982).

36) S.H.Glarum and J.H.Marshall, J.Chem.Phys. **46**, 55 (1967).

37) C.C.Wang and R.Pecora, J.Chem.Phys. **72**, 5333 (1980).

38) L.S.Selwyn,R.R.Vold and R.L.Vold, J.Chem.Phys. **80**, 5418 (1984).

39) P.L.Nordio,G.Rigatti and U.Segre, J.Chem.Phys. **56**, 2117 (1972).

40) J.M.Bernassau,E.P.Black and D.M.Grant, J.Chem.Phys. **76**, 253 (1982).

41) I.Haller, J.Chem.Phys. **57**, 1400 (1972).

42) S.Meiboom and R.C.Hewitt, Phys.Rev.Lett. **30**, 261 (1973).

43) I.Zupancic,J.Pirs,M.Luzar,R.Blinc and J.W.Doane, Solid State Comm. **15**, 227 (1974).

44) R.Y.Dong, E.Tomchuk, C.G.Wade, J.J.Visintainer and E.Bock, J.Chem. Phys. **66**, 4121 (1977).

45) P.R.Luyten,R.R.Vold and R.L.Vold, J.Phys.Chem. **89**, 545 (1985).

PBB, Vol. 2
Advanced Magnetic Resonance Techniques
in Systems of High Molecular Complexity
© 1986 Birkhäuser Boston, Inc.

DYNAMICAL AND STRUCTURAL CHARACTERISTICS OF BIAXIAL DISCOTIC MESOPHASES BY DEUTERIUM NMR

Z.Luz,D.Goldfarb and E.Lifshitz

The Weizmann Institute of Science

Rehovot 76100, Israel

and

H.Zimmermann

Max-Planck-Institut fuer medizinische Forschung

D-6900 Heidelberg, West Germany.

INTRODUCTION

Discotic liquid crystalline mesophases were discovered about eight years ago, almost simultaneously in India (1) and France (2). Since then several hundreds of compounds and at least seven structurally different classes of such mesophases were identified by optical and X-ray tech-niques (3,4). In recent years we extended these studies using in par-ticular deuterium NMR of labelled mesogens or dissolved probe molecules. These studies provided information on orientational order, conformati-onal equilibria and solute-solvent interaction in the various discotic mesophases. More recently we concentrated our attention on biaxial discotics and in the present contribution we describe results related to their structural and dynamic characteristics. Before describing these results in more detail we briefly review the chemical characteristics and classification of discotic liquid crystals (4).

CENTRAL CORES:

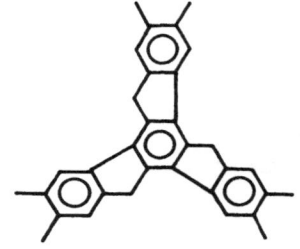

BENZENE **NAPHTALENE** **ANTHRAQUINONE**

TRIPHENYLENE

TRUXENE

SIDE CHAINS

$-C_nH_{2n+1}$, $-OC_nH_{2n+1}$, $-OC(O)C_nH_{2n+1}$,

$-OC(O)\hexagon C_nH_{2n+1}$, $-OC(O)\hexagon OC_nH_{2n+1}$,

$-CH_2O\hexagon OC_nH_{2n+1}$, $-(CH_2)_m OC(O)C_nH_{2n+1}$,

$-SO_2-C_nH_{2n+1}$, $-SC_nH_{2n+1}$

Figure 1 – Building blocks of discotic molecules. In the upper part of the figure are shown examples of typical cores, while in the lower part commonly occurring side chains and bridging groups.

Discotic mesophases are usually exhibited by compounds whose molecules consist of a rigid flat core to which a number of flexible side chains are bonded via appropriate bridging groups. Depending on the nature of the central core, the bridging groups, and the length of the side chains, discotic mesophases with different polymorphic properties are obtained. Examples of typical "cores" and side chains are shown in figure 1.

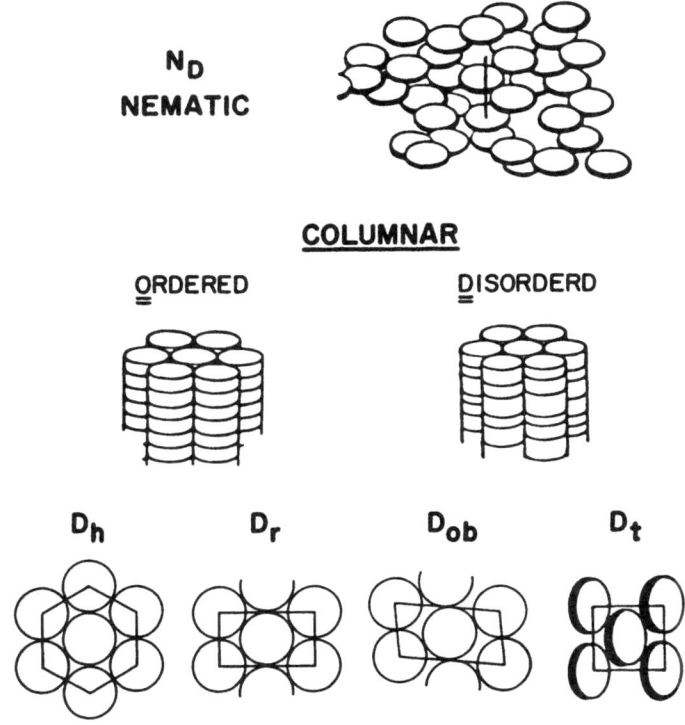

Figure 2 – Classification and schematic representation of discotic mesophases

Using X-ray and optical microscopy a classification of discotic meso-phases has emerged, analogous to the classification of the conventional calamitic liquid crystals (see figure 2). As for the latter there is a nematic discotic phase (N_D) in which the molecules are randomly distri-buted in space but tend to align with their symmetry axis parallel to some locally fixed orientation (director). Nematic discotic are however relatively rare; the more common discotics are columnar in which the mo-lecules are stacked into columns, which in turn form two dimensional ar-rays of well defined symmetries. These symmetries provide the basis for the classification of the columnar phases; thus D_h, D_r and D_{ob} corre-spond to columnar discotics with hexagonal, rectangular and oblique ar-rangements. A second subscript is often added to indicate whether the molecules are ordered or disordered within the phase (e.g. D_{ho} or D_{hd}). Finally D_t in fig.2 refers to a tilted discotic in which the molecules are uniformly tilted relative to the columnar axis. We note that some of the discotic mesophases are uniaxial (e.g. N_D, D_{ho}) while others are bi-axial (e.g. D_{rd}, D_{obd}, D_t). In the present paper we discuss several aspects related to the biaxial D_{rd} phase which appears in the phase dia-gram of two homologous series of discotic mesogens, i.e. truxenehexa-alkanoates and triphenylenehexaalkanoates. The phase diagrams of these series are summarized in figures 3 and 4. Using deuterium NMR we show that in this phase the unit cell consists of two types of columns tilted in opposite direction (5), and that jump diffusion of molecules between differently tilted columns is several orders of magnitude slower than between similarly tilted columns (6).

THE $D_{rd} \rightarrow D_{ho}$ TRANSITION IN TRUXENEHEXAALKANOATES

As shown in Fig.3 homologues of truxenehexaalkanoates exhibit highly polymorphic phase diagrams. It may be seen that for side chains having between 7 and 14 methylene groups the following phase diagram occurs:

$$\text{Solid} \rightarrow N_D \rightarrow D_{rd} \rightarrow D_{ho} \rightarrow \text{Iso.}$$

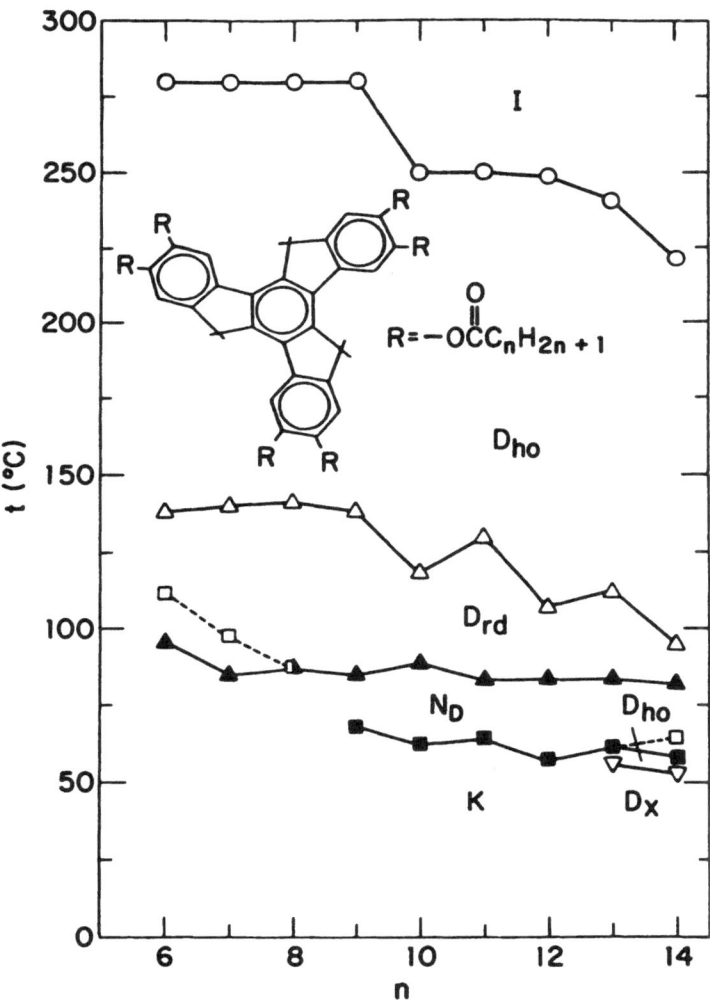

Figure 3 – Summary of polymorphism in the homologous series of truxenehexaalkanoates.

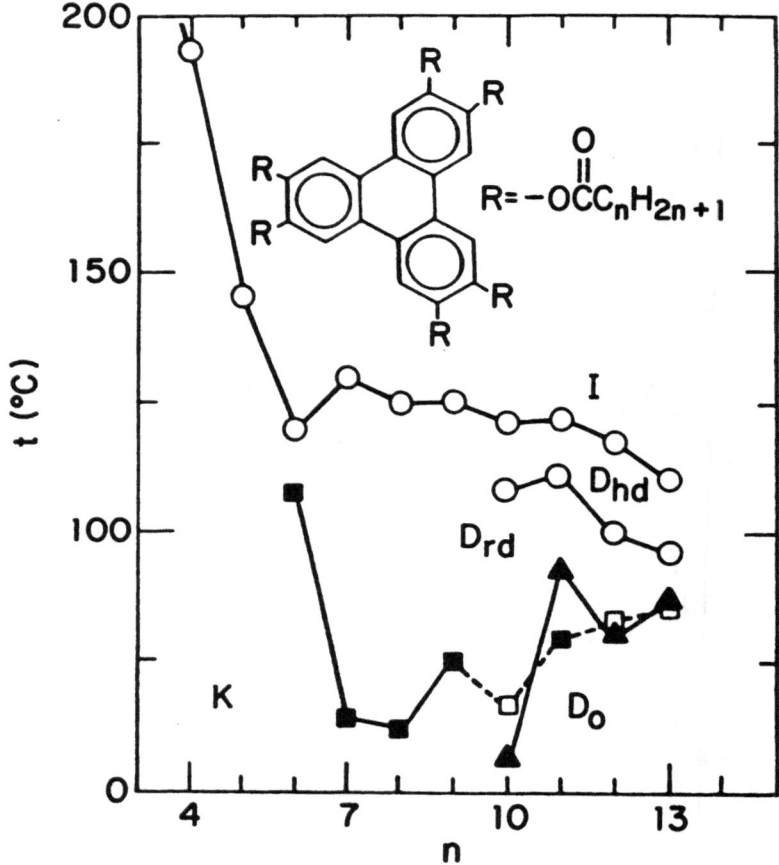

Figure 4 – Summary of polymorphism in the homologous series of triphenylenehexaalkanoates

When a solution of C_6D_6 in the various mesogens of this sequence is
introduced into an NMR probe the deuterium NMR spectra shown in Figure 5

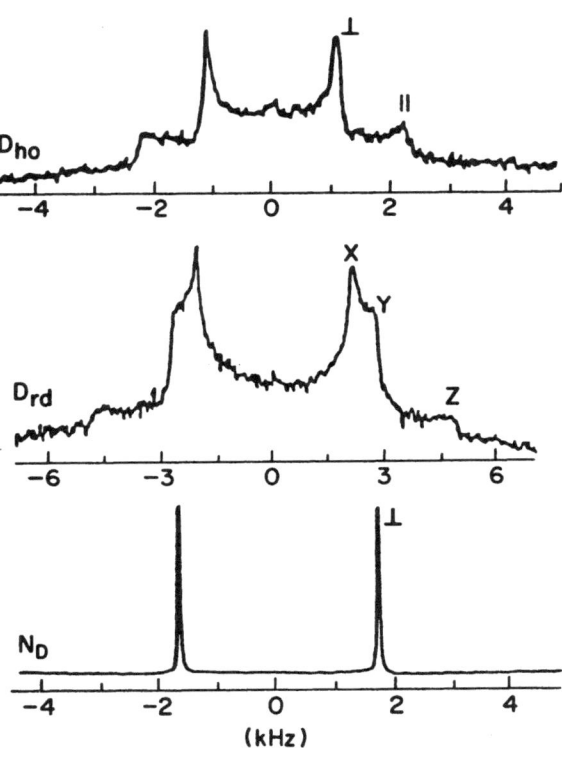

Figure 5 - Deuterium NMR spectra at 41.45 MHz of a 1.7 Wt.%C_6D_6 solution in truxenehexatridecanoate in the three mesophases indicated in the figure

are observed. The spectra indicate quite clearly that the nematic phase is effectively oriented by the magnetic field, while the D_{rd} and D_{ho} phases are not. The spectra due to the latter two samples exhibit powder patterns typical of biaxial and uniaxial phases respectively. The parallel features in the spectrum of the D_{ho} sample evidently correspond to domains whose columns are oriented parallel to the field direction, while the perpendicular features are due to domains in which the columns lie perpendicular to the field. We cannot make similar statements about the principal axes of the D_{rd} spectrum: if the D_{ho} to D_{rd} transition involves only a small perturbation of the phase geometry the z feature will most likely correspond to domains whose columns lie parallel to the magnetic field, while if the phase transition involves major geometrical changes, in particular if strong molecular tilts relative to the colum- nar axes take place, the assignment of the axes will interchange.

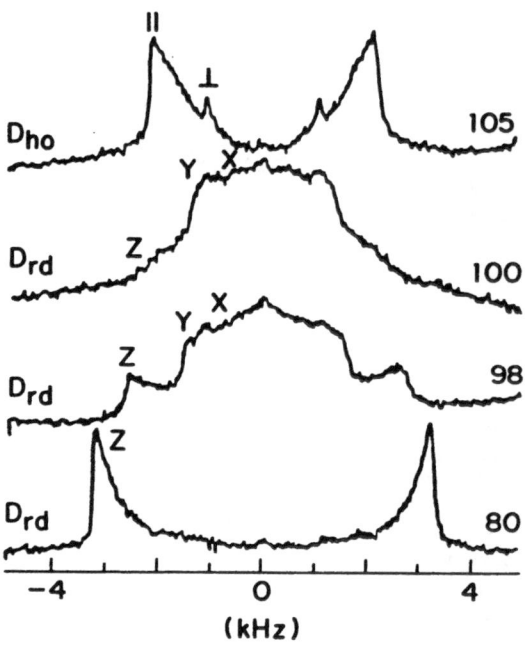

Figure 6 - Deuterium NMR spectra of the same solution as in Fig.5. The bottom trace was obtained after introducing a single domain sample in the D_{rd} phase at 80°C into the NMR spectrometer with its z axis parallel to the field direction. The other spectra were obtained after heating the sample to the indi- cated temperatures

To check on this point an experiment was performed in which a single domain D_{rd} sample oriented with its z axis along the magnetic field was heated within the NMR probe to the D_{ho} phase. The purpose of the experiment was to determine which of the principal axes of D_{ho} emerges from the z axis of D_{rd} (7). The results are shown in Figure 6. It may be seen that upon approaching the $D_{rd} \longrightarrow D_{ho}$ transition reorientation of the domains take place so that around 100°C almost no z feature remains, while significant intensity is observed at the x and y regions. However on transversing to the D_{ho} phase the dominant NMR intensity consists of the parallel features indicating that the parallel axis of D_{ho} corresponds to either the x or y direction of D_{rd}.

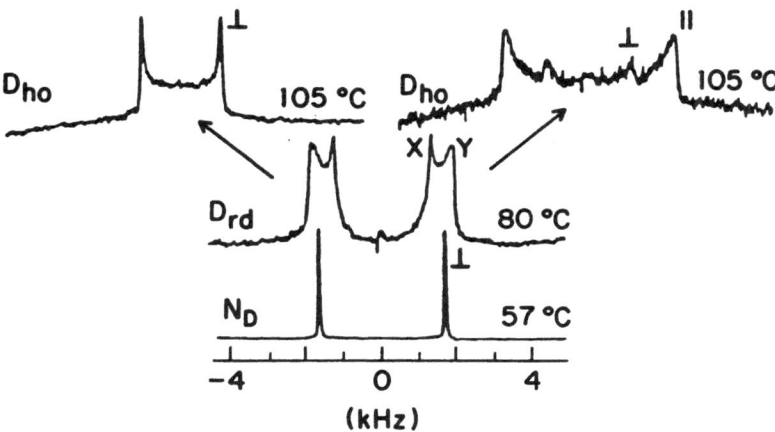

Figure 7 – Deuterium NMR spectra of the same solution as in Figs.5 and 6. The bottom spectrum is of the nematic phase at 57°C. The middle spectrum was obtained after heating to the D_{rd} phase at 80°C and corresponds to a planar distribution of domains with all z axes in a plane perpendicular to the field direction. The upper spectrum on the right was obtained by removing the D_{rd} sample from the magnetic field at 80°C, heating it to the D_{ho} phase and reinserting in the NMR spectrometer at 105°C. The upper spectrum on the left was obtained by heating the same sample from the D_{rd} to the D_{ho} phase inside the magnetic field.

The same conclusion was reached by another experiment described in figure 7. Here a nematic sample was heated up inside the magnetic field to the D_{rd} phase resulting in a distribution of domains in which all z directions lie in a plane perpendicular to the field direction and thus only features corresponding to the magnetic field in the x-y plane are observed (second trace from bottom). The sample was then heated to the D_{ho} phase, however to avoid domain reorientation, it was removed from the magnetic field during the phase transition. The NMR spectra so obtained are shown on the right hand side of Fig.7. For comparison spectra recorded from a sample heated inside the magnetic field are shown on the left side of the figure. The latter spectra clearly indicate that domain orientation took place during the transition, while those for the sample heated outside the field confirm the correlation determined above between the parallel direction of D_{ho} and the x or y direction of D_{rd}. More detailed experiments indicated that in fact $D_{ho}(\parallel) \longrightarrow D_{rd}(x)$ while $D_{ho}(\perp) \longrightarrow D_{rd}(y,z)$ (see figure 8). This result indicates very firmly that the molecules in the D_{rd} phase are strongly tilted with respect to the columnar axes. In order to preserve rectangular symmetry a structure as shown in figure 9 is proposed. It consists of a unit cell with two inequivalent, but symmetry related, columns with nearly opposite tilts. Such a structure was suggested by an X-ray study for the D_{rd} phase in homologues of triphenylenehexaalkanoates (8).

JUMP DIFFUSION IN THE D_{rd} PHASE OF TRIPHENYLENEHEXAALKANOATES

The results described in the previous section were obtained on a C_6D_6 probe dissolved in the mesophases of truxenehexaalkanoates. To substantiate the conclusions, measurements are underway on neat deuterated mesogens of truxene and other series which exhibit D_{rd} phases. In the course of these studies we observed features in the NMR spectra which provided information on the molecular diffusion between the two sub-lattices (i.e. between columns inclined in opposite direction) of the D_{rd} phase. The effects were observed in the triphenylenehexaalkanoate homologues and are described below.

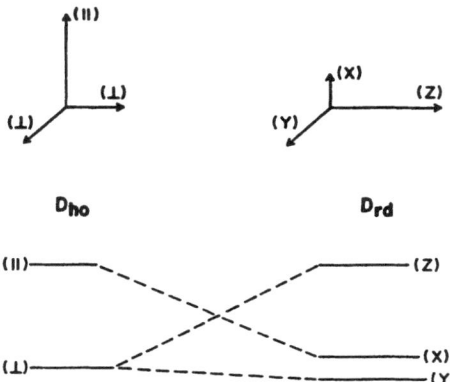

Figure 8 – Correlation of the principal axes of the D_{ho} and D_{rd} phases as determined by NMR experiments

Figure 9 – Schematic diagram of the arrangement of molecules in the columns and of the two dimensional array of columns in the D_{ho} and D_{rd} phases.

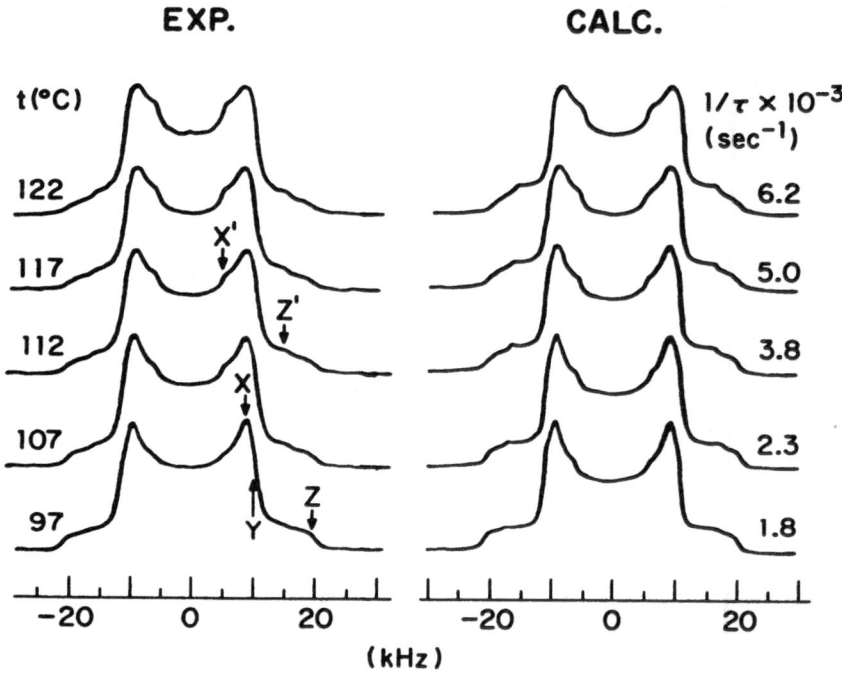

Figure 10 - Deuterium NMR spectra of triphenylenehexaoctanoate-αd_{12}. The spectra on the left are experimental and correspond to the D_{rd} phase at the indicated temperatures. The spectra on the right are calculated and were computed assuming a jump process as shown in Fig.11, with ß=69°, η=0.07 e^2qQ/h =26.5 kHz, and jump rates as indicated in the figure

In figure 10 are shown deuterium NMR spectra of triphenylenehexa-octanoate, deuterated in the α-methylenes, in the D_{rd} phase. The spectrum at 82°C is typical of a biaxial phase with a small asymmetry parameter, however on raising the temperature additional features appear in the spectrum (indicated by x' and z') whose intensity grows upon heating. We associate these features with the onset of a dynamic process which modulates the average quadrupolar interaction of a particular column. Specifically a jump process as described in figure 11 is consi-

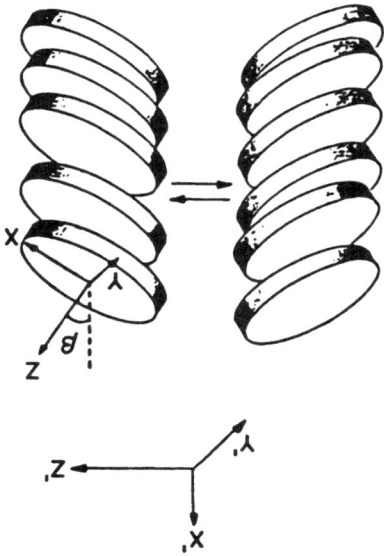

Figure 11 - Schematic representation of columns belonging to two sub-
lattices undergoing mutual molecular exchange. The unpri-
med and primed axes correspond to the quadrupole principal
axes in a single column and averaged over the two sub-
lattices respectively

stent with the experimental results. The process involves exchange of
molecules between oppositely oriented columns. In fact the characteristic dynamic features can readily be simulated by this model as
shown on the right hand side of fig.10, and by comparison of such simulated spectra with the experimental results the jump rates and activation parameters may be derived. For the hexaoctanoate homologue the
results are:

$$1/\tau \ (100°C) \sim 2x10^3 sec^{-1} \qquad\qquad \Delta E \sim 15 \ kcal/mol$$

This result may be related to the lateral diffusion constant, D, due to
jumps between oppositely inclined columns via the diffusion equation $\overline{\ell^2}$ =

$2D\tau$ where ℓ is the average distance between columns ($\sim$20-30 $\overset{\bullet}{A}$). This gives $D \sim 10^{-14} m^2/sec$ which is significantly lower (by a factor $\sim 10^3$) than the self diffusion constants determined in the D_{rd} phase of the undecanoate homologue by the NMR field gradient spin echo technique (9). We believe that the discrepancy indicates that the dominant translational diffusion in these systems occurs within a given sub-lattice of columns (i.e. between columns with the same molecular inclination) while diffusion between one sub-lattice and another is much slower in particular when the tilt angle is large.

SUMMARY

Most of the previous applications of deuterium NMR in the field of liquid crystals involved studies of molecular order and molecular structures of the mesogen or dissolved probe molecules. Here we have demonstrated that this technique can be extended to studies of phase structure and dynamic characteristics as well. In particular we showed how large tilt angles and jump diffusion in biaxial columnar discotics are reflected in the deuterium spectrum. More details will be given in forthcoming publications.

ACKNOWLEDGEMENTS

This work was supported by the United States-Israel Binational Science Foundation, Jerusalem and by the Israel Academy of Sciences.

REFERENCES

1)   S.Chandrasekhar,B.K.Sadashiva and K.A.Suresh,Pramana **9**, 471 (1977).

2) J.Billard, J.C.Dubois, N.H.Tinh and A.Zann, Nouv.J.Chim. 2, 535 (1978).

3) J.C.Dubois and J.Billard in "Liquid Crystals and Ordered Fluids", (A.C.Griffin and J.F.Johnson eds.), Plenum Publishing Corporation, 1984, p.1043.

4) D.Goldfarb,Z.Luz and H.Zimmermann, Israel J.Chem. **23**, 341 (1983); Z.Luz,D.Goldfarb and H.Zimmermann in "Nuclear Magnetic Resonance of Liquid Crystals" (J.W. Emsley ed), D.Reidel Publishing Co., 1984, p.000.

5) D.Goldfarb,I.Belsky,Z.Luz and H.Zimmermann, J.Chem.Phys. **79**, 6203 (1983).

6) D.Goldfarb,E.Lifshitz,H.Zimmermann and Z.Luz, J.Chem.Phys. **82**, 5155 (1985).

7) The single domain sample was obtained by spinning the container containing the liquid crystal in the nematic phase about an axis perpendicular to the field direction while slowly heating it into the D_{rd} phase (see D.Goldfarb,Z.Luz and H.Zimmermann, J.Physique **42**, 1303 (1981)).

8) A.M.Levelut, Proceedings of the International Liquid Crystal Conference, Bangalore (Heyden, London, 1979) p. 21; M.Takabatake and S. Iwayanagi, Jpn.J. Appl.Phys. **21**, L685 (1982).

9) R.Y.Dong, D.Goldfarb, M.E.Moseley, Z.Luz and H.Zimmermann, J.Phys. Chem. **88**, 3148 (1984).

PBB, Vol. 2
Advanced Magnetic Resonance Techniques
in Systems of High Molecular Complexity
© 1986 Birkhäuser Boston, Inc.

STOCHASTIC DESCRIPTION OF HINDERED MOTIONS

G.Moro and P.L.Nordio

Department of Physical Chemistry

University of Padova, Padova, Italy

The theories of the stochastic processes (1) provide models for describing molecular hindered motions at different level of generality. The
simplest one is certainly the random walk process where one considers
only jumps between adjacent sites, defined as the minima locations of
the hindering potential. In systems with equivalent sites, the time evolution of the site probability $p_j(t)$ is determined by just one phenomenological parameter, the transition rate w, according to the master equation:

$$\partial/\partial t \; p_j(t) = w \; (p_{j+1} + p_{j-1} - 2p_j) \qquad (1)$$

With the continuous diffusion equation, a more general picture of hindered motions is obtained, comprehensive of the small amplitude oscilla-
tions within the potential wells. For processes specified by a single
stochastic variable x, the time evolution of the probability $P(x,t)$ is
evaluated according to the equation:

$$\partial/\partial t \; P(x,t) = -\hat{R} \; P(x,t) \qquad (2a)$$

$$\hat{R} = -\partial/\partial x \; D(x) \; P(x) \; \partial/\partial x \; P(x)^{-1} \qquad (2b)$$

where $D(x)$ is the diffusion coefficient which in general is position dependent, and the equilibrium distribution function $P(x)$ is given in terms of the hindering potential $V(x)$ as:

$$P(x) = \exp\left|-V(x)/kT\right|/Z \qquad (2c)$$

We shall consider the following examples of one-dimensional hindered motions:

A) Dipole reorientation in uniaxial liquid crystals (2):

$x=\cos\beta$; β=orientation of the molecular electric dipole with respect to the director axis

$D=D^R(1-x^2)$; D^R=rotational diffusion coefficient

$V(x)/kT=-\lambda(3x^2-1)/2$; Maier-Saupe potential.

B) Translational diffusion in smectics (3):

x=displacement along the direction orthogonal to the smectic planes

$D=D^T$=molecular diffusion coefficient for the translational motion

$V(x)=-(\Delta/2)\cos(2\pi x/d)$; d=spacing between smectic layers.

C) Hindered internal motions (4):

x=relative orientation of a mobile group in a molecule

$V=(V_n/2)(1-\cos nx)$; n=number of equivalent conformers.

D) Chemical kinetics (5):

x=reaction coordinate

$V(x)=(V_o/a^4)(x^2-a^2)^2$; the region around $x=a$ ($x=-a$) defines the species A (B) of the unimolecular reaction A B.

In all these cases the diffusion equation can be solved numerically in order to calculate the relevant observables, in particular the relaxation time defined as the (normalized) time integral of the correlation function according to the following equation:

$$\tau = \int_0^\infty dt\ \overline{f(t)*f(0)}/\overline{f*f} = <f|\hat{R}^{-1}|fP>/<f|P|f> \qquad (3)$$

where the bracket denotes the integration over the stochastic variable, and $f(x)$ is the dynamic variable with null average probed by a specific experiment.

The two models, i.e. the continuous diffusion equation and the discrete random walk equation, are expected to be equivalent in the case of large potential barriers. In fact, under these limiting conditions, the transitions among potential wells are characterized by a much longer time scale than any other type of motion, e.g. the libration around the potential minima. The jump motion can be therefore considered uncoupled from the other motions, and the only one that affects the short frequency behaviour of the observables. Formally, this equivalence can be shown following two routes:

a) by comparing the correlation times obtained from the random walk model with the corresponding quantities calculated from the diffusion equation. In the second case an exact formula for τ can be derived from the solution of the second order differential equation $\hat{R}\ hP=fP$ with respect to the unknown function $h(x)$ (6). For reflecting boundary conditions, τ is given by the following equation:

$$\tau = <F|(PD)^{-1}|F>/<f|P|f> \qquad (4a)$$

where

$$F(x) = \int_a^x dy\ f(y)\ P(y) \qquad (4b)$$

with a being the lower limit of x.

b) By direct derivation of the random walk master equation from the more general diffusion equation. For each potential well centered at x_j, we define the localized function $g_j(x)$ which is non-vanishing only in the interval $x_{j-1} < x < x_{j+1}$, where it is given as:

$$g_j(x) = \int_x^{x_j+1} dy\, P(y)^{-1} / \int_{x_j}^{x_j+1} dy\, P(y)^{-1}$$ (5)

the positive (negative) sign holding for x greater (less) than x_j. The symmetrized diffusion operator can be now projected onto the subspace of the linear combinations of $g(x)P(x)^{1/2}$. This is equivalent to approximate the probability P(x,t) of the continuous variable x as a linear combination of site probabilities $p_j(t)$ according to the relation:

$$P(x,t) \approx \sum_j p_j(r) g_j(x)$$ (6)

The random walk master equation (1) is recovered with the transition rate w given as (6):

$$w = -\langle g_j | \hat{R}P | g_{j+1} \rangle = \langle dg_j/dx | PD | dg_{j+1}/dx \rangle$$ (7)

The route b) provides a molecular definition of the transition rate w which usually constitutes the main outcome of experimental measurements on hindered systems. Therefore it is possible to study in detail the dependence of w on the shape of the potential V(x), the Stokes–Einstein relations giving usually a fairly accurate estimate of the diffusion coefficient. In particular, with very large potential barriers, the Arrhenius-type behaviour for the transition rate w is recovered, as predicted by the Kramers theory of activated processes (7). As an example, in the case of dipole reorientation the following asymptotic relation is obtained:

$$w = 2\pi^{-1/2}\, D^R\, (\Delta E/kT)^{3/2}\, \exp(-\Delta E/kT)$$ (8)

where ΔE is the difference of the potential energy V(x) between a minimum (x=0,π) and the maximum (x=$\pi/2$). However it should be emphasized that good estimates of the transition rate are obtained from eq.(7) also when $\Delta E/kT$ is of the order of the unity, well outside the range of validity of the asymptotic relation.

REFERENCES

1)   N.G.van Kampen, "Stochastic Processes in Physics and Chemistry", North-Holland, 1981.

2)   P.L.Nordio,G.Rigatti and U.Segre, Mol.Phys. **25**, 129 (1973).

3)   G.Moro,P.L.Nordio and U.Segre, Chem.Phys.Letters **105**, 440 (1984); Mol.Cryst.Liq.Cryst. **114**, 113 (1984).

4)   D.Chandler, J.Chem.Phys. **68**, 2959 (1978); D.C.Knauss and G.T.Evans, J.Chem.Phys. **73**, 3423 (1980); J.Chem.Phys. **74**, 4627 (1981).

5)   J.L.Skinner and P.G.Wolyness, J.Chem.Phys. **69**, 2143 (1978); R.S. Larson and M.D.Kostin, J.Chem.Phys. **69**, 4821 (1978).

6)   G.Moro and P.L.Nordio, Mol.Phys. in press.

7)   H.A.Kramers, Physica (The Hague) **7**, 284 (1940).

PBB, Vol. 2
Advanced Magnetic Resonance Techniques
in Systems of High Molecular Complexity
© 1986 Birkhäuser Boston, Inc.

DEUTERIUM RELAXATION AND MOLECULAR MOTION IN CHANNEL CLATHRATES

Regitze R.Vold, Robert L.Vold, Michael S.Greenfield and Alan Ronemus

Department of Chemistry, University of California

San Diego, La Jolla, California 92093, USA

INTRODUCTION

The use of magnetic resonance techniques in studies of structure and molecular dynamics of inclusion compounds has increased dramatically in recent years. Considering the importance of many types of clathrates in catalysis this is to be expected. Considering the inherent charming physical chemical properties of many clathrates, it is surprising that physical chemists among the magnetic resonance community have waited so long to take an in depth look at the properties of these materials. It turns out that deuterium NMR is very well suited for investigating the dynamics of clathrates.

Our own interest in these materials has its roots in our work with thermotropic liquid crystals (1,2) which continue to present a challenge as far as developing a consistent, detailed picture of the alkyl chain dynamics is concerned. Alkyl chain dynamics in liquid crystals involve overall molecular tumbling and spinning, internal rotation and libra-tion, rotamer interconversion, and quasi-coherent fluctuations of the nematic director. Given these complexities, the suggestion by Geoffrey Luckhurst and Jim Emsley of the University of Southampton that urea and thiourea clathrates might present a simpler environment for studying alkyl chains is most welcome. Accordingly, we have begun detailed investigations of the deuteron relaxation behavior in two types of channel clathrates, biphenyl-d_{10} in ß-cyclodextrin and $\underline{n}$-nonadecane-d_{40} in urea, using relaxation measurements for determination of the spectral densities of motion as well as lineshape analysis of deuterium powder patterns.

CYCLODEXTRIN CLATHRATES

The cyclodextrins are a family of cyclic oligosaccharides containing from six to as many as eleven α-1,4-linked D- glucose molecules. α-, ß- and γ-cyclodextrin refer to the 6-, 7- and 8-membered species which are produced in the largest quantities from enzymatic degradation of starch and are the ones which form stable clathrates with a large variety of organic compounds. Interest has arisen in these clathrates because they serve as biomimetic model systems and because they have potential for drug delivery (3). Both cage-type and channel-type clathrates are formed, and molecules of different sizes can be included in α-, ß- and γ-cyclodextrins because of the different diameters of the channels.

It is known (4) from numerous X-ray investigations that guests in the cyclodextrins channel clathrates frequently are "statistically disordered" – that is, "mobile" – and ESR (5) studies and ^{13}C CP-MAS spectroscopy (6) have been used to obtain dynamic information. The channels of ß-cyclodextrin are large enough, d = 7.5 Å, that biphenyl should be able to wobble slightly about its long axis as well as undergo rotation about this axis. As shown in Fig. 1, this is indeed the case. The top spectrum is a room temperature deuterium NMR spectrum of biphenyl itself, obtained using quadrupole echo pulse sequences. Aside from effects of limited pulse power ($\pi/2$-pulse = 2.5 sec) the spectrum is typical of a "rigid" powder pattern with $<e^2qQ/h> = 171$ kHz. The bottom spectrum, of biphenyl-d_{10} in ß-cyclodextrin, shows narrowing characteristic of jump-like rotation of the o,m-deuterons about the long

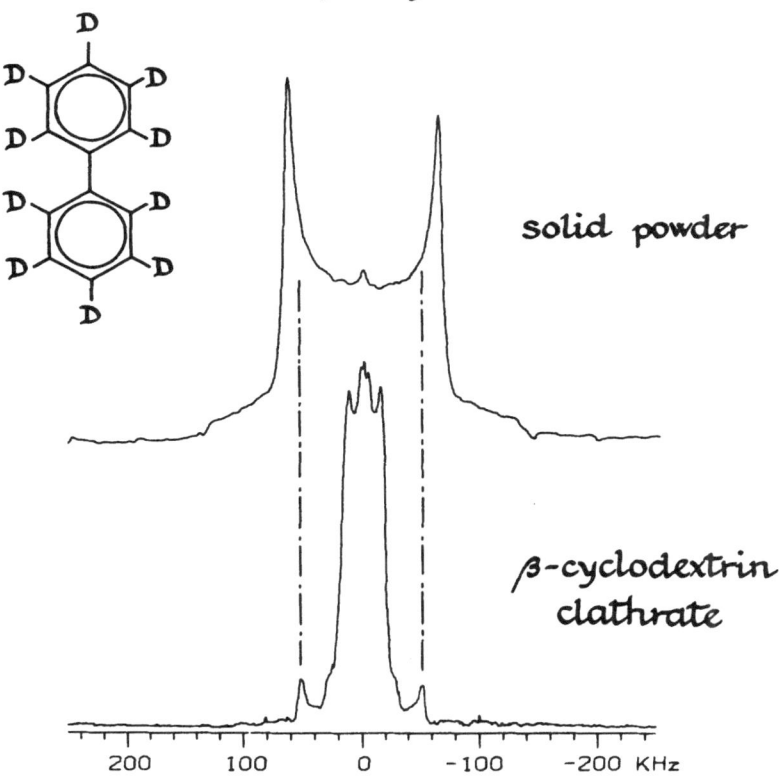

Figure 1 – Deuterium NMR spectra obtained at 38.4 MHz and 25°C of biphenyl-d$_{10}$ as solid powder (top) and in the ß-cyclo-dextrin channel clathrate (bottom). The spectra were obta-ined using phase cycled quadrupole echo pulse sequences with $\pi/2$ pulse equal to 2.5 μs and pulse intervals of 20 μs

molecular axis combined with a slightly narrowed "Pake pattern" from the p-deuterons undergoing a restricted ($\vartheta \simeq 25°$) wobbling motion. The effect of lowering the temperature is shown in Fig. 2. The molecular reorientation is slowed down almost completely near -165°C, although the spectrum suggests that the wobbling motion still occurs at this temperature.

BIPHENYL-d_{10} /β-CYCLODEXTRIN

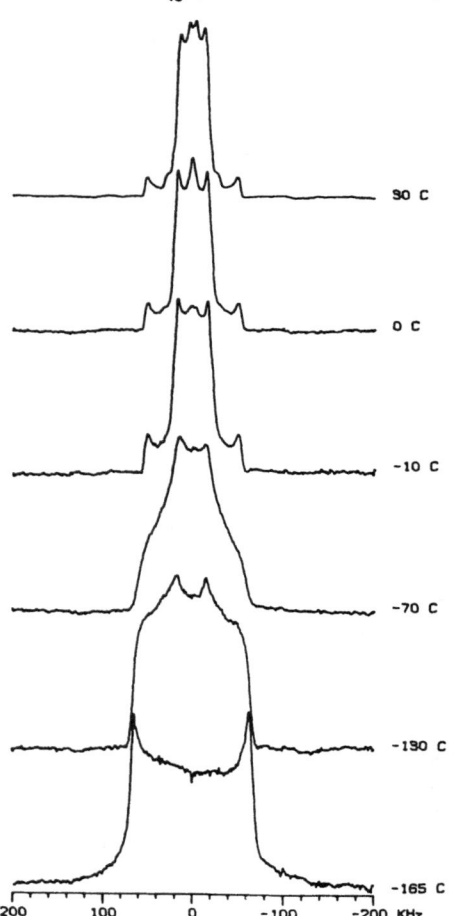

Figure 2 - Temperature dependence of the deuterium quadrupole echo powder patterns of biphenyl/ß-cyclodextrin. The lower temperatures may be off by $\pm 5°$

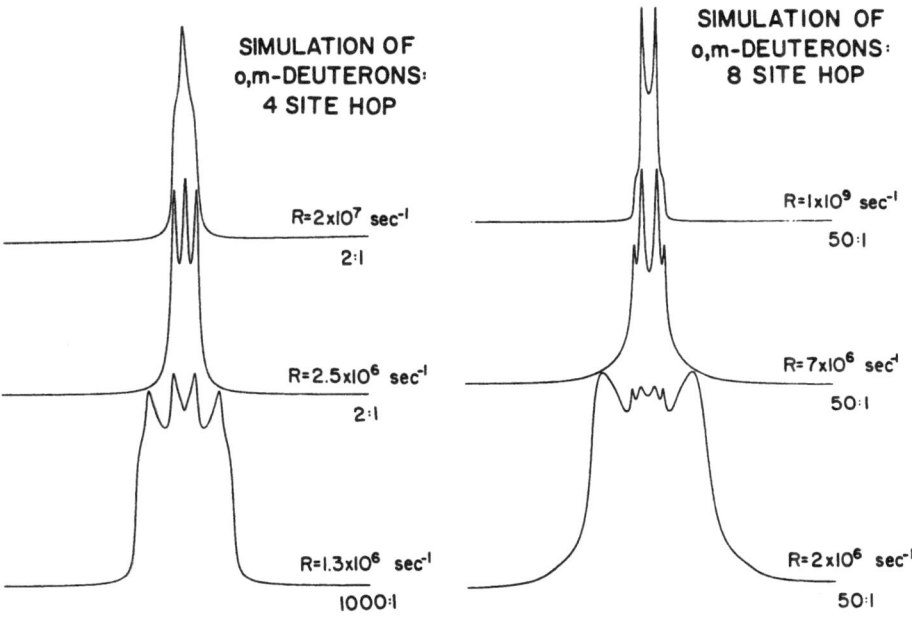

SIMULATION OF
o,m-DEUTERONS:
4 SITE HOP

R=2x10⁷ sec⁻¹

2:1

R=2.5x10⁶ sec⁻¹

2:1

R=1.3x10⁶ sec⁻¹

1000:1

SIMULATION OF
o,m-DEUTERONS:
8 SITE HOP

R=1x10⁹ sec⁻¹

50:1

R=7x10⁶ sec⁻¹

50:1

R=2x10⁶ sec⁻¹

50:1

Figure 3 — Calculated spectra corresponding to two different jump
models for the motion of biphenyl in the ß-cyclodextrin
clathrate

Fig. 3 shows our early attempts to simulate the spectra in Fig. 2 using
a lineshape calculation program developed in our laboratory (7). Our
initial approach has been to focus on the fact that the spectrum is characteristic of jump-like motion rather than continuous diffusion of the
molecule about its long axis, which would lead to a narrowed, but
axially symmetric powder pattern (8). The jumps may be associated with
internal, rather than overall rotation. Internal rotation in byphenyl is
thought (9) to be characterized by two distinct barriers: one at

dihedral angle $\varphi = 0$ with height of 4-5 kcal/mole arising from steric repulsion between the o-deuterons, the other of height ca. 2 kcal/mole at $\varphi = 90°$ where the degree of conjugation between the phenyl rings is minimal. We have assumed that the equilibrium conformation of the enclathrated byphenlyl is like that found in the gas (10) and liquid (11) phases with a dihedral angle of 42°. In one model one phenyl ring is assumed to jump among 4 sites, slowly over the high barrier ($\Delta\varphi_1 = 84°$) and fast over the low barrier ($\Delta\varphi_2 = 96°$). This model leads to the lineshapes on the left in Fig. 3. In the other model the two phenyl rings are jumping in concert across the barriers ($\Delta\varphi_1 = 42°$, $\Delta\varphi_2 = 48°$); this picture leads to 8 distinct sites for the o,m-deuterons of one ring and the simulated spectra on the right in Fig. 3. The curves in Fig. 3 show that there are characteristic spectral differences associated with 4-site and 8-site jumps, but as yet we are not close to an appropriate model for the motion. Further simulations will hopefully allow better understanding of the details, and we expect lineshape analysis to play a major role in the elucidation of molecular dynamics of these systems.

UREA CLATHRATES

Another interesting group of channel clathrates are the urea and thiourea inclusion compounds with alkanes and alkane derivatives. These inclusion compounds were discovered in 1949 (12) and received a great deal of attention because of their ability to specifically enclathrate alkane derivatives with different branching characteristics (13). When combined in solution with long (n > 6) n-alkanes or n-alkyl derivatives, urea forms structures with 'infinitely' long channels filled with extended all-trans conformers of the alkyl derivatives. The inner diameter of the channels is ca. 5 Å and X-ray measurements (14) have shown that above ca. -100°C the alkanes are free to spin about their long axis. NMR studies of the molecular dynamics of urea clathrates were first reported by Gilson and McDowell (15) who used proton wide-line spectroscopy to determine that the interaction between alkanes and urea was slight, allowing the included species considerable degree of motional freedom. A deuterium powder study on n-nonadecane (16) confirmed the observation that the alkane is spinning rapidly about its long axis at room temperature and that the spinning stops near the phase transition to the low temperature orthorhombic phase (17).

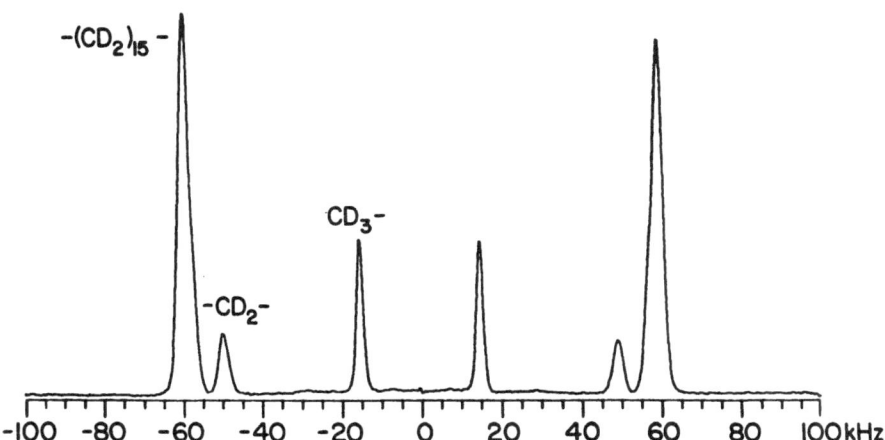

Figure 4 – Deuterium NMR spectrum of n-nonadecane-d$_{40}$ in urea channel clathrate. The spectrum was obtained at room temperature from a single crystal oriented with the long crystal axis parallel to the magnetic field

In order to provide quantitative information about the rates of motion in these materials, we prepared a single crystal of n-nonadecane/urea clathrate by controlled, slow cooling of a 1:55 mixture of nonadecane-d$_{40}$ and urea in a 1:1 mixture of acetone and methanol. The 38.4

MHz room temperature spectrum of this crystal oriented with the channel axes parallel to the magnetic field is shown in Fig. 4. The magnitude of the observed splittings show that the alkane is rotating rapidly about the long molecular axis, and the assignments indicated in the figure agree with those made earlier for the deuterium powder pattern by Casal et al. (16). It appears from this spectrum that it is reasonable to think of the included molecule as " α, ω -diethyl-rod", since conformational flexibility scarcely affects the spectrum of the central 15 methylene groups. A classic rotation experiment in which the crystal was rotated about an axis perpendicular to the field shows, after accounting for the fast spinning motion, that the effective quadrupole coupling constants (QCC) for the three types of deuterons are 42 kHz (CD_3), 136 kHz (outermost CD_2 groups) and 163 kHz (bulk CD_2 groups). The small effective QCC found for the methyl groups include the effect of rapid spinning about the C-C axis, while the values obtained for the methylene groups can be accounted for by different degrees of librational motion as discussed further below.

At 115 K and $\beta = 90°$ we obtain the spectrum shown in Fig. 5. As expected (15,16) the spinning has stopped and we observe an essentially rigid pattern which can be explained with reference to low temperature X-ray studies by Chatani et al. (17). An interpretation of their results is presented in Fig. 6, from which it can be seen that the channels now are distorted from hexagonal symmetry and that two unique orientations of the all-<u>trans</u> conformer exist relative to the magnetic field direction.

In order to learn something about the rates of rotational and librational motion of the included alkane we have used modified (18) Jeener-Broekaert (19) experiments to simultaneously measure the deuteron spin-lattice relaxation rate R_{1Z} and the rate of decay of quadrupolar order R_{1Q} by monitoring the evolution of the sum and difference magneti-

^{2}H NMR SPECTRA OF NONADECANE/UREA

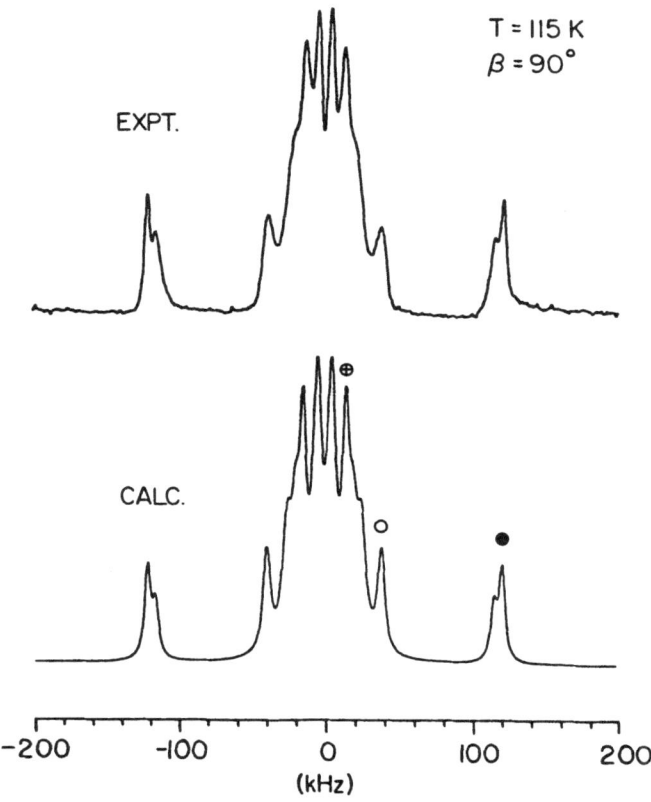

Figure 5 – Deuterium NMR spectrum (top) from a nonadecane/urea crystal oriented perpendicular to the magnetic field at 115 K. The bottom spectrum represents a composite of the spectra calculated for each of the three different orientations of the nonadecane deuterons indicated in fig.6. The relative intensities of the individual experimental spectra are distorted by the use of too short repetition times between quadrupole echo pulse sequences and this has been taken into account in the simulations

zations. The experiment allows the simultaneous determination of the spectral densities $J_1(\omega_o)$ and $J_2(2\omega_o)$ from the relations

$$J_1(\omega_o) = R_{1Q}/3C \qquad (1a)$$

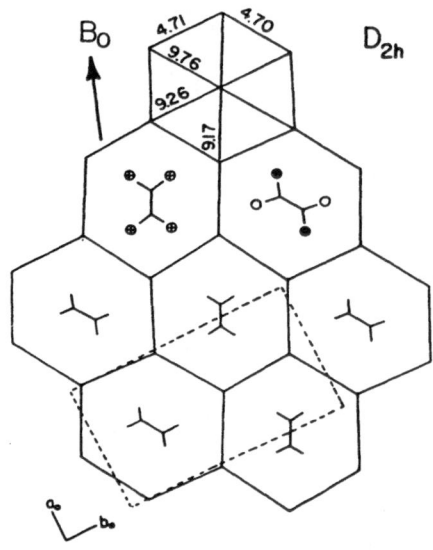

Figure 6 – Diagram of the low temperature orthorhombic phase of nona-
decane/urea clathrate (constructed from results presented
by Chatani et al.(17)) with circles indicating the three
distinct types of C-D bonds present at this orientation of
the crystal relative to the magnetic field

$$J_2(2\omega_o) = (R_{1Z} - \tfrac{1}{3} R_{1Q})/4C \qquad (1b)$$

where $C = 3\pi^2/2(e^2qQ/h)^2$. The results for the bulk methylenes are
plotted as a function of inverse temperature in Fig. 7 for two orientations of the crystal with respect to the magnetic field. One sees
that when the crystal is oriented along the field $J_2(2\omega_o)$ is significantly larger than $J_1(\omega_o)$, while the opposite is true near the perpendicular orientation. Spectral densities of motion determined for the
methyl groups are presented in Fig. 8, and the relative magnitude of the
J's is seen to be opposite to that observed for the methylene groups.

This reflects the presence of an additional deuteron oriented at a different angle as well as an additional spinning motion being available to the methyl deuterons.

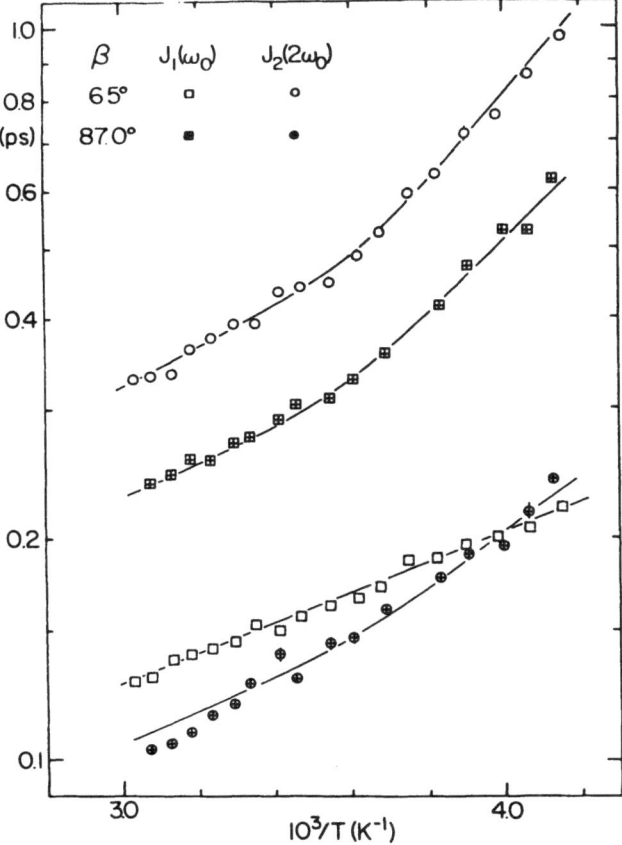

Figure 7 – Spectral densities of motion $J_1(\omega_0)$ and $J_2(2\omega_0)$ determined for nonadecane/urea using Jeener–Broekaert pulse sequences and eqs.(1a) and (1b) as a function of temperature at two orientations of the crystal with respect to the magnetic field. The solid lines represent fits to the experimental data using $<e^2qQ/h>_0$=195 kHz, a libration angle ϑ_0=24° and the motional rate constants described in the text and Fig.9

SPECTRAL DENSITIES OF MOTION
FOR NONADECANE METHYL GROUP

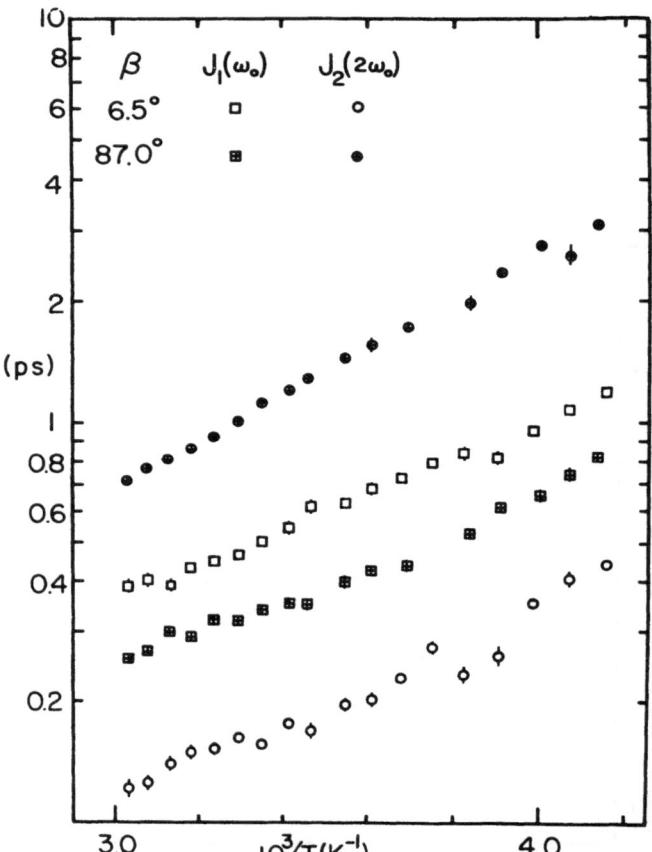

Figure 8 – Spectral densities of motion for the CD_3 groups in nonadecane/urea as a function of temperature at two orientations of the crystal with respect to the magnetic field. The spectral densities were determined as described for the methylene groups in the caption to Fig.7

Since additional experiments carried out at 15.4 MHz revealed no frequency dependence of the spectral densities in the range from 15.4 to 76.8 MHz, it is clear that all motions contributing to the relaxation are fast. It is also immediately clear, that if spinning motion was all that contributed to the relaxation of nonadecane/urea we should observe no relaxation at all for the methylene groups when the channels are oriented parallel to the field, because this motion does not change the angle of the C-D bonds relative to the magnetic field. A wagging motion of the methylene groups combined with spinning about the long axis can, however, be used to quantitatively account for the observed spectral densities. A sketch of the model is provided in Fig. 9. This particular choice of model involves the assumption of small step rotational diffusion about the long molecular axis, combined with a wagging motion

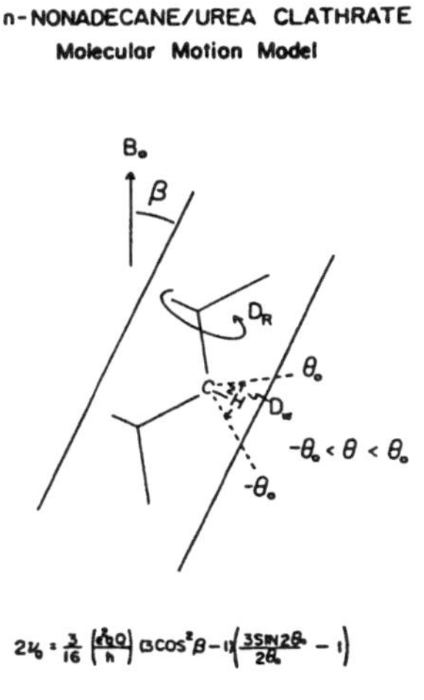

n-NONADECANE/UREA CLATHRATE
Molecular Motion Model

$$2J_1 = \frac{3}{16}\left(\frac{e^2qQ}{\hbar}\right)\left(3\cos^2\beta - 1\right)\left(\frac{3\sin2\theta_o}{2\theta_o} - 1\right)$$

Figure 9 - Diagram of the molecular motion model used to interpret the spectral densities for the bulk methylene deuterons in nonadecane/urea. In addition to spinning about the long molecular axis, the deuterons are assumed to "wag" in a plane containing the long axis. The deuterons are assumed to have uniform distribution within the interval $-\vartheta_o \leq \vartheta \leq \vartheta_o$ and the spinning and wagging motions are assumed to be uncorrelated.

of the C-D bond in a plane containing the long axis. If we further assume a uniform probability of finding a methylene deuteron along the arc of libration, we can write the following expression for the methylene doublet splittings observed as a function of the angle ß between the channel axis and B_0:

$$2\nu_Q = \tfrac{3}{4} S_{zz} <e^2qQ/h>_0 \tfrac{1}{2} \{3\cos^2 ß - 1\} \qquad (2a)$$

Here $S_{zz} \equiv \tfrac{1}{2} <3\cos^2\vartheta -1>$ is the order parameter for the principal component of the quadrupole coupling tensor in a molecular frame defined with the z-axis along the equilibrium direction of the CD bond, and the factor 3/4 (instead of 3/2) accounts for the spinning of the alkane about its long axis. The order parameter S_{zz} is given by

$$S_{zz} = \tfrac{1}{2} (3(\sin 2\vartheta_0/2\vartheta_0) - 1) \qquad (2b)$$

where ϑ_0 is the maximum librational angle shown in Fig. 6. We emphasize the importance of using the proper value of the quadrupole coupling constant in Eq.(2) by the use of $<...>_0$, but note that so far we do not know what value to use. Similar remarks can be made regardind the choice of QCC in the relaxation expression (1a) and (1b).

On the basis of earlier, simpler expressions given by London and Avitabile (20) and by Wittebort and Szabo (21) we have derived an expression for the spectral densities of motion for the methylene deuterons undergoing the motion described above. The model includes the assumption that the spinning motion is uncorrelated with the wagging motion, and leads to the following rather formidable result:

$$J_1(\omega) = 0.023 \; A(1-\cos^4 ß)+1.5 \; B(4\cos^4 ß-3\cos^2 ß+1)+1.69 \; C\sin^2 2ß \qquad (3a)$$

$$J_2(\omega) = 0.0058 \ A(\cos^4\beta + 6\cos^2\beta + 1) + 1.5 \ B(1 - \cos^4\beta) + 1.69 \ C\sin^4\beta \qquad (3b)$$

where

$$A = \left[(\sin^2 2\vartheta_0/2\vartheta_0^2) + (2\sin 2\vartheta_0/\vartheta_0) + 2\right] \tau_{2,0}/(1+\omega^2\tau_{2,0}^2) +$$

$$+ \ 16\vartheta_0^2\sin^2 2\vartheta_0 \sum_{n=1}^{\infty} 1/(4\vartheta_0^2 - n^2\pi^2)^2 \cdot \tau_{2,2n}/(1+\omega^2\tau_{2,2n}^2) \qquad (4a)$$

$$B = \vartheta_0^2\cos^2 2\vartheta_0 \sum_{n=1}^{\infty} 1/\left[4\vartheta_0^2 - 0.25(2n-1)^2\pi^2\right]^2 \cdot \tau_{1,2n-1}/(1+\omega^2\tau_{1,2n-1}^2) \ \ (4b)$$

$$C = \vartheta_0^2\sin^2 2\vartheta_0 \sum_{n=1}^{\infty} 1/(4\vartheta_0^2 - n^2\pi^2)^2 \cdot \tau_{0,2n}/(1+\omega^2\tau_{0,2n}^2) \qquad (4c)$$

and

$$1/\tau_{m,n} = m^2 D_R + n^2\pi^2 D_W/4\vartheta_0^2 \qquad (5)$$

The rotational rate is given here in terms of the rotational diffusion constant D_R and the librational rate by D_W. Eqs. (3a) and (3b) summarize the effect of sample orientation, β, with respect to the magnetic field, while Eqs. (4a)-(4b) have the form of sums of exponentials. The sums converge within 0.1% after 4 terms, and Eqs. (4a)-(4c) further show how the amplitudes of the Lorentzians are a function of the librational angle ϑ_0.

We have fitted the data presented in Fig. 5 for the bulk methylenes to Eqs. (3)-(5) and the solid lines in Fig. 5 correspond to calculations of the spectral densities of motion $J_1(\omega_0)$ and $J_2(2\omega_0)$ using $<e^2qQ/h> = 195$ kHz, $\vartheta_0 = 24°$ and rotational rate constants D_R and D_W shown in Fig. 10. D_R is observed to be roughly ten times larger than D_W, the precise ratio being a function of the value chosen for the QCC. ϑ_0 may be obtained

RATES OF ROTATION (D_R) AND LIBRATION (D_W) IN NONADECANE/UREA CLATHRATE

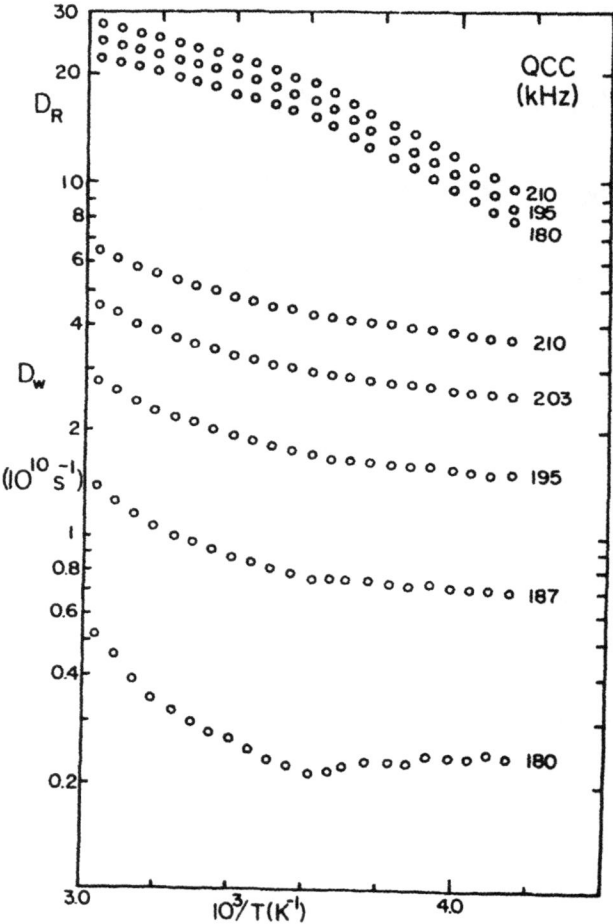

Figure 10 – Rotational and librational rate constants obtained from fits of Eqs.(2)-(5) to the experimental data for different values of the quadrupole coupling constant appropriate for use in the relaxation expressions, Eqs.(1a) and (1b)

from Eq.(2) and the (essentially temperature independent) observed value for the quadrupolar splitting, $2\nu_Q = 122\pm1$ kHz for any particular choice of the QCC. As seen in Fig. 10 an unphysical minimum in the rate of libration, D_W, appears as a result of the fits when values of $e^2qQ/h <$ 185 kHz are used. It is not possible to fit the observed relaxation rates to spectral densities given by Eqs.(3)-(5) using values in the range 160-170 kHz which is often quoted for deuterium QCC's in alkanes. The use of a QCC >185 kHz is consistent with estimates provided by Ragle and Sherk (22) and theoretical calculations (23, 24) for methane and ethane. More significantly, it agrees with the value 191.5 kHz determined by Wofsy et al. (25) for CD_4 in a cold molecular beam where only the lowest 2-3 rotational states were populated. The beam measurement thus provides a value for the deuterium quadrupole coupling constant in alkanes, which is affected only by the 3N-6 = 9 v = 0 vibrations in methane. The values 160-170 kHz typically measured by NMR for alkanes and lipids are affected by numerous lower frequency modes as well; since it is these very modes which are responsible for the relaxation in nonadecane/urea we need to use a minimally vibrationally averaged QCC, $<e^2qQ/h>_o$, as the interaction strength in Eqs.(1a) and (1b). It is gratifying that a very recent normal mode analysis by Henry and Szabo (26) has confirmed our conclusions. Their calculations show that when low frequency modes in the vibrational spectrum of n-octane are included in the averaging, one obtains a 15% reduction of the deuterium quadrupole coupling constant relative to the value calculated for a very short alkane.

The spectral densities for the methyl deuterons presented in Fig. 8 can be interpreted in a similar fashion by including the spinning of the methyl group about its C_3 axis. The expression for the spectral density functions is quite cumbersome, and will not be reproduced here, but a quantitative analysis yields rotational rate constants which are consistent with the data obtained from the methylene data.

SUMMARIZING REMARKS

Deuterium NMR spectroscopy, in particular relaxation measurements, have been found by many research groups to be extremely valuable in the studies of molecular dynamics of the solid state as well as of numerous order mesophases. The data presented here, while preliminary, confirms the utility of deuterium NMR measurements for solid samples. When molecular motion takes place at rates comparable to the width of the static spectrum lineshape analysis is undoubtedly the method of choice. For fast motion, when the spectrum is in the narrowed limit, determination of individual spectral densities provides the most information one can hope to get from NMR relaxation measurements. If single crystals are available, it is comparatively simple to obtain such information through the use of techniques developed first for molecular motion studies of thermotropic liquid crystals (1). To get the same kind of detailed information for powders, a complete analysis of the recovery of the full lineshape is required. We particularly wish to emphasize that in either case it is necessary to ensure that the value used for the deuteron QCC is not effected by vibrational averaging over modes which are responsible for the relaxation.

ACKNOWLEDGEMENTS

This work was supported by grant CHE81-22097 from the National Science Foundation.

293

REFERENCES

1) R.R.Vold and R.L.Vold, in "Liquid Crystals and Ordered Fluids" (A.C.Griffin and J.F.Johnson eds.), Vol. 4, 561, 1984.

2) R.L.Vold,R.R.Vold,J.F.Martin,B.C.Nishida and L.Selwyn, this same book.

3) J.Szejtli, "Cyclodextrins and their Inclusion Complexes", Akademiai Kaodo, Budapest (1982).

4) R.K.McMullan,W.Saenfer,J.Fayos and D.Mootz, Carbohyd.Res. **31**, 371 (1973).

5) K.Flohr,R.M.Patton and E.T.Kaiser, J.Amer.Chem.Soc. **97**, 1209 (1975).

6) H.Ueda and T.Nagai, Chem.Pharm.Bull. **29**, 2710 (1981).

7) M.S.Greenfield,A.D.Ronemus,P.D.Ellis,R.L.Vold and R.R.Vold, to be published.

8) M.Mehring, in "Principles of High Resolution NMR in Solids", Springer-Verlag, Berlin, 1983.

9) J.C.Rayez and J.J.Dannenberg, Chem.Phys. Letters **41**, 492 (1976).

10) O.Bastiansen, Acta Chem.Scand. **3**, 408 (1949).

11) A.d'Annibale,L.Lunazzi,A.C.Boicelli and D.Macciantelli, J.Chem.Soc. Perkin II, 1396 (1973).

12) M.F.Bengen, Ann.Chem. **565**, 204 (1949).

13) R.M.Barrer, in "Non-Stoichiometric Compounds" (L.Mandelcorn Ed.), Ch. 6, Academic Press, New York, 1964.

14) A.E.Smith, Acta Cryst. **5**, 224 (1952).

15) D.F.R.Gilson and C.A.McDowell, Mol.Phys. **4**, 125 (1961).

16) H.L.Casal,D.G.Cameron and E.Kelushky, J.Chem.Phys. **80**, 1407 (1984).

17) Y.Chatani,H.Anraku and Y.Taki, Mol.Cryst.Liq.Cryst. **48**, 219 (1978).

18) R.L.Vold,W.H.Dickerson and R.R.Vold, J.Magn.Reson. **43**, 213 (1981).

19) J.Jeener and P.Broekaert, Phys.Rev. **157**, 232 (1967).

20) R.E.London and J.Avitabile, J.Amer.Chem.Soc. **100**, 7159 (1978).

21) R.J.Wittebort and A.Szabo, J.Chem.Phys. **69**, 1722 (1978).

22) J.Ragle and K.Sherk, J.Chem.Phys. **50**, 3553 (1969).

23) T.Caves and M.Karplus, J.Chem.Phys. **45**, 1670 (1966).

24) J.F.Harrison, J.Chem.Phys. **48**, 2379 (1968).

25) S.C.Wofsy,J.S.Muenter and W.Klemperer, J.Chem.Phys. **53**, 4005 (1970).

26) E.R.Henry and A.Szabo, J.Chem.Phys. **82**, 4753 (1985).

PBB, Vol. 2
Advanced Magnetic Resonance Techniques
in Systems of High Molecular Complexity
© 1986 Birkhäuser Boston, Inc.

LINE SHAPE IN PMR SPECTRA OF MOLECULES DISSOLVED IN NEMATIC SOLVENTS

G.Chidichimo, D.Imbardelli, M.Longeri

Dipartimento di Chimica, Università della Calabria

A.Saupe

Liquid Crystal Institute, Kent State University, USA

INTRODUCTION

Line width in high resolution NMR spectra can be affected by several relaxation mechanisms.

Following Redfield theory it is possible to calculate the contributions to the line width coming from these different mechanisms, discovering by comparison with experimental data, what is the perturbation responsible for the spin system relaxation.

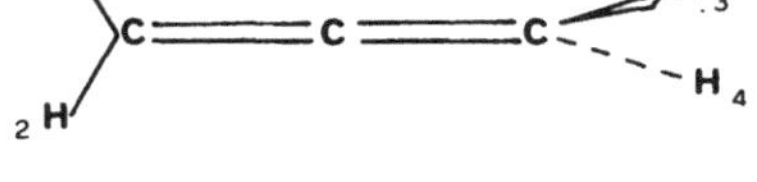

Figure 1 - Notation allene 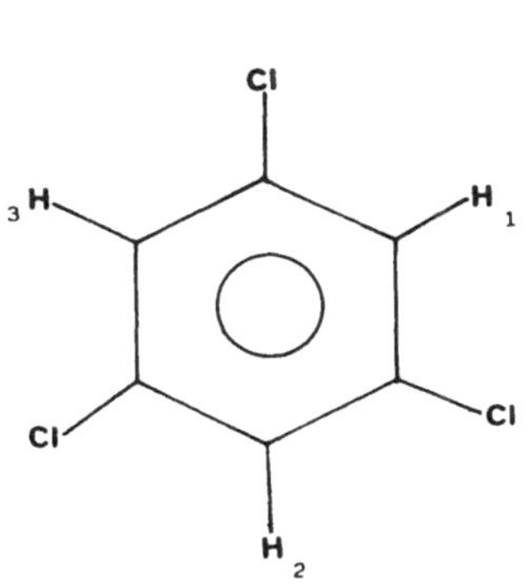   Figure 2 - Notation trichlorobenzene

In this work the contribution to the line width due to intramolecular and intermolecular dipolar interactions were calculated for very simple spin systems, such as allene (figure 1) and 1,3,5-trichlorobenzene (figure 2) dissolved in a nematic mesophase.

THEORY

For a given NMR transition the half width at half height is given by the Redfield matrix element

$$-1/T_2 = R_{\alpha\alpha'\beta\beta'} = \left[J_{\alpha\beta\alpha'\beta'}(\omega_{\alpha\beta}) + J_{\alpha\beta\alpha'\beta'}(\omega_{\alpha'\beta'}) - \delta_{\alpha'\beta'} \sum_\gamma{}' J_{\gamma\beta\gamma\alpha}(\omega_{\gamma\beta}) + \right.$$

$$\left. - \delta_{\alpha\beta} \sum_\gamma J_{\gamma\alpha'\gamma\beta'}(\omega_{\gamma\beta'}) \right]$$

where

$$J_{\alpha\beta\alpha'\beta'}(\omega) = \int_{-\infty}^{+\infty} G(\tau)e^{-i\omega\tau}d\tau =$$

$$= \int_{-\infty}^{+\infty} \overline{<\alpha|\mathcal{H}_1(\tau)|\beta><\beta'|\mathcal{H}_1(0)|\alpha'>} - \overline{<\alpha|\mathcal{H}_1|\beta>}\,\overline{<\beta'|\mathcal{H}_1|\alpha'>}\, e^{-i\omega\tau}d\tau$$

while $\mathcal{H}_1(\tau)$ is the perturbing Hamiltonian.

The considered relaxation mechanisms were:

1) INTRAMOLECULAR DIPOLAR RELAXATION

In this case the perturbing Hamiltonian may be written as (1):

$$\mathcal{H}_1(\tau) = -K \sum_{p<q} \sum_{l,\blacksquare} (-1)^m F_{-m}^{2(p,q)} \left| D_{1,m}^2 \left[\Omega(\tau) \right] - <D_{1,m}^2>_{av} \right| A_1^{2(p,q)}$$

where $F_{-m}^{2(p,q)}$ are the components of the second rank tensor which describes dipolar interaction between two nuclei p and q, in the molecular frame; $A_1^{2(p,q)}$ are the components of a second rank spin tensor in the laboratory frame; $K = \gamma_p \gamma_q \hbar^2$.

The correlation function of the perturbation is then:

$$G(\tau) = \overline{<\mathcal{H}_1(0)_{\alpha B}\,\mathcal{H}_1^*(\tau)_{\alpha'B'}>} - <\mathcal{H}_1>_{\alpha B} <\mathcal{H}_1^*>_{\alpha'B'}$$

$$= \sum_{\substack{11' a<b \\ \blacksquare\blacksquare'p<q}} K^2 (-1)^{m+m'} F_{-m}^{2(a,b)} F_{-m'}^{2*(p,q)} g(\tau) <\alpha|A_1^{2(a,b)}|B><B'|A_1^{2(p,q)}|\alpha'>$$

where

$$g(\tau) = \left[\overline{<D_{1,m}^2(0)D_{1'm'}^2{}^*(\tau)>} - <D_{1,m}^2>_{av} <D_{1',m'}^2{}^*>_{av} \right]$$

$$= \int d\Omega_o P(\Omega_o) D_{1,m}^2(\Omega_o) \int d\Omega P(\Omega_o/\Omega i\tau) D_{1',m'}^2{}^*(\Omega) -$$

$$- \int d\Omega_o P(\Omega_o) D_{1,m}^2(\Omega_o) \int d\Omega P(\Omega) D_{1',m'}^2{}^*(\Omega)$$

$g(\tau)$ can be evaluated if the equation for rotational diffusion equation is solved (3,4). The rotational diffusion equation, as proposed by Nordio, is

$$\partial/\partial\tau \; P(\Omega_o/\Omega i\tau) = D_\perp \mathcal{D}_R P(\Omega_o/\Omega i\tau)$$

where $\mathcal{D}_R = \left\{ \nabla^2 + (1/2KT)\left[(\nabla^2 V_N) + \nabla^2 V_N - V_N \nabla^2 \right] \right\}$, is the rotational-diffusion operator. ∇^2 is the Laplace operator, whose eigenfunctions are the Wigner matrices

$$\nabla^2 D_{1,m}^J(\Omega) = -\left\{ J(J+1) + (D_{\parallel}/D_\perp - 1)m^2 \right\} D_{1,m}^J(\Omega)$$

$D_{\parallel}$ and $D_\perp$ are respectively parallel and perpendicular components of rotational diffusion tensor. V_N/KT is the orienting potential, which can be expressed as a series of Legendre polynomials

$$V_N/KT = \sum_n \lambda_n D_{oo}^{\ n}(\Omega)$$

If $\mathscr{D}_R$ is diagonalized in Wigner matrices space with eigenvalues a_{lm}^K and eigenvectors $\chi_{JK}^{(1,m)}$, the spectral densities which contribute to Redfield matrix have the expression

$$J_{\alpha\beta\alpha'\beta'}(\omega) = \sum_{1,\blacksquare} \sum_{\substack{a<b \\ p<q}} K^2 \left\{ \left[F_{-m}^{2(a,b)} F_{+m}^{2(p,q)} \right] + \left[F_{-m}^{2(p,q)} F_{+m}^{2(a,b)} \right] \right\}$$

$$\cdot \left\{ \left[<\alpha | A_1^{2(a,b)} | \beta><\beta' | A_1^{2(p,q)} | \alpha'> \right] + \right.$$

$$\left. + \left[<\alpha | A_{-1}^{2(a,b)} | \beta'><\beta' | A_{-1}^{2(p,q)} | \alpha'> \right] \right\}$$

$$\cdot 0.2 \left| \chi_{00}^{-1(00)} \sum_J \chi_{J0}^{(00)} \sum_{i,k} \chi_{ki}^{-1(1,m)} C(J,2,i;011) C(J,2,i;0mm) \right.$$

$$\left. \cdot D_\perp a_{lm}^k / [(D_\perp a_{lm}^k)^2 + (\omega_{\alpha\beta})^2] \right|$$

2) INTERMOLECULAR DIPOLAR RELAXATION

The perturbing Hamiltonian for a N spin system with a local fluctuating field may be written as (2):

$$\mathcal{H}_1^{(\tau)} = -\gamma h \sum_{p=1}^N \left[I_{zp} H_z + \tfrac{1}{2}(I_p^- H^+ + I_p^+ H^-) \right]$$

with $H^+ = H_x + iH_y$, $H^- = H_x - iH_y$; (H_x, H_y, H_z are the components of the fluctuating field).

In the hypothesis that all the components fluctuate independently:

$$\overline{H_i(0)H_j(\tau)} = 0$$

$$\overline{H_i(0)H_i(\tau)} = \overline{H_i^2}e^{-\tau/\tau_c}$$

The expression for spectral densities which contribute to Redfield matrix elements are:

$$J_{\alpha\alpha'\alpha'}(0) = \gamma^2\hbar^2\sum_{p=1}^{N}<\alpha|I_{zp}|\alpha><\alpha'|I_{zp}|\alpha'>\overline{H_z^2}\tau_c$$

$$J_{\alpha\beta\alpha'\beta'}(\ ) = \gamma^2\hbar^2\sum_{p=1}^{N}(<\alpha|0.5(I_p^{+}+I_p^{-})|\beta>)(<\beta'|0.5(I_p^{+}+I_p^{-})|\alpha'>)(\overline{H_x^2}+\overline{H_y^2})$$

$$\cdot\tau_c/(1+\omega^2\tau_c^2)$$

RESULTS AND DISCUSSION

The geometry and direct coupling constants of the analysed systems are reported in Table I and Table II.

Table I

Parameters of the spin Hamiltonian and internuclear

distances for allene (5)

Atom pair ij	D_{ij}	J_{ij}	r_{ij}
$(H_1;H_2)$	678-35	---	1.8532
$(H_1;H_3)$	-124.0	-7	3.8672

Theoretical line shapes obtained by using the intramolecular dipolar relaxation model are shown in figures 3 and 4 for allene and trichlorobenzene respectively, while Tables III and IV show the connection among line widths and rotational diffusion motion parameters in the analysed cases.

Table II

Parameters of the spin Hamiltonian and internuclear

distances for 1,3,5-trichlorobenzene

Atom pair ij	D_{ij}	J_{ij}	r_{ij}
$(H_1;H_2)$	101.0	-1.35	2.54

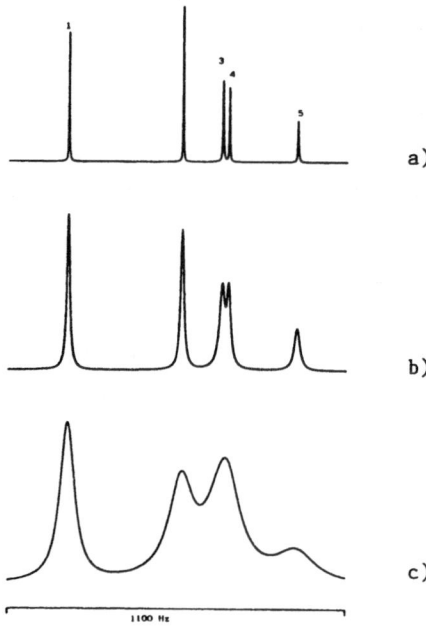

Figure 3 – Contribution to the linewidth of high field part of allene spectrum due to intramolecular dipolar interactions. a) $D_{\parallel}=D_{\perp}=10^9$ rad/sec; b) $D_{\parallel}=D_{\perp}=10^8$ rad/sec; c) $D_{\parallel}=D_{\perp}=10^7$ rad/sec.

It can be observed that the geometry of the spin system is a critical

Table III

Connection between line widths (Hz) of allene spectra

(high field part) and rotational diffusion tensor elements (rad/sec)

	$D_{\parallel} = D_{\perp} = 10^9$	$D_{\parallel} = D_{\perp} = 10^8$	$D_{\parallel} = D_{\perp} = 10^7$
Line n.1	3.31	21.58	151.95
Line n.2	2.14	14.79	95.72
Line n.3	2.98	20.21	130.67
Line n.4	1.65	12.52	95.17
Line n.5	2.05	11.31	56.32

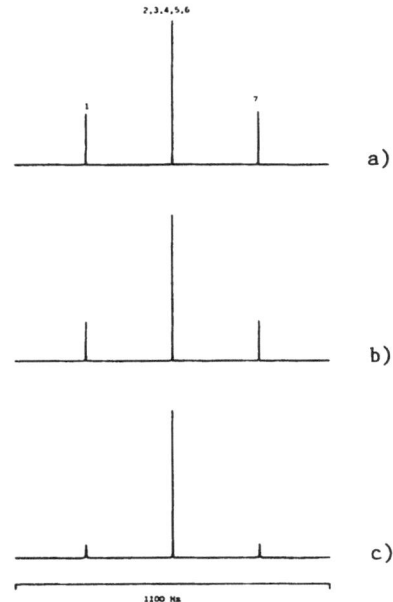

Figure 4 – Contribution to the line width of 1,3,5-trichlorobenzene
spectrum due to intramolecular dipolar interactions. a)
$D_{\parallel} = D_{\perp} = 10^9$ rad/sec; b) $D_{\parallel} = D_{\perp} = 10^8$ rad/sec; c) $D_{\parallel} = D_{\perp} = 10^7$ rad/sec.

point in intramolecular dipolar relaxation. In fact dipolar perturbation between two nuclei p and q depends on r_{pq}^{-3}. Then a spin system having very close spin pairs will be more affected by such a relaxation mechanism than a spin system which exhibits larger internuclear distances. That's why the intramolecular contribution to the line width in the case of allene is much bigger than in the case of 1,3,5-trichlorobenzene.

Table IV

Connection between line widths (Hz) of 1,3,5-trichlorobenzene spectra and rotational diffusion elements

	$D_{\parallel} = D_{\perp} = 10^9$	$D_{\parallel} = D_{\perp} = 10^8$	$D_{\parallel} = D_{\perp} = 10^7$
Line n.1	0.032	0.238	1.804
Line n.2 *	0.015	0.082	0.229
Line n.3	0.003	0.018	0.083
Line n.4	0.001	0.006	0.028
Line n.5	0.008	0.041	0.194
Line n.6	0.005	0.029	0.139
Line n.7	0.032	0.238	1.804

*The central line of the spectrum consists of five overlapping lines.

Table V shows the sensitivity of the line width, in the case of allene spectra, to the variation of $D_{\parallel}$ and $D_{\perp}$. It is a noticeable point that the variation of $D_{\perp}$ influences line widths much more than $D_{\parallel}$ does.

The intermolecular contributions to the line width (see Tables VI and

VII) which can be obtained for allene spectra are comparable to that calculated for 1,3,5-trichlorobenzene spectra.

Table V

Connection between line widths (Hz) of allene spectra and rotational diffusion tensor elements (rad/sec) asymmetry

	$D_\| = 10^9$ $D_\perp = 10^8$	$D_\| = 10^9$ $D_\perp = 10^7$	$D_\| = 10^8$ $D_\perp = 10^9$	$D_\| = 10^7$ $D_\perp = 10^9$
Line n.1	16.55	124.71	5.38	5.90
Line n.2	11.17	67.36	3.21	3.52
Line n.3	16.24	102.36	4.21	4.55
Line n.4	9.03	66.85	2.67	2.96
Line n.5	7.64	28.54	3.34	3.67

Table VI

Correlation between line widths (Hz) of allene spectrum (high field part) and the correlation time (sec) of fluctuating fields

	$\tau_c = 1.2 \times 10^{-10}$	$\tau_c = 1.2 \times 10^{-9}$	$\tau_c = 8 \times 10^{-9}$
Line n.1	0.30	3.00	12.00
Line n.2	0.74	7.40	29.60
Line n.3	0.86	8.60	34.40
Line n.4	0.30	3.00	12.00
Line n.5	0.70	7.00	28.00

Table VII

Correlation between line widths (Hz) of 1,3,5-trichloro-

benzene spectrum and the correlation time (sec)

of the fluctuating fields

	$\tau_c = 1.2 \times 10^{-10}$	$\tau_c = 1.2 \times 10^{-9}$	$\tau_c = 8 \times 10^{-9}$
Line n.1	0.50	5.00	20.00
Line n.2 *	0.10	1.00	4.00
Line n.3	0.10	1.00	4.00
Line n.4	0.10	1.00	4.00
Line n.5	0.10	1.00	4.00
Line n.6	0.10	1.00	4.00
Line n.7	0.50	5.00	20.00

*The central line of the spectrum consists of five overlapping lines

CONCLUSIONS

The results of our theoretical calculations indicate that the relative importance of intramolecular and intermolecular relaxation mechanisms on PMR spectra obtained from liquid crystalline solutions depends on:

-The rotational diffusion tensor elements,

-The molecular geometry,

-The solute solvent interactions.

It has also been observed that in the case of molecules that have a small orientation in the nematic mesophase, the degree of order of liquid crystalline solution has only a negligible influence.

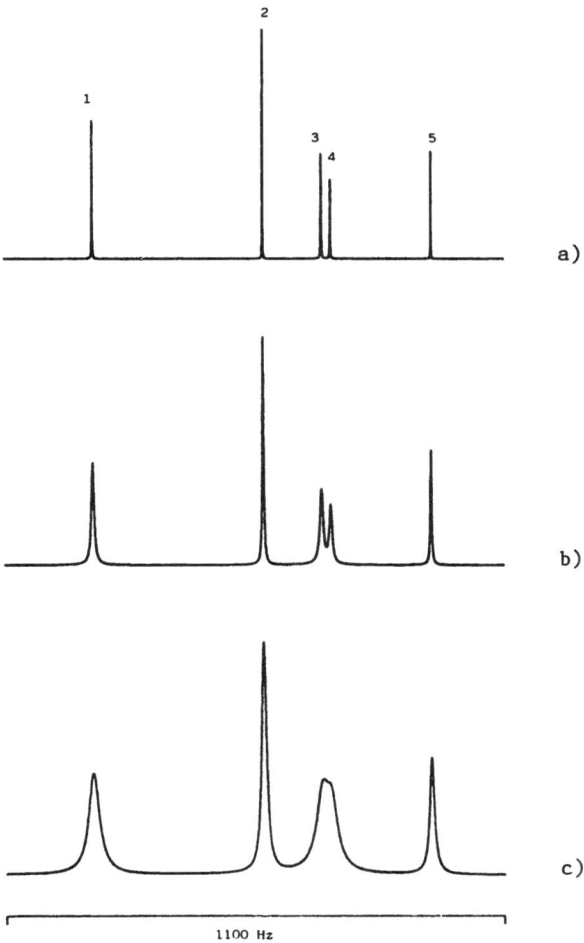

1100 Hz

Figure 5 - Contribution to the linewidth of high field part of allene spectrum coming from intermolecular dipolar interactions. a) $\tau_c = 1.2 \times 10^{-10}$ sec; b) $\tau_c = 1.2 \times 10^{-9}$ sec; c) $\tau_c = 8 \times 10^{-9}$ sec.

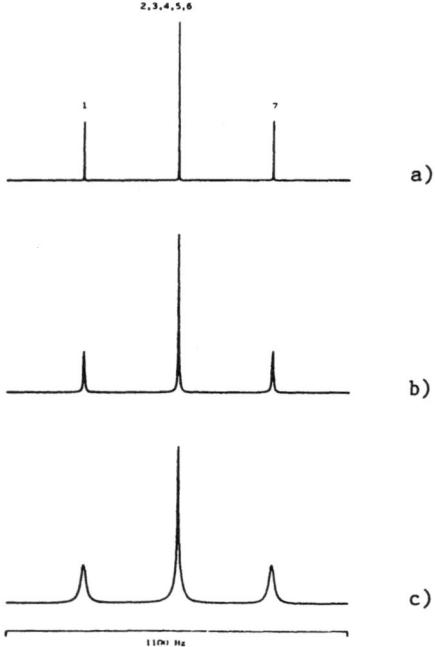

Figure 6 - Contribution to the line width of 1,3,5,-trichlorobenzene
spectrum coming from intermolecular dipolar interactions.
a) τ_c =1.2x10^{-10} sec; b) τ_c =1.2x10^{-9} sec; c) τ_c =8x10^{-9} sec.

REFERENCES

1) J.H.Freed and G.K.Fraenkel, J.Chem.Phys. **39**, 326 (1963).

2) B.B.Sharma,G.Chidichimo,A.Saupe,H.Chang and G.H.Brown, J.Magn.Reson.
    49, 287 (1982).

3) P.L.Nordio,G.Rigatti and U.Segre, Mol.Phys. **25**, 129 (1973).

4) P.L.Nordio and U.Segre, in "The Molecular Physics of Liquid
    Crystals" (G.R.Luckhurst and G.W.Gray eds.), Academic Press, New
    York, 1979, Chapt.19.

5) E.Sackmann, J.Chem.Phys. **51**, 2984 (1969).

PBB, Vol. 2
Advanced Magnetic Resonance Techniques
in Systems of High Molecular Complexity
© 1986 Birkhäuser Boston, Inc.

EXTENDED-TIME EXCITATION ELECTRON SPIN ECHO SPECTROSCOPY

L.Braunschweiler, A.Schweiger, J.-M.Fauth and R.R.Ernst

Laboratorium für Physikalische Chemie

Eidgenössische Technische Hochschule

CH-8092 Zürich, Switzerland

In electron spin resonance (ESR), refocusing experiments (1) led to the development of electron spin echo (ESE) spectroscopy which developed into a powerful tool for the study of relaxation and of hyperfine and quadrupole interactions (2). The main interest in electron spin echo experiments relies on the modulation of the echo envelope as a function of the pulse separation in a two- or three-pulse sequence. This modulation is caused by interactions which are not fully refocused by a pulse (2-4). To record an ESE modulation trace, the time between the microwave pulses is incremented from experiment to experiment, each providing one point of the trace. The measurement of a full echo envelope may therefore be time-consuming. The use of pulse trains for the simultaneous measurement of the entire echo envelope in analogy to Carr-Purcell experiments (5) is, unfortunately, often not feasible because of the extended dead-time disqualifying sampling between the pulses.

In this contribution we present an alternative approach to refocusing which permits the entire echo envelope modulation to be recorded in a single experiment whitout any pulses during detection (6). In effect the experiment provides a "continuous refocusing" of the interactions by means of a particular extended-time preparation of the spin system prior to detection.

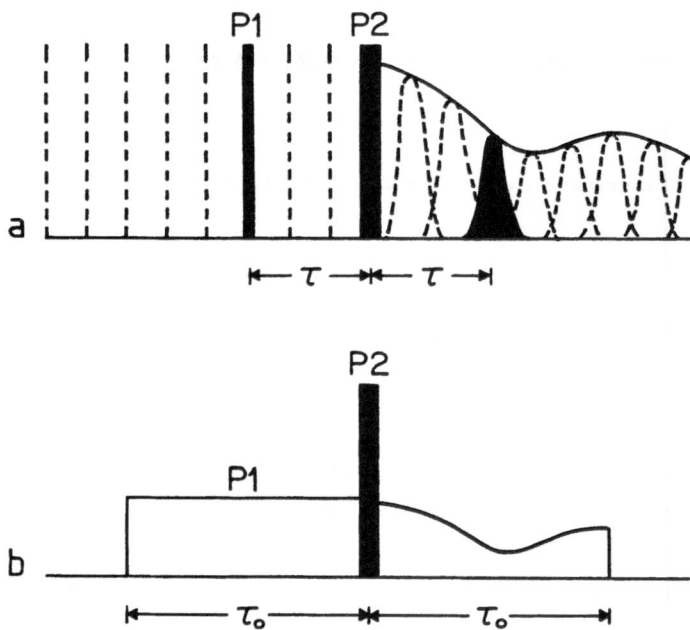

Figure 1 – a) Conventional two-pulse Hahn-echo experiment. The echo
              amplitude is recorded point-wise by stepping τ from ex-
              periment to experiment.
           b) Echo experiment with soft-pulse excitation followed by a
              short refocusing π pulse producing an entire echo decay
              in a single experiment.

Instead of applying an initial $\pi/2$ pulse followed by a π refocusing
pulse as in Fig. 1a, we excite the system by a low level irradiation
$V(t)$ for an extended time τ_o prior to the pulse as illustrated in Fig.
1b. In the linear response approximation, each time interval within the
excitation period τ_o causes an echo in the corresponding time interval
in symmetric position after the π pulse. The superposition of all echoes
leads then to a continuous echo envelope which can be measured in a single experiment. The irradiation $V(t)$ can either be a continuous wave

microwave field of constant amplitude with a frequency placed in the center of an ESR transition, a broadband stochastic noise irradiation, or a burst of small flip angle microwave pulses.

Extended-time excitation experiments which replicate the shape of a low level excitation sequence have been discussed previously (7-13). In an early paper on nuclear magnetic resonance, Fernbach and Proctor first showed that by means of a short powerful "reading pulse" an applied sequence of events can be recalled (7). In coherent optics, corresponding information storage effects have recently been described (10-13). In contrast to the mentioned references, the technique presented in this contribution is applied to map interactions inherent within the spin system itself. No information storage is attempted.

For the basic two-pulse experiment (Fig. 1a), the density operator $\sigma(t)$ at the echo maximum $t = \tau$ is given by

$$\sigma(t=\tau) = -\exp(-i\mathcal{H}\tau)\exp(-i\pi S_x)\exp(-i\mathcal{H}\tau)S_y\exp(i\mathcal{H}\tau)\exp(i\pi S_x)\exp(i\mathcal{H}\tau)$$

$$= \exp(-i\mathcal{H}\tau)\exp(-i\tilde{\mathcal{H}}\tau)S_y\exp(i\tilde{\mathcal{H}}\tau)\exp(i\mathcal{H}\tau) \qquad (1)$$

with the transformed Hamiltonian $\tilde{\mathcal{H}} = \exp(-i\pi S_x)\mathcal{H}\exp(i\pi S_x)$. The propagator $\exp(-i\mathcal{H}\tau)\exp(-i\tilde{\mathcal{H}}\tau)$ is responsible for the echo amplitude modulation. Of particular importance are the non-commuting parts of $\mathcal{H}$ and $\tilde{\mathcal{H}}$ involving the hyperfine and quadrupole interactions.

To derive a similar relation for the case of an extended-time low level excitation $V(t)$, we can solve the density operator equation

$$\dot{\sigma}(t) = -i\left[\mathcal{H} + V(t), \sigma(t)\right] \qquad (2)$$

by expanding $\sigma(t)$ in terms of the perturbation $V(t)$,

$$\sigma(t) = \sigma^{(0)}(t) + \sigma^{(1)}(t) + \ldots\ldots \qquad (3)$$

Solving Eq.(2) for the linear response density operator $\sigma^{(1)}(t)$ we find immediately before the π pulse

$$\sigma^{(1)}(0^-) = -i \int_{-\tau_0}^{0} \exp(i\mathcal{H}x) \left[V(x), \sigma_0 \right] \exp(-i\mathcal{H}x)dx \qquad (4)$$

with the equilibrium density operator σ_0. The evolution of the density operator after the π pulse is then determined by

$$\sigma^{(1)}(t) = i \int_{-\tau}^{0} \exp(i\mathcal{H}t)\exp(i\tilde{\mathcal{H}}x) \left[\tilde{V}(x), \sigma_0 \right] \exp(-i\tilde{\mathcal{H}}x)\exp(i\mathcal{H}t)dx \qquad (5)$$

For a sufficiently broad inhomogeneous frequency distribution $g(\omega)$ of the ESR line, one can show (14) that only the integrand for $x = -t$ contributes to the integral and one obtains

$$\sigma^{(1)}(t) = ig(0)\exp(-i\mathcal{H}t)\exp(-i\tilde{\mathcal{H}}t) \left[\tilde{V}(-t), \sigma_0 \right] \exp(i\tilde{\mathcal{H}}t)\exp(i\mathcal{H}t) \qquad (6)$$

$$\text{for } 0 \leq t \leq \tau_0$$

Eq.(6) leads to a refocusing for any $t \leq \tau_0$ and generates a continuous echo envelope which depends on the properties of the spin system and on the excitation $V(-t)$. Schenzle et al. (15) showed already that the response of an inhomogeneously broadened system to an excitation of length τ_0 can last no longer than for an additional time τ_0.

The response becomes particularly simple for a cw soft-pulse with $V(t) = \omega_{MW}S_x$ for $-\tau_0 \leq t \leq 0$. Full equivalence to Eq.(1) is established when $\sigma_0 = S_z$ is inserted:

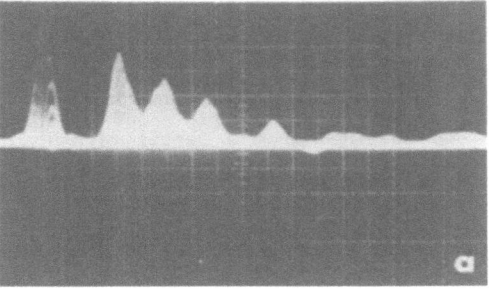

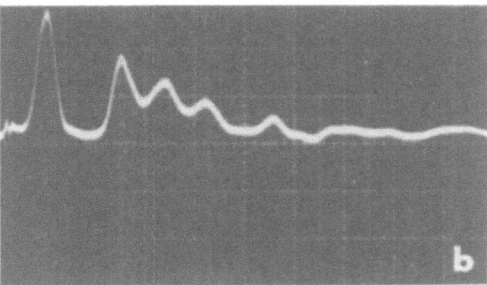

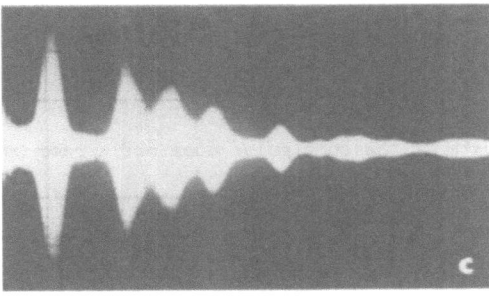

Figure 2 – Echo modulation amplitude of Coacacen diluted in a single
crystal of Niacacen measured for an arbitrary crystal orientation at 9.15 GHz and a temperature of 4.2 K. The hori-
zontal deflection is $0.2\,\mu$s per division. The vertical axis
is arbitrary.

a) Echo modulation obtained by multiple-exposure to 200 two-
pulse experiments with $\tau_{\pi/2}$=10 ns and τ_{π}=20 ns. Delay τ is
incremented in steps of 10 ns.

b) Echo modulation obtained from a single event using 2 μs
soft pulse excitation (pulse rotation angle $\simeq 30°$) and a 20
ns π refocusing pulse.

c) Echo modulation obtained by multiple exposure to 10^4 stochastic response signals using a 2 μs noise excitation followed by a 20 ns π refocusing pulse.

$$\sigma^{(1)}(t) = g(0)\ \omega_{MW}\exp(-i\mathcal{H}t)\exp(-i\widetilde{\mathcal{H}}t)S_y\exp(i\widetilde{\mathcal{H}}t)\exp(i\mathcal{H}t) \qquad (7)$$

for $0 \leq t \leq \tau_0$. Thus, the soft-pulse single-experiment echo envelope is identical to the one obtained from a series of basic two-pulse experiments as long as the linear regime is valid for the low level pulse.

Figure 2 shows an experimental verification of this prediction. The echo envelope obtained from a series of two-pulse electron spin echo experiments on N,N'-ethylenebis (acetylacetonatiminato)Co(II) (Coacacen) diluted in a single crystal of Niacacen is reproduced in Fig. 2a. The photograph is a multi-exposure of the echoes of 200 τ -values incremented in steps of 10 ns. The deep modulations are due to the hyperfine and quadrupole couplings of the two nitrogen ligands of Coacacen with nuclear frequencies lying in the range between 1-7 MHz (16).

The trace shown in Fig. 2b has been obtained with a single experiment of the type shown in Fig. 1b with a soft excitation pulse of 2 μs length. It is apparent that the echo envelope is faithfully reproduced. It has been found that no visible distortions are encountered as long as the total rotation angle of the soft pulse does not exceed 30°.

The same echo envelope can also be reproduced by stochastic excitation (17) of equal duration τ_0 prior to the application of the short π pulse. To eliminate the random character of a single response, a series of stochastic response experiments may be combined. Figure 2c shows the multi-exposure photograph of 10^4 stochastic response experiments with a repetition rate of 1 kHz. Again a perfect echo envelope matching the one of Fig. 2a is obtained.

Comparable results as shown in Fig. 2 can also been obtained by use of a rapid repetitive burst of weak microwave pulses for excitation. Then a sampled record of the echo envelope is produced.

We should mention that the frequency range of excitation of the long low-level pulse is usually much narrower than the hyperfine couplings, without affecting the survival of the modulation pattern. In contrast to this finding, it has previously been stated in the literature (18, 19) that the frequency range of excitation of all pulses must exceed the maximum envelope modulation frequency to be observed. We have found by theory and experiment that for a system with a dominant inhomogeneous broadening mechanism it is sufficient that the refocusing pulse fulfills this condition (20).

Experimental results are summarized in Fig. 3. Figure 3a gives the echo envelope modulation for a Cu (II)-aquo complex at 4.2 K using short microwave pulses both for excitation (10ns) and refocusing (20ns) with nominal flip angles of $\pi/2$ and π, respectively. The Fourier transform of this pattern consists of two peaks at 14 and 28 MHz corresponding to the proton Zeeman frequency and its second harmonics.

Replacing the short excitation pulse by a long (210ns) weak pulse, no change in the modulation pattern is observed (Fig. 3b). However, a loss of signal amplitude occurs caused by the reduced bandwidth of excitation, not visible in Fig. 3b due to normalized amplitudes. In Fig. 3c, the short refocusing pulse is replaced by a long (210ns) and weak pulse. In contrast to the first two experiments the echo modulation pattern disappears completely.

By increasing the amplitudes of the two pulses part of the echo modulation may be recalled (Fig. 3d). The pattern observed, however, does not fully correspond to the one found in Fig. 3a and 3b using short refocusing pulses. This is because the origin of the time reversal is

314

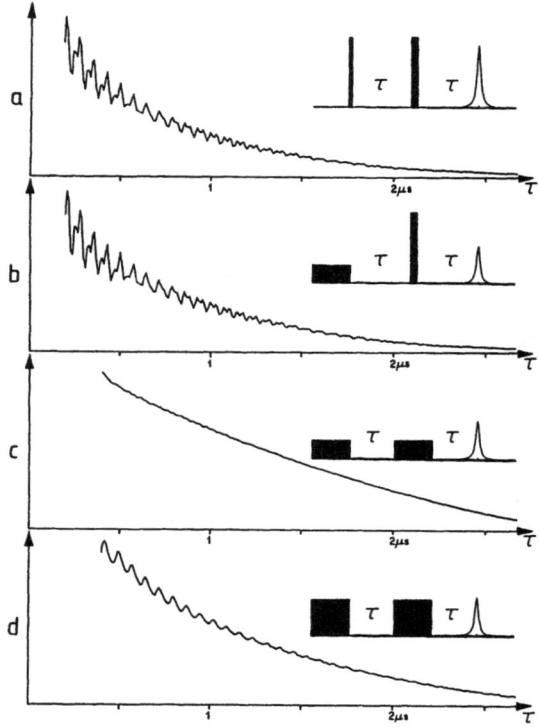

Figure 3 – Two pulse electron spin echo envelope modulation traces of
copper(II)chloride in water {(0.01 M),pH=1,diluted with an
equal volume of glycerol}.All traces are recorded at 4.2 K
a) Short microwave pulses for both excitation (10ns) and refocusing (20ns) with nominal flip angles of $\pi/2$ and π, respectively.
b) Excitation pulse replaced by a long (210ns) weak pulse,
nominally $< \pi/2$.
c) Both excitation and refocusing pulses are long (210 ns)
and weak.
d) Selective excitation and refocusing pulses of increased
amplitude.

not well defined causing a smearing and low-pass filtering of the modulation pattern. Thus, for distortionless modulation a short refocusing pulse seems to be a prerequisite, whereas no such condition applies to the excitation pulse.

The equivalence of selective and non-selective excitation can be proved for inhomogeneously broadened sytems where the inhomogeneous broadening is much wider than the spread of the relevant ESR frequencies as visualized in Fig. 4 for an electron nuclear two-spin system with S=I=½ (20).

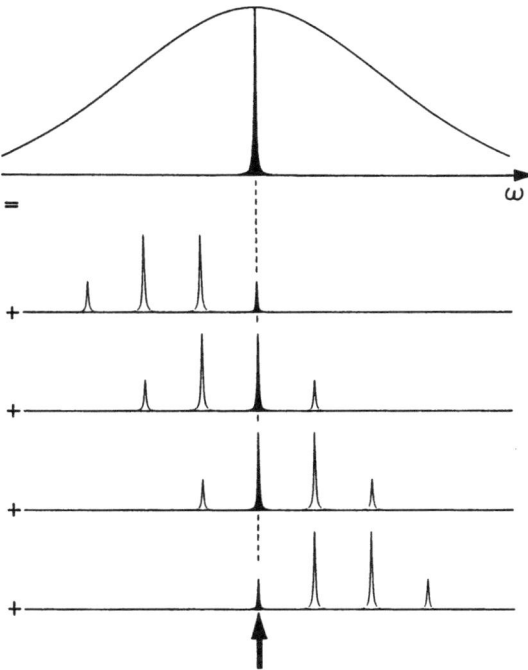

Figure 4 – Effect of inhomogeneous line broadening: the underlying spectra of the individual spin systems are shifted by different amounts with respect to the center of irradiation. A selective pulse excites all resonance lines present though only one line in each spin system

The inhomogeneous broadening in effect shifts the underlying ESR spectrum by different amounts such that a selective perturbation excites all resonance lines, though only one line in each spin system. The initial density operator for selective excitation, when averaged over the inhomogeneous distribution, becomes proportional to that reached by a non-selective pulse,

$$\sigma_{sel}(0) \propto \sigma_{non-sel}(0) \qquad (8)$$

This can be shown more formally as follows: in the case of a non-selective excitation by an x-pulse with rotation angle β, we have for the relevant part of the density operator,

$$\sigma(0) = \exp(-i\beta S_x)S_z\exp(i\beta S_x) \propto -S_y\sin(\beta) \qquad (9)$$

For a selective excitation of the transition (kl), on the other hand, we find

$$\sigma^{(kl)}(0) = -S_y^{(kl)}\sin(\beta^{(kl)}) \qquad (10)$$

where $S_Y^{(kl)}$ is a single transition operator of transition (kl) (21, 22) and the effective rotation angle $\beta^{(kl)}$ is given by

$$\beta^{(kl)} = \beta 2 <k|S_x|1> \qquad (11)$$

In other words, the effective rotation angle is weighted by the transition moment of the particular transition and varies from transition to transition. For a large rotation angle, therefore the excitation will be non-uniform. However for small β, we may expand $\sin(\beta^{(kl)})$, and find

$$\sigma^{(kl)}(0) \simeq -S_y^{(kl)} <k|S_x|1> 2\beta \qquad (12)$$

leading to the average initial density operator

$$\sigma^{(0)} \propto \sum_{(kl)} \sigma^{(kl)}(0) = -\beta \, S_y \, 2B \tag{13}$$

Taking into account in addition the linear character of the equation of motion,

$$\dot{\sigma} = -i\left[\mathcal{H}, \sigma\right] \tag{14}$$

the echo modulation resulting for non-selective and selective excitation with small rotation angle will be identical except for a uniform scaling . factor.

The above arguments can easily be generalized to any arbitrary form of excitation, irrespective whether broad- or narrow-band, provided the excitation is sufficiently weak.

We should like to point out that the essential phenomena responsible for the echo modulation are caused by the refocusing pulse π. The echo modulation may be understood as a coherence transfer induced by the π pulse between the so called allowed and forbidden ESR transitions (4, 20, 23).

A quantititative comparision of the relative sensitivities of the different excitation schemes in extended-time excitation spectroscopy will be given at another place (14). We note here that in preliminary measurements the amplitude of a single soft-pulse echo decay is approximately 1/10 of a Hahn two-pulse echo while the maximum excursions of the stochastic response can reach up to 1/3 of a Hahn echo, without causing appreciable envelope distortion. Stochastic and pulse burst excitation can lead to enhanced sensitivity due to the wider bandwidth of irradiation (24).

The soft pulse, stochastic and pulse burst excitations can easily be extended to more complicated experiments such as the three-pulse stimulated echo experiment where the first $\pi/2$ pulse is replaced by an extended-time excitation which is immediately followed by the second (short) $\pi/2$ pulse. The response after the third $\pi/2$ pulse is then of the same length than the extended-time excitation. A single experiment of this type exhibits an envelope modulation which corresponds to a full set of three-pulse sequences with a variable time τ between the first and second pulse and thus drastically simplifies the data acquisition in two-dimensional electron spin echo experiments (25).

The experiments have been performed on a home-made electron-spin-echo spectrometer described at another place (26, 27). The microwave pulses have been amplified by a 1 KW travelling wave tube amplifier. The stochastic excitation used the noise of a second travelling wave tube amplifier. The transient signals have been visualized on an oscilloscope. A loop gap resonator with three loops and two gaps was employed.

The extension of the proposed extended-time excitation experiment to other coherent spectroscopies like nuclear magnetic resonance and optical spectroscopy is straightforward and may for inhomogeneously broadened systems offer similar advantages in comparison to standard echo experiments.

ACKNOWLEDGEMENT

This research has been supported by the Swiss National Science Foundation. The authors are grateful to Mr. J. Forrer for technical support with the spectrometer and to Mr. G. Grassi for the preparation of the samples. Furthermore, we acknowledge the lending of a microwave power amplifier by the Bundesamt für Militärflugplätze.

REFERENCES

1) R.J.Blume, Phys.Rev. **109**, 1867 (1958).

2) L.Kevan and R.N.Schwartz, Eds. "Time Domain Electron Spin Resonance", Wiley-Interscience, New York, 1979.

3) L.G.Rowan,E.L.Hahn and W.B.Mims, Phys.Rev. **137**, 61 (1965).

4) W.B.Mims, Phys.Rev. **B5**, 2409 (1972).

5) H.Y.Carr and E.M.Purcell, Phys.Rev. **94**, 630 (1954).

6) A.Schweiger,L.Braunschweiler,J.-M.Fauth and R.R.Ernst, Phys.Rev. Lett. **54**, 1241 (1985).

7) S. Fernbach and W.G. Proctor, J. Appl. Phys. **26**, 170 (1955).

8) P.Mansfield,A.A.Maudsley,P.G.Morris and I.L.Pyckett, J.Magn.Reson. **33**, 261 (1979).

9) D.F.Hoult, J.Magn.Reson. **35**, 69 (1979).

10) S.O.Elyutin,S.M.Zakharov and E.A.Manykin, Sov.Phys.JETP **49**, 421 (1979).

11) V.A.Zuikov,V.V.Samartsev and R.G.Usmanov, JETP Lett. **32**, 270 (1980).

12) N.W.Carlson,L.J.Rothberg,A.G.Yodh,W.R.Babbitt and T.W.Mossberg, Opt. Lett. **8**, 483 (1983).

13) N.W.Carlson,Y.S.Bai,W.R.Babbitt and T.W.Mossberg, Phys.Rev. **A30**, 1572 (1984).

14) L.Braunschweiler,A.Schweiger and R.R.Ernst, to be published.

15) A.Schenzle,N.C.Wong and R.G.Brewer, Phys.Rev. **A22**, 635 (1980).

16) M.Rudin,A.Schweiger and Hs.H.Günthard, Mol.Phys. **46**, 1027 (1982).

17) R.R.Ernst, J.Magn.Reson. **3**, 10 (1970).

18) L.Kevan,M.K.Bowman,P.A.Narayana,R.K.Boeckman,V.F.Yudanov and Y.D. Tsvetkov, J.Chem.Phys. **63**, 409 (1975).

19) R.M.Macfarlane,R.M.Shelby and R.L.Shoemaker, Phys.Rev.Lett. **43**, 1726 (1979).

20) L.Braunschweiler,A.Schweiger,J.-M.Fauth and R.R.Ernst, J.Magn. Reson., in press.

21) A.Wokaun and R.R.Ernst, J.Chem.Phys. **67**, 1752 (1972).

22) S.Vega, J.Chem.Phys. **68**, 5518 (1978).

23) G.Bodenhausen,H.Kogler and R.R.Ernst, J.Magn.Reson. **24**, 425 (1976).

24) R.Beach and S.R.Hartmann, Phys.Rev.Lett. **53**, 663 (1984).

25) R.P.J.Merks, R.de Beer, J.Phys.Chem. **83**, 3319 (1979).

26) J.Forrer, M.Müri,J.-M.Fauth,A.Schweiger and R.R.Ernst, Proc. XXIInd Congress Ampere, p. 623 (1984).

27) M.Müri,J.Forrer,J.-M.Fauth,L.Braunschweiler,A.Schweiger   and   R.R. Ernst, to be published.

PBB, Vol. 2
Advanced Magnetic Resonance Techniques
in Systems of High Molecular Complexity
© 1986 Birkhäuser Boston. Inc.

STRUCTURE AND DYNAMICS OF EXCIMERS IN ORGANIC MOLECULAR

CRYSTALS. ENDOR AND PICOSECOND STUDIES

H.C.Wolf

Physikalisches Institut, Teil 3, Universität Stuttgart

Pfaffenwaldring 57, 7000 Stuttgart 80, West Germany

1 - INTRODUCTION

In the organic solid state photochemistry excimers play an important
role as intermediates of photochemical reactions. As photochemical
reaction products they can give us a better understanding of photochemical reactions which are governed by the specific crystal structure.
This is the reason for the great interest in excimers during the last 20
years. In the following I want to discuss some new experimental results
which are able to help to a better understanding of this interesting
solid state photoproduct.

2 - WHAT ARE EXCIMERS?

Excimers are molecules which absorb light as monomers, dimerize in the
excited state and emit light as dimers. The main characteristic is the
broad optical emission which is shifted to lower energies with respect
to the monomer emission, and also the longer decay time of singlet emission as compared to the monomer decay time. In organic molecules, they
first were discovered and described in this way by Förster et al. (1954)
in concentrated solutions of pyrene in the liquid phase. A kinetic
analysis gave evidence for the dimeric character of the emitting spe-

cies. For reviews see (1-3).

In the following years it has been shown that excimer emission can be observed in many solutions of aromatic molecules at high enough concentrations, also in molten organic substances. It was concluded that excimer formation is possible only if the two molecules of the dimer are able to come into a characteristic geometrical configuration allowing for a large overlap of the π-orbitals.

In the solid state, the oldest known excimer is that of pyrene. In pyrene crystals the molecules are arranged pairwise in a sandwich like configuration. So the molecules can come into a configuration in which their planes are parallel to each other and the interplanar distance is small. It was observed, in addition, that the intensity of excimer emission has usually a characteristic temperature dependence with a minimum both at very low and at very high temperature. From this it was concluded that often excimers need a thermal activation energy to be formed, and with increasing temperature they can again dissociate before emitting light.

Excimers are observed in two types of single crystals. Those like pyrene where the molecules in the crystal are arranged in sandwich-like pairs (another example is perylene) and those where the molecules are arranged in stacks of molecules with the planes parallel to each other (examples are Dichloro-Anthracene (DCA) and Hexamethyl-Benzene (HMB)).

In addition, excimer emission is observed in crystals where the structure does not permit sandwich-like dimerisation if the crystals contain specific defect sites. By deformation of the crystals such defect sites can be created.

3 - WHAT ARE THE INTERESTING PROBLEMS?

Among the open questions in the physics of excimers, I want to address mainly the following two problems:

- How is excimer formation possible in a long stack of parallel molecules? Is this possible as a self-trapping process in an ideal single crystal, or does one need defect sites in order to make excimer formation possible? Or is an exciton a perturbation strong enough in order to break the stack into dimers?

- How fast is the relaxation process from the monomeric absorbing into the dimeric emitting state? Is this a one-step process or are there intermediate or precursor states between the monomer and the excimer which live long enough in order to be observable?

The structure of excimers can be studied by electron spin resonance techniques if the excimers are in the triplet state. In order to contribute to the first question, we have made ESR and ENDOR experiments on the only known triplet excimer system in the solid state, hexachlorobenzene. The relaxation processes, as mentioned as the second main problem, can be studied using time resolved optical spectroscopy in the singlet state. A picosecond time resolution is needed. After earlier measurements on perylene excimers we present here new results on the prototype excimer crystal, that is pyrene.

4 - TRIPLET EXCIMERS IN HCB (HEXACHLOROBENZENE (4))

4.1 Crystal structure

The crystal structure of HCB is shown in Fig 1. There are two molecules in the unit cell, and two stacks (A and B) of molecules with almost parallel planes in the crystal. The interplanar distance is 3.52 Å.

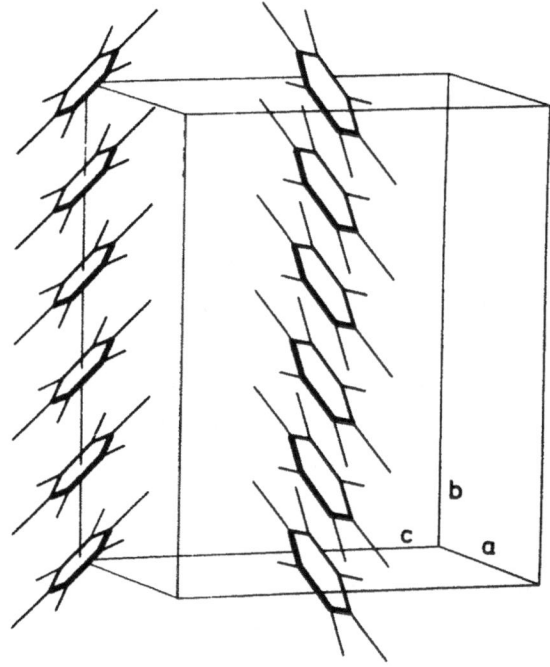

Figure 1 - Crystal structure of HCB (4)

4.2 Optical phosphorescence emission spectrum

In pure, undoped HCB the phosphorescence spectrum is a typical X-trap
spectrum with relatively narrow lines and 0.0 at 389 nm. This spectrum
is superimposed on a broad structureless background which is perhaps the
excimer spectrum.

Using ODMR, the fine structure parameters of the X-trap are measured as $D=0.21$ cm^{-1}, $E=0.03$ cm^{-1}. If the crystals are doped with tri-chlorobenzene (TCB), the phosphorescence spectrum changes drastically. One observes a broad structureless emission with the maximum at 460 nm and no corresponding absorption; this means a typical excimer spectrum. The fine structure parameters as measured by ODMR are 0.14 and 0.017 cm^{-1} for D and E, respectively, independent of the dopant, much smaller than those of the X-traps. The lifetime is 60 msec. If the temperature increases above 12 K, the excimer spectrum disappears and one observes the X-trap spectrum.

4.3 ESR spectra

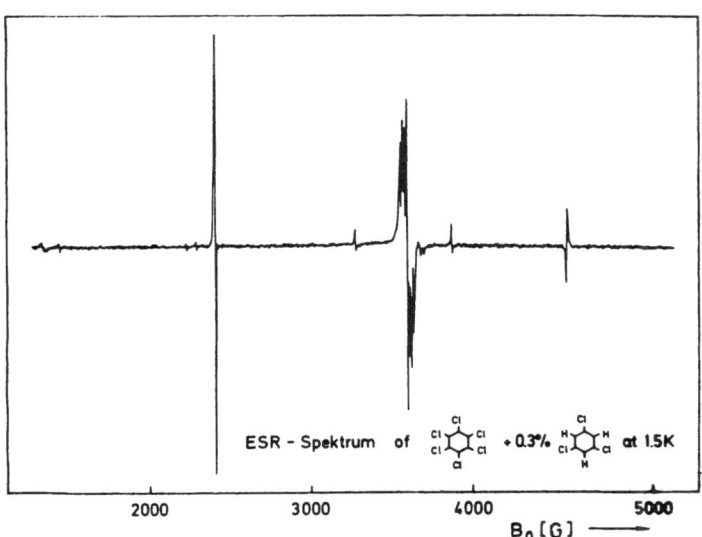

Figure 2 - ESR spectrum of HCB doped with 0.3% of TCB at 1.5 K, for B_0 parallel to the x axis of the fine structure tensor. The two excimer signals are spin polarized. The g=2 signal is due to non-metastable photochemical defects, the two small signals close to g=2 are stable and not identified (4).

The ESR spectrum of the excimer state (Fig. 2) is a typical triplet spectrum with two lines, which are spin polarized. From the angular dependence one can conclude that the excimers belong to one of the two stacks only, and that they are misoriented by 20° as compared to the orientation of molecules in the original crystal. From this angular dependence we can also conclude that the excimers are of type AA or BB, not AB.

The shape of the ESR lines is purely Gaussian with a halfwidth of 6.5 G. It is characteristic for unresolved hyperfine structure, can be saturated easily, thus showing that ENDOR should be possible.

The temperature dependence of the ESR spectrum is the same as that of the phosphorescence. It disappears with increasing temperature, the activation energy being 16 cm^{-1}.

4.4 PMDR

By measuring the PMDR (phosphorescence-microwave double resonance) spectrum, one gets clear evidence for a complete correlation between excimer emission and ESR. This means that the ESR spectrum really belongs to the excimer emitting species.

4.5 Proton ENDOR

HCB has only Cl-atoms as substituents. Thus from HCB we expect Cl-ENDOR only. TCB as partner or neighbour of the excimer, on the other hand, should give both proton and Cl-ENDOR.

H-ENDOR signals between 9 and 11 MHz and Cl-ENDOR signals between 0 and

3 and between 37 and 45 MHz are observed. An example is shown in Fig. 3. The angular dependence of all these lines has been measured. In addition, triple resonance measurements were performed where not only the ESR transition but also one of the possible nuclear transitions have been saturated in order to measure the influence of this saturation on the intensity of a different nuclear transition.

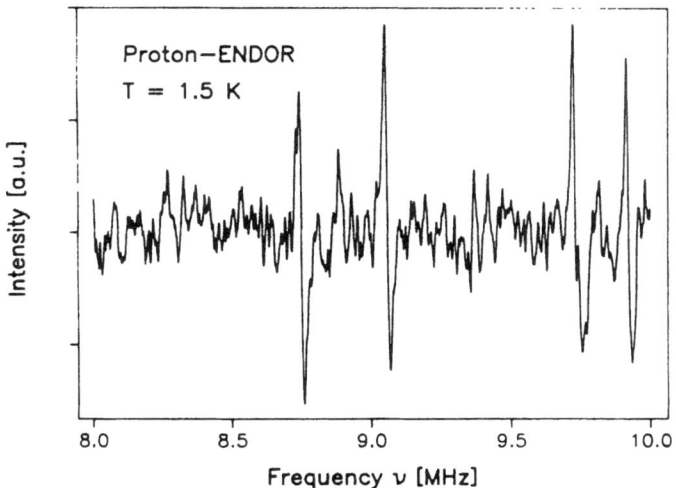

Figure 3 - Part of the ENDOR spectrum of the triplet excimer in HCB at 1.5 K. The singlets are due to three inequivalent protons. The free proton frequency is ν_H=9.76 MHz (4)

The proton ENDOR spectrum shows 4 lines (3 x 2, but 3 are those of free protons) for the three different H-atoms on TCB.

The isotropic interaction is zero, so there is no contact interaction; this means no spin density on TCB, so we can conclude that TCB is not a partner of the excimer.

The measured anisotropic interaction, on the other hand, is consistent with a model where TCB is nearest neighbour to HCB, according to a simple point-dipole calculation. So the molecular structure of the excimer must be (HCB-HCB-TCB). The orientation of the proton Hfs-tensor has molecular symmetry, wherefrom we conclude that TCB is trans-lationally equivalent to HCB.

4.6 Cl-ENDOR

Analysis and discussion of Cl-ENDOR is more complicated by the presence of a large quadrupole interaction of the Cl-atom.

The quadrupole interaction is of the order of 35 MHz, much larger than the nuclear spin interaction. This large interaction determines the main characteristics of the ENDOR spectrum. This results also into an angular dependence that is different from what one expects if the spins were quantized in B_0, or in bonding direction.

One can analyse the ENDOR spectra of 6 non-equivalent Cl-atoms, 3 of them (1,2,3) on HCB (pairwise equivalence within the 6 Cl-atoms) and 3 of them (1',2',3') on TCB.

For the Cl-atoms 1,2,3 we find, by analyzing the ENDOR angular depen-dence, an isotropic interaction which corresponds to a spin density $\varrho = 0.077$. If one multiplies 0.007 by 12, one gets almost 1. This means: the spin is totally localized on two HCB molecules with 12 Cl-atoms; both HCB molecules are completely equivalent; the measured values of the anisotropic interaction of Cl-atoms 1,2,3 are also consistent with the values calculated using a point-dipole model for direct neighbours in the stack.

From the angular dependence we get the quadrupole tensor with the result that the symmetry axes of the molecule are parallel to the main axes of the fine structure tensor.

For the Cl-atoms 1',2',3', on the other hand, the isotropic hyperfine interaction is zero, so the spin density $\varrho = 0$. This is consistent with the model derived above that TCB is not partner of the excimer, but neighbour, and the neighbour HCB on the other side must have the same orientation as TCB. Otherwise additional ENDOR lines would be expected.

The orientation of the quadrupole tensor is identical to that of the excimer. So HCB and TCB have an orientation which is identical to that of the excimer. Finally, the measured quadrupole interaction frequency is 2.5 MHz lower than that measured for the atoms 1,2,3. This is explained by the fact that the triplet state is localized on the excimer. Thus one measures at the same time the quadrupole interaction in the excited triplet state (HCB, excimer) and in the ground state (TCB, neighbour).

4.7 Conclusions for HCB excimers

By analyzing the ENDOR spectra, we come to the following model, as shown in Fig. 4:

                TCB (HCB-HCB) HCB   with a 20° misorientation

The excimer binding energy must be of charge-transfer type, since the reduction of the excimer fine structure parameters, as compared to those of the monomer, is inconsistent with excitonic interaction only.

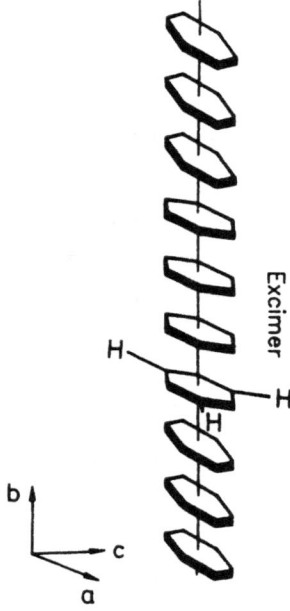

Figure 4 - Model of the excimer structure. Within one stack, two HCB
molecules are misoriented and constitute the excimer. The
dopant molecule TCB, the H atoms of which are indicated, is
translationally equivalent. The HCB neighbour on the other
side of the excimer is also misoriented. The Cl atoms are
not shown, for simplicity (4).

5 - SINGLET EXCIMERS IN PYRENE; RISE AND DECAY TIME MEASUREMENTS (5)

5.1 Crystal structure

In the pyrene crystal the molecules are arranged in sandwich pairs both
in the high temperature phase (HT, > 120 K) and in the low temperature
phase (LT), Fig.5. The overlap of the two partners of the sandwich
changes a little at the phase transition. When cooling the crystals very
slowly, one can transform the HT into the LT phase (and vice versa)
without cracking the crystals.

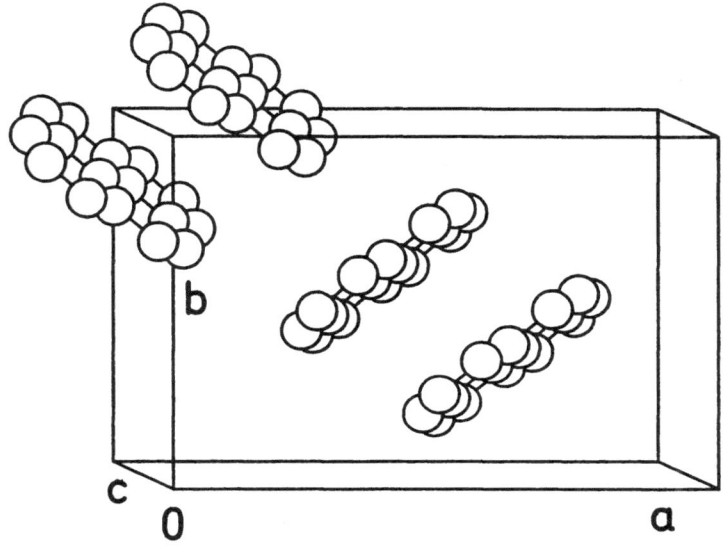

Figure 5 - Crystal structure of pyrene - High temperature phase (5)

5.2 Experimental technique

The measurements in the picosecond time domain were done using fre-
quency-doubled (527 nm) single pulses of a mode locked Nd glass laser
for excitation and a streak camera for detection. So we used low density
two photon excitation of pyrene in order to avoid surface and annihilation effects. The time resolution was 10 picoseconds.

5.3 Fluorescence spectrum

The optical emission spectrum (Fig. 6) as a function of temperature is a typical excimer spectrum with little changes between 7 and 290 K. With increasing temperature the excimer emission intensity decreases.

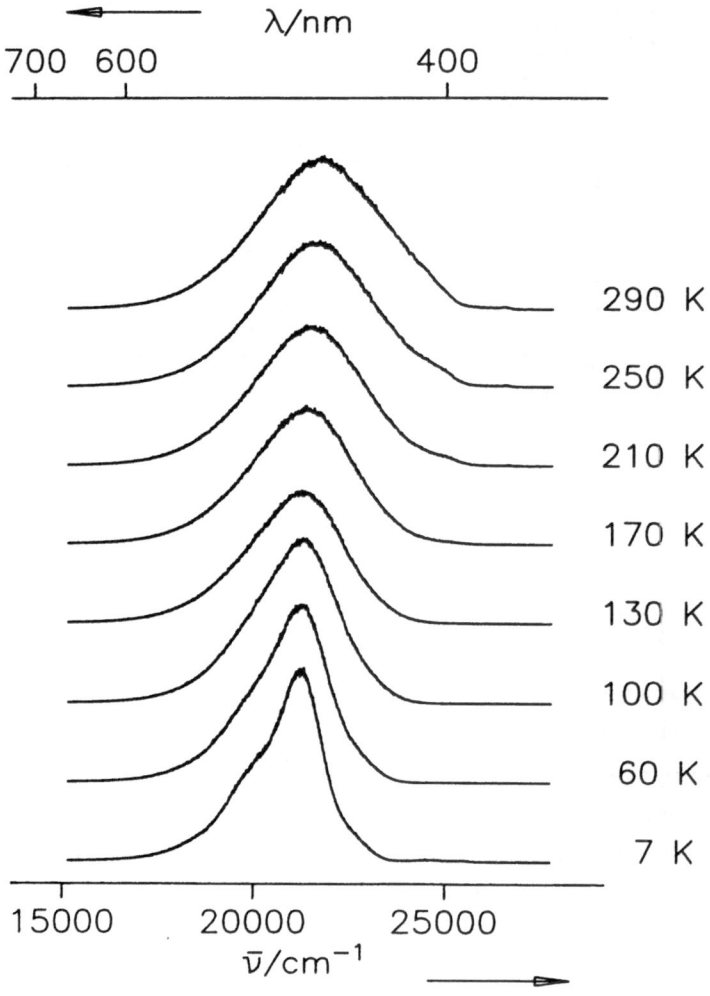

Figure 6 - Temperature dependent fluorescence spectra of pyrene (5)

5.4 Rise and decay of excimer emission

Under the conditions of low excitation densities, both build up and decay of the excimer emission follow single exponentials.

The decay times τ_E decrease with increasing temperature from 185 ns below 30 K to 115 ns at 300 K with an abrupt change at the phase transition, with a small hysteresis (see Fig. 7). The curves can be fitted using a thermally activated excimer quenching process.

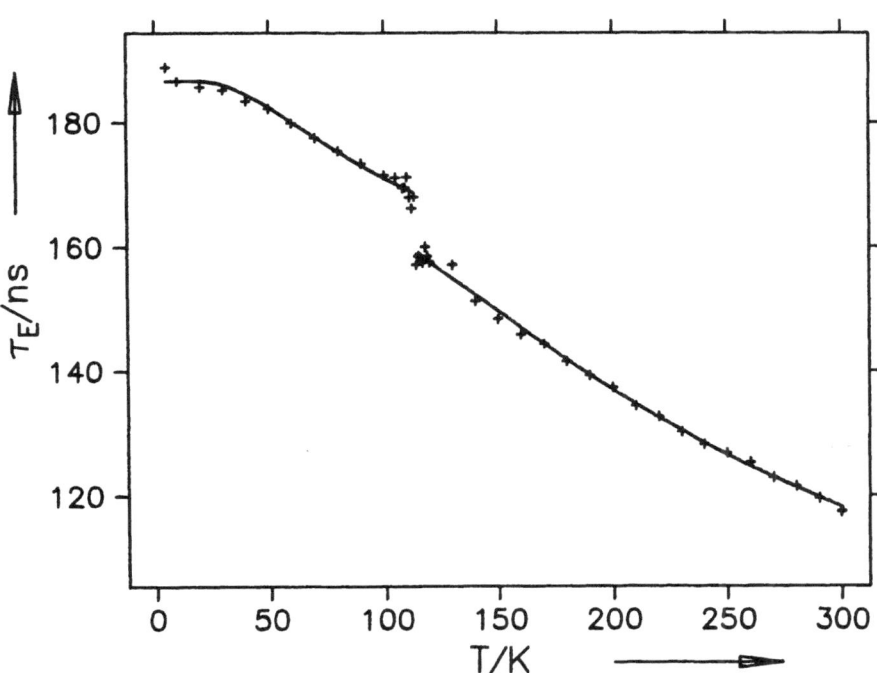

Figure 7 - Decay time τ_E of the excimer fluorescence in pyrene as a function of temperature. Experimental values are given together with theoretical curves (5)

The build-up of the excimer fluorescence after two photon ps excitation is much faster (τ_E^r = 85 ps in the LT phase), increases by a factor of three within less than one degree at the phase transition temperature, and then it decreases monotonically up to the room temperature value of roughly 100 ps (see Fig. 8).

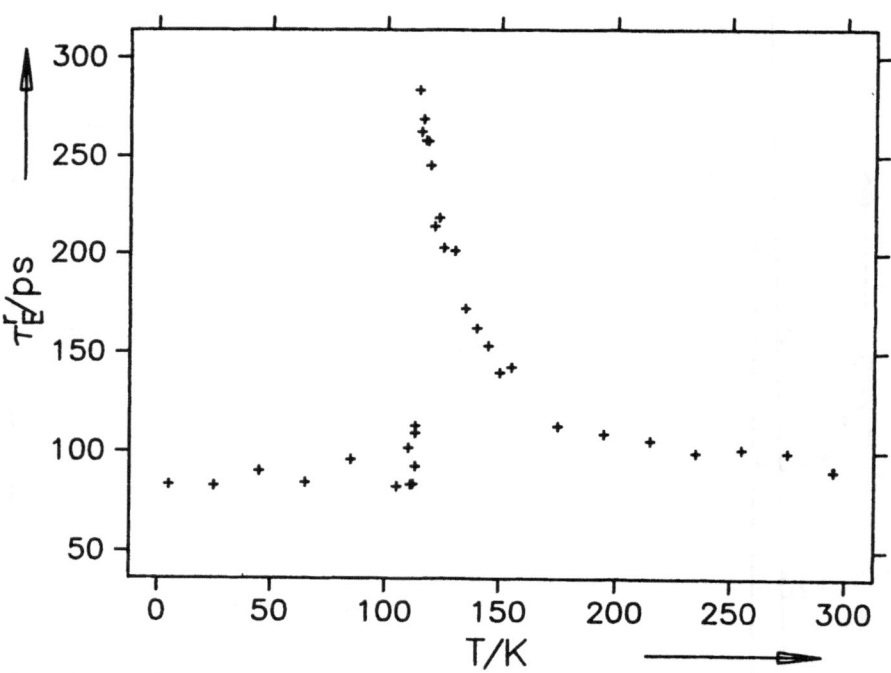

Figure 8 – Rise time τ_E^r of the excimer fluorescence in pyrene as a function of temperature (5)

Measurements on deuterated crystals give similar results.

5.5 Additional emission: B emission

In the time and spectrally resolved experiments a fast fluorescence component, which we call B, is identified in the high temperature phase only.

While no change in the nanosec. transients with detecting wavelength could be identified at room temperature, a pronounced change is observed close to the phase transition temperature.

The intensity of this B emission, which is in the wavelength region below 420 nm, is extremely weak.

At 113 K, rise and decay of the B-fluorescence occur at comparable time scales. With increasing temperature, the build-up is getting considerably faster (35 ps at 190 K). The intensity at this temperature is already very weak. The decay time of the B-fluorescence decreases with increasing temperature.

5.6 Discussion of the pyrene excimer formation

In the low temperature phase, the time constant for excimer build-up is independent of temperature. This means that excimer formation does not need thermal activation. There seems to be a direct relaxation pathway from the monomeric to the excimeric state without a precursor. The measured τ_E^r might still contain contributions from both internal conversion and lattice relaxation.

In the high temperature phase, on the other hand, the excimer formation might well be a two step process similar to that previously observed in perylene (6).

The B-fluorescence might originate from a precursor state at $T > T_p$. The decay of B-fluorescence and the rise of excimer emission are comparable in the vicinity of the phase transition where both can be measured. The B-state is thermally activated and transformed into the E-state.

5.7 Conclusions on pyrene

We get, from our measurements, the impression that, in the low temperature phase, excimer formation in pyrene is a direct relaxation process from the originally excited monomeric state. In the high temperature phase, on the other hand, the formation seems to follow a two step scheme similarly to that observed in perylene.

ACKNOWLEDGEMENTS

This work has been performed in collaboration with Dr. H. Port, Dr. J.U. von Schütz, E. Betz, W. Schrof, R. Seyfang and J. Waldmann.

REFERENCES

1) Th.Förster, Angew.Chem.Int. Ed.8, 333 (1969).
2) J.B.Birks,"Photophysics of Aromatic Molecules", Wiley (1970).
3) V.Yakhot,Z.Ludmer and M.Cohen, Adv.Photochem. 11, 489 (1979).
4) For more details and for references see
   J.Waldmann,J.U.von Schütz and H.C.Wolf, Chem.Phys. 92, 1 (1985).
5) For more details and for references see
   R.Seyfang,E.Betz,H.Port,W.Schrof and H.C.Wolf, J.Luminesc. 1985 (in press).
6) B.Walker,H.Port and H.C.Wolf, Chem.Phys. 92, 177 (1985).

PBB, Vol. 2
Advanced Magnetic Resonance Techniques
in Systems of High Molecular Complexity
© 1986 Birkhäuser Boston, Inc.

METHYL DYNAMICS STUDIED BY ENDOR SPECTROSCOPY: A NEW METHOD

M.Brustolon,T.Cassol,L.Micheletti and U.Segre

Dipartimento di Chimica Fisica, Università di Padova

Via Loredan 2, 35131 Padova (Italy)

We present a new method to study classical methyl group rotation in radicals by exploiting the ENDOR line intensities and their temperature
variation. The proposed method works for low barrier as well as high
barrier systems. The method lies on the fact that the methyl rotation
modulates the proton contact coupling constant. This gives rise to a
specific relaxation path which affects in a dramatic way the EPR desatu-
ration determining the ENDOR line intensity.

The spin hamiltonian for methyl protons, in β-position with respect to
the unpaired electron, is given by:

$$\mathcal{H} = \omega_e S_z + \sum_{i=1}^{3} (-\omega_n I_{iz} + A_i(\vartheta)\underline{S}\cdot\underline{I}_i + \underline{S}\ \underline{D}_i(\vartheta,\chi)\underline{I}_i)$$

A_i, $\underline{D}_i$ are the contact and dipolar interaction of the i-th proton. χ is
the angle between the external magnetic field and the C_3 methyl axis

$$A_i(\vartheta) = a + b\ \cos^2(\vartheta+\varphi_i) \qquad \text{with } \varphi_i=2\pi i/3$$

At sufficiently high RF fields, the transition moments are ineffective
in determining the ENDOR enhancements for radicals trapped in molecular
crystals. It is therefore possible to analyze them in term of the spin
relaxation properties by using the methods developed by Freed for radi-

cals in solutions.

The modulation of the h.f.c.c. due to methyl group rotation is a very effective source of relaxation for the electron level populations, and therefore it influences ENDOR line intensities. While contact coupling modulation brings about only flip-flop $(\Delta M_S + \Delta M_I) = 0$ transitions, all kinds of transitions are induced by the dipolar term. The dipolar induced transition probabilities however are nearly two orders of magnitude less than the contact induced flip-flop.

Table I

transition probabilities	time-dependent interaction
– electron WE	hyperfine dipolar
– nuclear WN	"
– flop-flop WX2	"
– flip-flop WX1	" hyperfine isotropic

WX1 dominates the saturation pathways. WX1 is given by the spectral density for the contact coupling correlation function:

$$WX1 \cong j_A(\omega_e) = \text{Re} \int_0^\infty g_A(t) \exp(-i\omega_e t)dt$$

$$j_A(\omega_e) = (b^2/8)\, \tau\, (1+\omega^2\tau^2)^{-1}$$

τ is the average time between consecutive jumps, and its temperature dependence is given by the Arrhenius relation:

$$\tau^{-1} = \tau_{\infty}^{-1} \exp(-V/kT) \qquad (1)$$

then we have

$$E \propto (b^2/WE) \, \tau \, (1+\omega^2 \tau^2)^{-1} \qquad (2)$$

The experimental temperature dependence of E has been fitted by a non-linear-least-square procedure to Eqs.(1,2).

Two radicals have been studied:

$$CH_3-\overset{\bullet}{C}H-COOH \quad (I) \qquad\qquad CH_3-C_6H_4-\overset{\bullet}{O} \quad (II)$$

produced by γ-irradiation of 1-alanine and 4-methyl-2,6 di-t--butylphenol as examples of high (15 ± 0.8 kJ/m) and of low ($2.4-2.9$ kJ/m) barriers.

The best fit values for the Arrhenius parameters τ_{∞} and V are:

Table II

	V (kJ/m)	τ_{∞} (psec)
I (crystal)	17.1	0.031
I (powder)	17.6	0.028
II (crystal)	3.26	2.05
II (powder)	3.18	1.75

The dynamical parameters determined previously by different methos are:

radical I (1) 15 ± 0.8 kJ/m 0.12 ± 0.05 psec

radical II (2,3) $2.4-2.9$ kJ/m

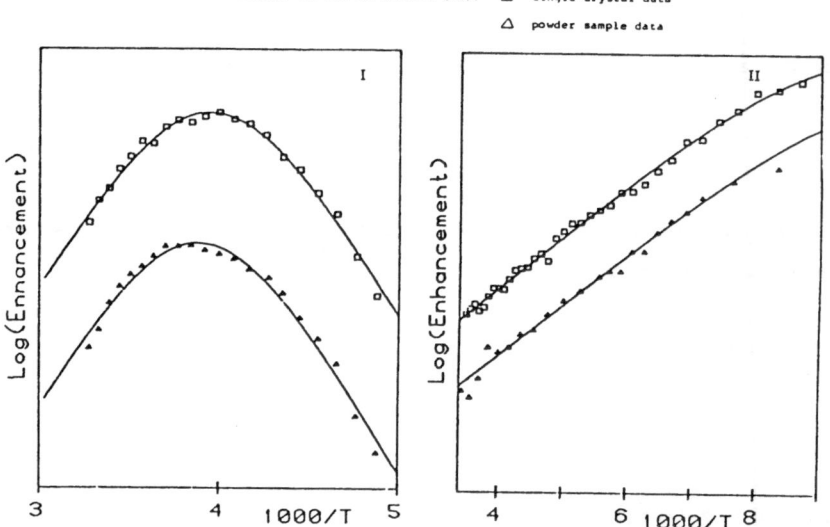

Figure 1 - ENDOR enhancements for radical (I) and for radical (II):
(□) single crystal data; (△) powder sample data

REFERENCES

1)   I.Miyagawa and K.Itoh, J.Chem.Phys. **36**, 2157 (1962).

2)   S.Clough and F.Poldy, J.Chem.Phys. **51**, 2076 (1969).

3)   S.Clough and B.J.Mulady, Phys.Rev. Lett. **30**, 161 (1973).

PBB, Vol. 2
Advanced Magnetic Resonance Techniques
in Systems of High Molecular Complexity
ⓒ 1986 Birkhäuser Boston, Inc.

PHOTOEXCITED TRIPLET STATES IN ZERO MAGNETIC FIELD

C.J.Winscom

Institut für Molekülphysik, Freie Universität Berlin

Arnimallee 14, D-1000 Berlin 33

INTRODUCTION

After the advent of EPR and its establishment as a technique for detecting paramagnetic molecular ground states, it was still some 15 years
before the photoexcited triplet state was observed. The first successful
attempt (1) was performed on a single crystal of durene, dilutely
substituted with naphtalene, and finally resolved a controversy which
had, at that time, already existed for some decades.

The failure of many previous attemps using conventional high field EPR
was attributed to the inability to establish a high enough equilibrium
concentration of the metastable triplet states, and the fact that glassy
sample were used. The many different orientations and the large anisotropy dipole-dipole interaction between the two unpaired electron spins
resulted in a "smearing" of the total EPR signal intensity over many
hundreds of Gauss.

The ensuing decade saw a wealth of EPR experiments-even in glassy samples, but it was nor until 1967 that magnetic resonance of phospho-
rescent triplet states was detected optically, not only in high field
(2), but also in zero field (3). Several salient features of both the
general characteristics of photoexcited triplet states and the optical
detection technique became apparent. These are illustrated with the help
of figure 1

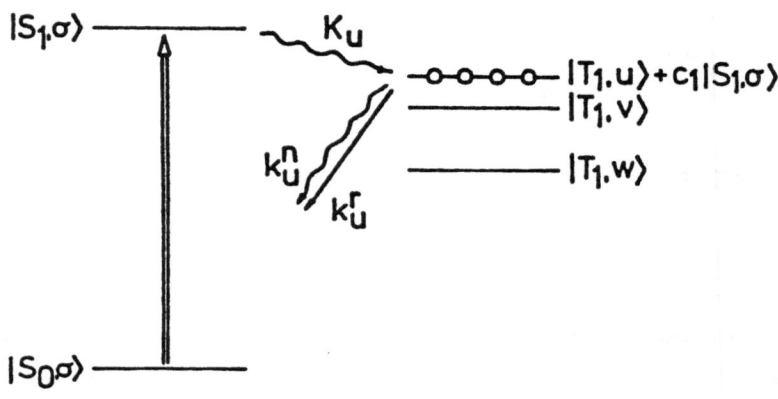

Figure 1. A simplified representation of the optical pumping cycle of an organic chromophore. For the singlet states S_0, S_1, etc. σ represents the electronic spin part of the wavefunction, and analogously u, v, w (= x, y, z) for T_1.

The relevant feature of photoexcited triplet states here may be concisely stated as follows: the chromophoric fragment of the molecule generally has enough local symmetry to permit selective spin–orbit singlet admixture in the zero field triplet space x spin wavefunctions. In group theoretical language, the spin orbit rule becomes:

$$\Gamma_{S1} \times \Gamma_{Tu} \supset \Gamma_{A1} \qquad (u=x,y,z) \qquad (1)$$

This rule, and its variations, governs the populating process, intersystem crossing (isc), and the non-radiative-(also isc) and radiative (phosphorescence) depopulating processes. In a very selective case, the typical photokinetic pathway of an optical pumping cycle might be that depicted in fig. 1

A second feature concerns the technique itself. Assuming that experiments are carried out at temperatures low enough to render spin-lattice relaxation (srl) negligible, it should be clear that any magnetic resonance transition involving T_u causes a change in the phosphorescence intensity. What is important, however, is the comparison with conventional magnetic resonance. For the two spin levels, u, v involved in the transition, the signal intensity depends on the following quantities:

microwave detection $\sim N_u - N_v$

optical detection $\sim k_u^r N_u - k_v^r N_v$

where N_u, N_v are the respective equilibrium populations, and k_u^r, k_v^r the corresponding radiative decay rates. Since, in general, $N_u \sim (k_u^r + k_u^n)^{-1}$ etc.(where k_u^n is the non-radiative decay rate), conventional EPR becomes less sensitive for short-lived triplets, whereas ODMR retains its capability. Furthermore, the quantum transformation from the microwave to the optical domain further promotes the detection sensitivity of ODMR.

Without forgetting that microwave detection methods have continued to make advances in their application to the study of metastable states in the past two decades, the forthcoming discussion will be restricted to optical detection methods. From the present point in time it is relevant to discuss ODMR from two standpoints, namely (i) zero magnetic field-, and (ii) high field experiments. Figure 2 illustrates the origins of the main differences.

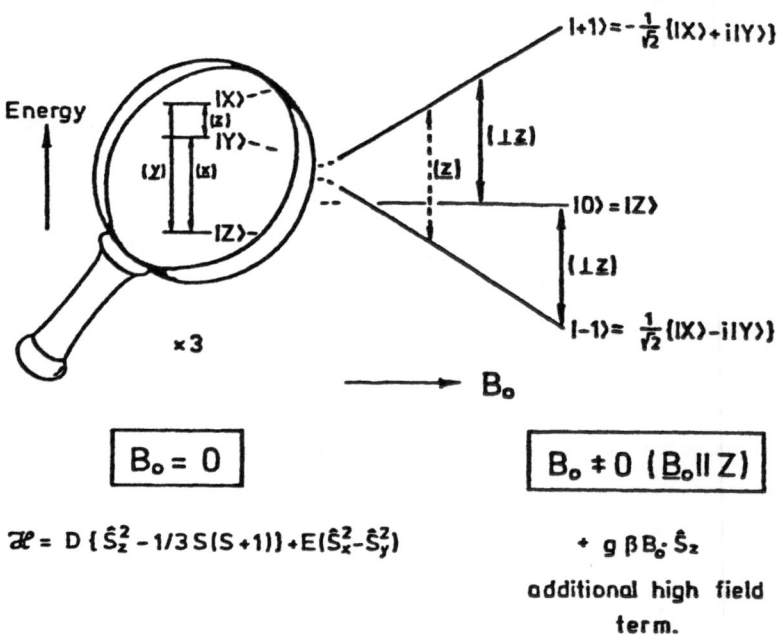

$$\mathcal{H} = D\{\hat{S}_z^2 - 1/3\,S(S+1)\} + E(\hat{S}_x^2 - \hat{S}_y^2)$$

$$+ g\,\beta B_o \hat{S}_z$$

additional high field
term.

Figure 2. The relationship between the T_1 spin-sublevel characteristic
and an external magnetic field, $\underline{B}_o$.

On the left hand side, the zero field situation is depicted,
illustrating the three spin sublevels and the possible allowed microwave
transitions. Ignoring hyperfine interactions, the Spin Hamiltonian
comprises only the dipolar interaction between the two unpaired electrons (zero field splitting (zfs)). Applying an increasing magnetic
field, the additional Zeeman term in the Hamiltonian gradually decouples
the electron spin from the molecular frame. (For simplicity, the
g-tensor is assumed to be isotropic). Two $\Delta m_s = 1$ allowed transitions
occur, as well as a partially allowed $\Delta m_s = 2$ transition. For the case
$\underline{B}_o//Z$ shown here, the respective polarisations of the transitions are
indicated in the brackets. The act of decoupling the electron spin from

molecular (zf) frame brings about a reduced selectivity in the population and depopulation of the spin sublevels, and also affects the influence of electron-nuclear dipole- and nuclear quadrupole hyperfine (dhf
and qhf, respectively) interactions on the electron and nuclear spin level energies.

Whilst it might be interesting to measure the spin selective photokinetics, the zf splitting, or the g-tensor to assess the coarser, more
global characteristics of the triplet state, detailed structural information from specific localised regions within a given chromophore will
be retrieved more readily via the smaller dhf and qhf interactions.
Considering this as a main experimental goal, and allowing the full
arena of single and double resonance variations, the main consequences
of the zero field vs. high field characteristics may be summarised as
follows.

Quite generally, zero field experiments suffer no loss of information
through using non-oriented samples (e.g. frozen solutions). In high
field, single crystal samples would seem to be mandatory to retain high
resolution, but with the advantage of gaining direct principal axis
information. In zero field the population and depopulation kinetics will
be optimally selective, whereas in high field this selectively will be
strongly dependent on the orientation of the sample. In single resonance
experiments, dipolar hyperfine splitting will occur only in 2nd order in
zero field: only the largest component will be observed, the remainder
will contribute to inhomogeneous line-broadening with concomitant loss
of resolution. In high field dhf will be observed as a first order effet
for the allowed transitions. Whilst qhf occurs as a first order effect
in zero field, it is only observable as a partially-allowed or forbidden
transition. A similar situation exists in high field.

In the domain of double resonance experiments, where nuclear spin transitions are probed, it is in principle possible to determine qhf splitting directly in zero field; in high field both dhf and qhf may be determined as first order interactions, however not directly – they are folded with various other Zeeman and dipolar interactions and require a fairly complex analysis procedure. Furthermore, the strong dependence of photokinetics on orientation restricts measurements to $\pm$ 20° about the principal axes of the zfs tensor!

Summarising, both options have advantages and disadvantages; where possible the two methods should be used to complement each other. Where this is not possible would seem to be where a lack of suitable single crystal hosts exists and in this respect the study of biological system is most affected.

DETERMINATION OF NUCLEAR QUADRUPOLE COUPLING IN PHOTOEXCITED TRIPLET STATES

With the foregoing in mind, our efforts have concentrated themselves in the zero field regime directed at the determination of nq coupling constants as a main goal. A typical candidate commonly occurring with 100% natural abundance in biomolecules is, of course, ^{14}N (I=1). The possibility of determining ^{14}N qhf in zero field is not new and was well-recognised in the original ODENDOR (4) experiments. However, before concentrating fully ^{14}N, it is useful to note some of the more graphic examples of direct determination of qhf on "easier" nuclei. In this respect the mono-substituted halo-naphtalenes studied by Kothandaraman and coworkers (5-7) are clear examples, where forbidden transitions symmetrically disposed about a central allowed ODMR transition allow direct determination of e^2qQ/h for $^{35,37}Cl$, $^{79,81}Br$ and ^{127}I. These situations are ideal because the qhf $\gg 2^{nd}$ order dhf; adequate microwave (mw) power can be applied withouth being frustrated by the inevitable broadening of the centrally-placed allowed transition.

By comparison the situation is not so ideal for ^{14}N. Qualitatively, the same sort of behaviour is expected, but relatively, the qhf is much smaller in magnitude. The situation may be illustrated by two examples. Figure 3 shows the 2E-ODMR transitions for acridine-d$_9$ (A-d$_9$) and phena- zine-d$_8$ (P-d$_8$), respectively. In both cases the triplet state is $\pi\pi^*$ in character; in these, and similar $^3\pi\pi^*$ aza-aromatics, the only important dhf component is the out-of-plane component, A_{zz}.

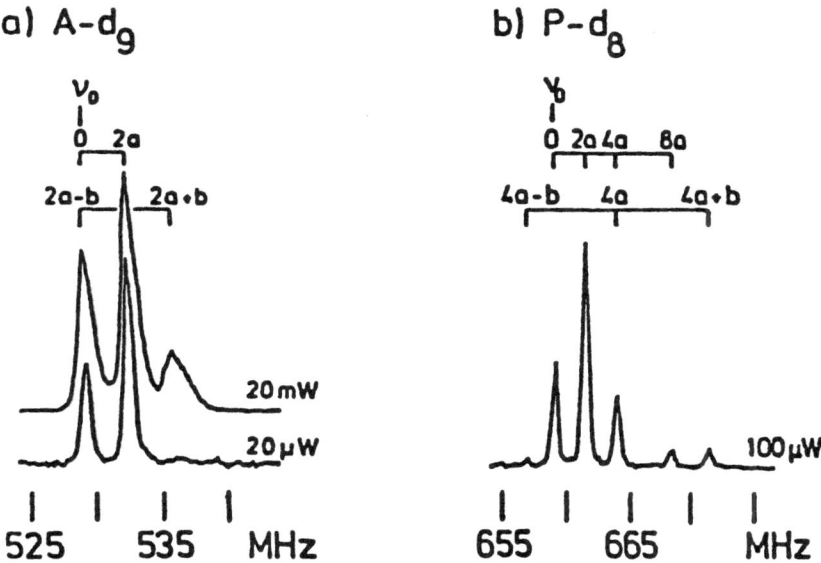

Figure 3 - The 2E ODMR spectra of a) acridine-d$_9$ (A-d$_9$) in n-heptane-h$_{16}$ at low (20μW) and saturating (20 mW) power levels, and b) phenazine-d$_8$ (P-d$_8$) in n-heptane-d$_{16}$ at low mw power level. For discussion of the spectral parameters a,b,ν_o, see text.

In the case of acridine, having just one ^{14}N nucleus, the transition exhibits two allowed lines whose splitting is governed by 2nd order dhf interaction, conveniently expressed in units of $a = A_{zz}^2/2E$. At higher powers partially allowed satellites occur symmetrically about the central "2a" line, and yield qhf information in the form of the difference $(\varepsilon_y - \varepsilon_x)$ of the x- and y-principal values. Phenazine, with two equivalent ^{14}N nuclei is somewhat similar; again the 2nd order dhf splitting is clearly resolved, this time with four allowed lines occurring at 0, 2a, 4a and 8a, relative to the $\nu_o = 2E$ frequency. The quadrupole satellites corresponding to those in acridine are also observed, symmetrically placed about the 4a line with a separation 2b. (b takes a slightly different functional form that for one ^{14}N, but renders the same $\varepsilon_y - \varepsilon_x$ information.)

At this stage it is useful to make two general observations. The first is that the corrisponding D$\pm$E transitions would also exhibit a 2nd order dhf splitting arising from A_{zz}, but reduced to exactly a half of that for the 2E transition. Secondly, in principle two other pairs of quadrupole satellites containing the $\varepsilon_y - \varepsilon_z$ and $\varepsilon_z - \varepsilon_x$ information should be present. Their "allowed-ness" would be governed by A_{xx}^2 and A_{yy}^2, respectively; ca. 2-3 orders of magnitude less than A_{zz}^2.

The spectra shown in figure 3 have been obtained under high resolution conditions (8), involving optical site-selection, and guest deuteration to minimise line-broadening effects as a result of 2nd order proton interaction.

Returning to our two examples, what is needed is an additional double resonance experiment to determine the remaining quadrupole information. An optically detected, but otherwise conventional, ENDOR experiment (ODENDOR) yields only strongly allowed (hyperfine enhanced) transitions

containing only redundant information. The weaker non-enhanced transitions of interest are masked at the rf power levels required to detect
them, by their enhanced, and consequently broadened, counterpart. Fortunately, however, well-resolved 2E ODMR spectra are to hand, whose allowed transitions can be easily assigned to pairs of specific (electron x
nuclear) spin levels. This allows a strategy (9) to be designed which
excludes the strongly-enhanced ENDOR transitions.

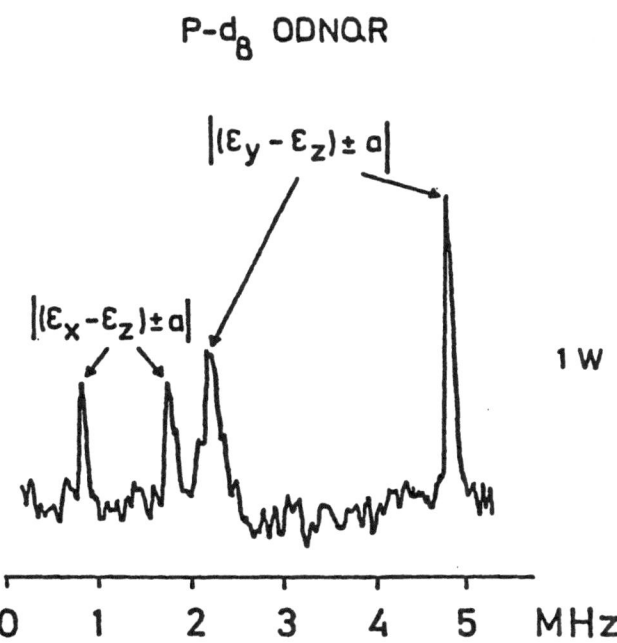

Figure 4. The ODNQR spectrum of P-d$_8$ in n-heptane-d$_{16}$. For definition
of the spectral parameter, see text.

Application of this strategy to phenazine yields the optically detected
nuclear quadrupole resonance (ODNQR) spectrum shown in figure 4. Four
transitions are observed; their relation to the quadrupole and second-
order hyperfine parameter, a, shown above, clearly involve the outstan-

ding quadrupolar information. Assuming that the N lone-pair direction is the prinipal axis of largest positive principal value, the quadrupole coupling constants can be determined. Acridine (10) may be studied using a similar strategy and yield almost identical coupling constants.

In contrast to the $\pi\pi^*$ triplet examples, pyrazine (Pz) is a smaller molecule having an $n\pi^*$ triplet state and two equivalent ^{14}N nuclei, and represents an extreme case. The goal of determining the ^{14}N qhf has been the subject of several previous ODMR/ODENDOR studies (11,12), their results were unfortunately at variance with one another and thus somewhat inconclusive. In the light of the foregoing, this is quite understandable because in pyrazine all the components of the ^{14}N and ^{1}H dhf tensors must be regarded as significant on a 2nd order mixing basis. Briefly stated both ODENDOR and ODNQR experiments will not serve as a particular help to solve the problem; in principle all the information will be contained in a highly resolved 2E-ODMR spectrum. The principal components and axes of the various hyperfine tensors are known (13); computer simulation to optimise the ^{14}N quadrupole tensor against experimental 2E-ODMR spectra was then method chosen. For simulation, a Hamiltonian containing all dipolar- and quadrupolar hf interactions of all the magnetic nuclei must be used. For photokinetic reasons, the 2E spectrum can only be adequately accessed by means of an optically detected EEDOR (ODEEDOR) experiment; a pulsed variation based on similar principles to the foregoing ODNQR strategy was employed (14). The spectra of a) fully deuterated - and b) fully protonated pyrazine are shown in figure 5.

Comparison shows that inhomogeneous broadening is not alleviated by deuteration, rather the opposite since the 2nd order proton dhf effects now become well-resolved in the protonated case. One ^{14}N quadrupole satellite (*) in figure 5 is resolved and may be cleary identified; it

yields direct information concernign <u>one</u> principal qhf value (ε_x).
Optimised fitting sets strict limits on the remaining pair.The result is
that the principal values of the quadrupole tensor are all very small
compared with other ^{14}N situations.

2E - ODEEDOR

a) Pz - d$_4$ b)   Pz-h$_4$

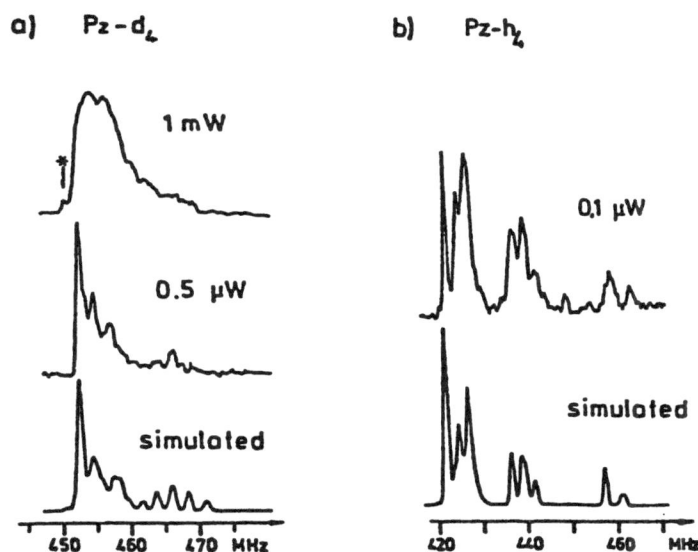

Figure 5 – The 2E ODEEDOR spectra of a) pyrazine-d$_4$ (Pz-d$_4$) in
benzene-d$_6$ at low (0,5 μW) and saturating (1 mW) mw power
levels, and b) Pz-h$_4$ in benzene-d$_6$ at low (0,1 μW) mw power
level

Table I summarises the results of the model $^3\pi\pi^*$ and $^3n\pi^*$
representatives in both their ground (15,16) and photoexcited states,
respectively. As expected the ground states of both molecules and the
$^3\pi\pi^*$ state of phenazine have similar ^{14}N quadrupole tensors, whereas in
the $^3n\pi^*$ a marked complementary change in the y- and z-axis field
gradients is observed, completely consistent with the n $\rightarrow \pi^*$ relocation
of electron density.

TABLE I

Comparison of the experimentally determined

^{14}N nq tensor principal values (MHz) in the singlet (S_o)

ground-and triplet (T_1) excited states of Phenazine

and Pyrazine, determined in polycrystalline hosts at 1,3 K

	Phenazine(S_o) (ref. 15)	Phenazine(T_1,$^3\pi\pi^*$) (ref. 9)	Pyrazine(S_o) (ref. 16a)	Pyrazine(T_1,$^3n\pi^*$) (ref. 14)
ε_x^N	(-)0.637	(-)0.86	(-)0.565	(-)0.6(2)
ε_y^N	(+)2.290	(+)2.18	(+)2.430	0(3)
ε_z^N	(-)1.653	(-)1.32	(-)1.865	(+)0.6(3)

For phenazine (T_1), sign choice is inferred from acridine (T_1) (10) which has almost identical magnitudes and where the sign choice may be determined. For phenazine (S_o) and pyrazine (S_o), the ordering of the levels is not expected to be changed. For pyrazine (S_o) ab initio calculations using an extended basis (16a) are in excellent agreement with experimental magnitudes and this sign choice.

For pyrazine (T_1), the sign choice has been inferred from INDO calculations (14) where good agreement is obtained.

APPLICATION TO PORPHYRIN SYSTEMS

From the model systems described above, the prospect for tackling more complex molecules might, at first sight, seem less hopeful. On the other hand, looking towards biomolecules, as an application, typical chromophores will tend to have a more extensive distribution of the electron spin density. The other extreme, exemplified by the pyrazine situation, will be rarely, if ever, met.

Typical chromophores often found embedded in protein environements are porphyrins and their near analogues. As a chromophoric class they have similar, but individually distinct optical characteristics. The possibility of using optical discrimination to specifically locate a triplet state spin probe is very attractive. However, within this structural class, ^{14}N quadrupole coupling determination, not only in their excited states but also in their ground states, is very sparsely reported.

The first zf ODMR experiment on a representative of this class, namely zinc porphin, was performed by Chan et al.[17] in 1971. Since that time zf ODMR has provided very much information on porphyrin and chlorin excited triplet states, as a whole [18].

The representative we have chosen for preliminary studies is zinc (II) octaethylporphyrin, which is very similar in its triplet state properties to the zinc porphin. The triplet state observed is Jahn-Teller stabilised against an orbitally degenerate partner. As a result, the four ^{14}N atoms are pairwise inequivalent, and high-field experiments [19] show that they distinguish themselves via their dhf principal values, the dominant out-of-plane (z) components of which are ca. 11.2 and 3.6 MHz, respectively. Other hyperfine components which might make a significant contribution to line-broadening in zf are the in-plane dhf principal values (ca. 10 MHz) of the methine bridge protons. In zero field (17,20), the D+E transitions have been readily observed, but exhibit unstructured single lines of width ca. 6 MHz. As one might anticipate, ODENDOR is little help; it comprises a broad poorly-structured line, extending from 1 to 5 MHz.

It is straightforward to experimentally show that Zn OEP has an almost identical optical site structure, zfs parameters and photokinetic characterisics. Accordingly, the axis assignment and ordering of the zf levels of ref. 19 are adopted. Using a 75 W Xe arc lamp, irradiatig continuously in the region 562 +10 nm, the D-E zf ODMR transition of the main site may be recorded. The spectrum shown in fig. 6 obtained under non-saturated conditions.

D-E ODMR of ZnOEP

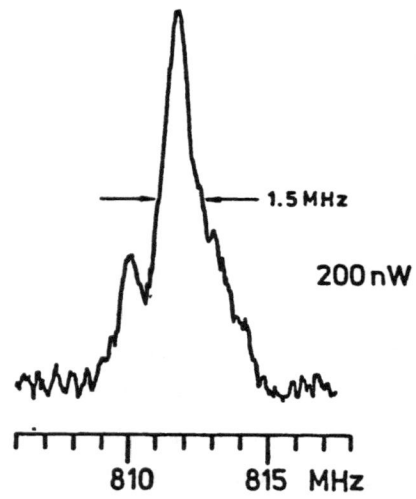

Figure 6. The D-E ODMR spectrum of zinc(II) octaethyl porphyrin (ZnOEP) in n-octane-h_{18}

The central line has a width of ca. 1.5 MHz and is bordered on each side by weaker features. Remembering the phenazine case, and the earlier remarks, it may be anticipated that the low frequency shoulder is either the outermost lying dhf component or the partly-allowed quadrupole satellite, and the high-frequency shoulder almost certainly the quadrupole counterpart. Driving the low-frequency shoulder with amplitude modulated microwaves, applying a slowly swept cw rf frequency and performing digital phase-sensitive detection of the phosphorescence, the spectrum shown in figure 7 is obtained.

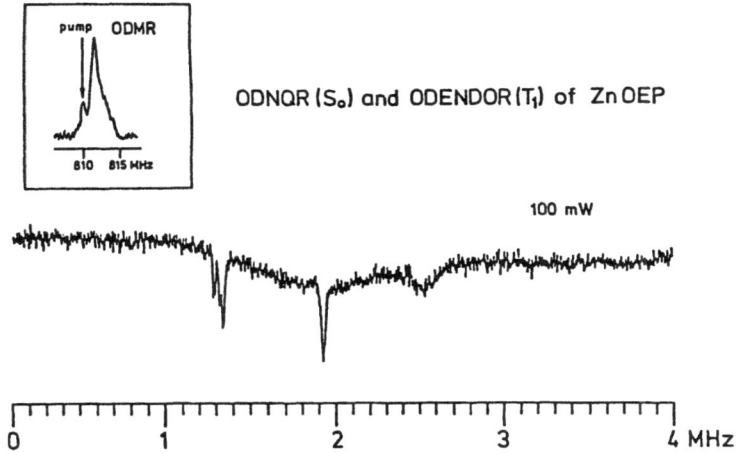

Figure 7 - The S_o (ODNQR) and T_1 (ODENDOR) spectra of ZnOEP in n-octane-h_{18}. The inset (top left) indicates the position of the mw pump frequency (809.7 MHz, 100 mW, am ½ Hz). The rf was slowly stepped at 1 kHz/2 sec. The spectrum is the result of a single sweep.

The complete trace comprises two different superimposed spectra. The first is the broad background with maxima at ca.1.9 MHz and 2.5 MHz, it returns to the baseline at ca.5 MHz, and is associated with a time constant corresponding to the triplet spin sublevel decay rates (ca. 200 ms). The second comprises two structured groups at ca. 1.3–1.35 MHz and 1.93 MHz; the individual linewidths are ca. 20 kHz. This spectrum is associated with a time constant of ca.10 sec. The explanation is as follows: the background spectrum is straightforward triplet ODENDOR, and corresponds to that observed by Van Noort (20). The sharp features arise from pure ground state (S_o) NQR. By driving an essentially forbidden line, the well-known solid-state effect (21) is called into play. The cw optical pumping cycle then "carries" the ensuing nuclear alignment throughout all electron spin states involved in the cycle (22). This effect has been observed for halogen nuclei in zero field (23), but to our knowledge this is the first time ^{14}N has been observed under these conditions. The region of the third transition may be ascertained by driving one of the transitions already observed and sweeping a second frequency through the 600 kHz and the 3.2 MHz regions; the 600 kHz region yields the third transition. Spectral simulation, using a Hamiltonian containing only the ^{14}N quadrupole interaction, and assuming a constant Lorenzian linewidth of ca.20 kHz, suggests that the spectrum arises from four almost equivalent ^{14}N nuclei. The average values of their coupling constants are: $|e^2qQ/h|$ = 2.16 MHz, η = 0.56. Axis assignment and sign determination are not directly determined; they may, however, be inferred by comparison with copper (II) tetraphenylporphyrin (CuTPP) (24) and aquometmyoglobin (myogl.) (25) where both these determinations have been made (see Table II)

An interesting extension now presents itself which seems promising for the extraction of the ^{14}N quadrupole data from the rather unresolved ODMR domain of the T_1 state. The nuclear alignment, so nicely prepared

TABLE II

The average ^{14}N nq principal values (MHz)

for ground state metal porphyrins.*

Principal axis	Cu TPP	Myogl.	Zn OEP
in-plane, ⊥ to N–Metal	+0.93	+1.04	(+)1.08
out-of-plane,	−0.31	−0.27	(−)0.24
in-plane, ∥ to N–Metal	−0.62	−0.77	(−)0.84

*see text for further explanation.

in the previous experiment, may now be destroyed in coded fashion. By pumping a specific S_o nqr frequency using a low frequency modulation, the T1 ODMR spectrum may be swept using cw irradiation, and the amplitude of the phosphorescence modulation recorded. The result is shown in lower part of Fig. 8; the normal D-E ODMR spectrum is shown above for comparison.

This type of spectroscopy records only that part of the ODMR transition leading to eventual nuclear alignment in S_o, i.e. the strongly-allowed transitions where nuclear spin is conserved, are decoupled. The relevant transitions will be those which give rise to quadrupole satellites, and those between levels which are substantially mixed as a result of accidental degeneracy. In both cases qhf information is directly involved. Clearly, microwave powers many orders of magnitude higher than the ODMR saturation level may now be applied – just that needed to bring out the weaker satellites. Examination of different combinations of S_o nqr and T_1 ODMR transitions give rise to different spectra. Furthermore, by driving hitherto undetected satellites, new transitions occur in the T_1

ODENDOR spectrum.

Decoupled D-E ODMR of ZnOEP

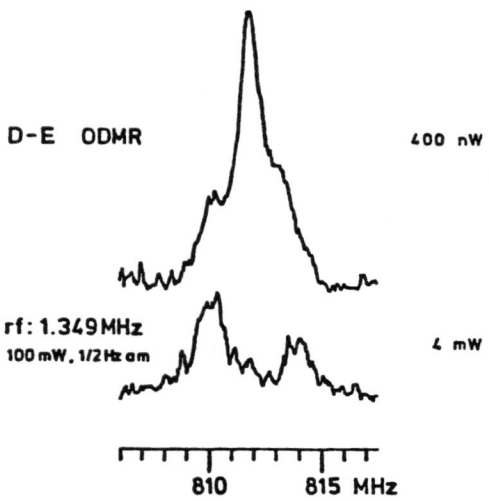

Figure 8 – The decoupled D-E ODMR spectrum of ZnOEP/n-octane. The lower part shows one possibility under the rf pumping conditions indicated at the left. The upper part shows the normal D-E ODMR spectrum under just-saturating (4oonW) mw power conditions (c.f. Figure 6).

For Zn OEP the experimental analysis is as yet incomplete; it also awaits fairly demanding spectral simulation techniques. Nevertheless, with the strategy outlined we believe the prospects are very promising for obtaining the quadrupole coupling constants for all four ^{14}N nuclei, in both the ground- and photoexcited triplet states. Furthermore, it would seem generally applicable to this chromophoric class - even the ODMR of biologically incorporated porphyrins or chlorins which do not phosphoresce may be obtained by monitoring either their S_1-S_0 fluorescence (26) or S_0-S_1 absorption (27) under conditions of cw optical pumping.

In conclusion, in complex systems specific local information is hard won by any spectroscopic method. Nuclear quadrupole tensors, which are primarily sensitive to local electronic charge distribution and the spacial relationships of the nearest neighbour nuclei to the quadrupolar probe, are good reporters of such information. zf ODMR allows several strategies which access these tensors in both their singlet ground - and photoexcited triplet states.

ACKNOWLEDGEMENTS

This work was supported by the Deutsche Forschungsgemeinschaft (Sfb 161). It is a pleasure to acknowledge the various collaborators and colleagues, namely Dr. W.Fröhling (AEG Berlin) Dr. K.P.Dinse (Universität Dortmund), Dr. M.Plato and Dr. K.Möbius (Freie Universität Berlin), and Dr. A.L.Maniero (University of Padua) who have contributed to various aspects of the work described. I am particularly indebted to H.Zimmermann (Max-Planck-Institut, Heidelberg) for providing isotopically substituted compounds.

REFERENCES

1) C.A.Hutchison,jr. and B.W.Mangum, J.Chem.Phys. **29**, 952 (1958).

2) M.Sharnoff, J.Chem.Phys. **46**, 3263 (1967).

3) J.Schmidt and J.H.an der Waals, Chem.Phys.Lett. **2**, 640 (1968).

4) a) C.B.Harris,D.S.Tinti,M.A.El Sayed and A.H. Maki, Chem.Phys. Lett. **4**, 409 (1969);

   b) I.Y.Chan,J.Schmidt and J.H.Van der Waals, Chem.Phys.Lett. **4**, 269 (1969).

5) D.W.Pratt in "Excited States", Vol. 4 (E.C. Lim ed.), Academic Press, New York, 1979. p. 137.

6) G.Kothandaraman,H.J.Yue and D.W.Pratt, J.Chem.Phys. **61**, 2102

(1974).

7) G.Kothandaraman,D.W.Pratt and D.S.Tinti, J.Chem.Phys. **63**, 337 (1975).

8) W.Fröhling,C.J.Winscom,K.P.Dinse and K.Möbius, Chem. Phys. **51**, 369 (1980).

9) K.P.Dinse and C.J.Winscom, J.Chem.Phys. **68**, 1337 (1978).

10) W.Fröhling,C.J.Winscom and K.Möbius, Chem. Phys. **60**, 301 (1981).

11) A.L.Kwiram, in "MTP International Review of Science", Physical Chemistry Series 1, Vol. 4, (A.D. Buckingham and C.A. McDowell eds.), Butterworths, London, 1972. p. 271.

12) M.J.Buckley, Ph.D. Thesis (Lawrence Radiation Lab.),Univ. of Calif., Berkeley (1971).

13) L.T.Cheng and A.L.Kwiram, Chem.Phys.Lett. **4**, 457 (1969).

14) W.Fröhling,C.J.Winscom and K.Möbius, Chem.Phys. **75**, 389 (1983).

15) A.M.Achlama,U.Harke,H.Zimmermann and K.P.Dinse, Chem.Phys.Lett. **85**, 339 (1982).

16) a) A.Peneau,M.Gourdji and L.Guibé, J.Chem.Phys. **60**, 4295 (1974); b)J.Almlöf,B.Roos,U.Wahlgren and H.Johansen, J.Electr.Spectros. Relat.Phenom. **2**, 51 (1973).

17) I.Y.Chan, W.G.Van Dorp, T.J.Schaafsma and J.H.Van der Waals, Mol. Phys. **22**, 741, 753 (1971).

18) R.E.Connors and W.E.Leenstra in "Triplet State ODMR", (R.H.Clarke ed.), John Wiley, New York, 1982. p. 257.

19) J.A.Kooter and J.H.Van der Waals, Mol Phys. **37**, 997 (1979).

20) H.M.Van Noort, Ph.D.Thesis (Univ. Leiden, 1981).

21) A.Abragam and W.G.Proctor, C.R.Acad.Sci. **246**, 1803 (1959).

22) K.P.Dinse and C.J.Winscom in "Triplet State ODMR", (R.H. Clarke ed.), John Wiley, New York, 1982. p. 114.

23) K.P.Dinse and C.Von Borczyskowski, Chem.Phys. **44**, 93 (1979).

24) T.G.Brown and B.M.Hoffman, Mol.Phys. **39**, 1073 (1980).

25) C.P.Scholes,A.Lapidot,R.Mascarenhas,T.Inubushi,R.A.Isaacson and

G.Feher, J.Amer.Chem.Soc. **104**, 2724 (1982).

26) W.G.Van Dorp,T.J.Schaafsma,M.Soma and J.H.Van der Waals, Chem.Phys.Lett. **21**, 221 (1973).

27) R.H.Clarke and R.E.Connors, Chem.Phys.Lett. **33**, 365 (1975).

PBB, Vol. 2
Advanced Magnetic Resonance Techniques
in Systems of High Molecular Complexity
© 1986 Birkhäuser Boston, Inc.

EPR DETERMINATION OF THE NUMBER OF NITROGENS COORDINATED TO

Cu IN SQUARE-PLANAR COMPLEXES

J.S.Hyde,W.E.Antholine,W.Froncisz*and R.Basosi**

National Biomedical ESR Center, Department of Radiology,

Medical College of Wisconsin, Milwaukee, Wisconsin 53226 U.S.A.

*On leave from the Department of Biophysics,

Institute of Molecular Biology

Jagiellonian University, Krakow, Poland

**Work performed while on leave from the

Department of Chemistry, University of Siena

Pian dei Mantellini 44, 53100, Siena, Italy

I. INTRODUCTION

In their article on g-strain and A-strain in non-blue Type II copper

complexes, Froncisz and Hyde (1) established that best resolution in the

$g_\parallel$ region of powder EPR spectra occurs generally for the turning point

assigned to the $M_I = -\frac{1}{2}$ copper nuclear quantum number using a microwave

frequency of about 2 GHz. Under these experimental conditions g-strain

and A-strain cancel. The concepts were further elaborated in the review

article by Hyde and Froncisz (2) on the role of microwave frequency in

EPR spectroscopy of copper complexes.

Of the likely copper ligands, only nitrogen has a nuclear spin. When

copper spectra are obtained at about 2 GHz under conditions of optimum

resolution, superhyperfine coupling to ligand nuclei can be observed. W.

E.Antholine and his colleagues in a series of papers (3-9) have examined the M_I=-½ line in the S-band region of a number of biological and model compounds with the goal of determining whether there are 0, 1, 2, 3 or 4 in-plane nitrogen ligands. Our purpose in the present paper is to review the present level of understanding of g- and A-strain in copper complexes, to summarize the more recent experimental work of Antholine, and finally to reach a determination of the reliability in using low microwave frequencies to answer the single question: How many nitrogens are there?

Focus is on the $g_\parallel$ region. In the $g_\perp$ region of powder spectra, nitrogen superhyperfine couplings are more readily detected than in the $g_\parallel$ region. But in the opinion of the authors, reliable answers to our primary question of the number of nitrogens have not been obtained by analysis of $g_\perp$ spectral features. The problems are the following:

1. Nitrogen superhyperfine couplings are similar to copper hyperfine couplings, causing confusion.

2. The nitrogen superhyperfine coupling tensor is somewhat axial with the principal axis along the Cu-N bond. As a consequence, in $g_\perp$, there is a powder pattern over both $A_\parallel$ and $A_\perp$ of the nitrogen. Typical values are $A_\parallel$ = 15.5 G, $A_\perp$ = 10.6 G (10).

3. Overshoot or extra absorption or hyperfine anomaly (all synonyms for the same effect) lines are seen (11,12).

4. There always is the possibility of rhombic distortion requiring the introduction of g_x, g_y and also A_x, A_y for the copper hyperfine interaction.

It is believed that a rather general computer based optimization of the spin Hamiltonian input parameters required to achieve best fit to spectra obtained over a range of microwave frequencies (i.e., multifrequency EPR) should provide a unique solution and permit the use of $g_\perp$ region of the spectrum, but this must remain for the future.

II. THEORETICAL BACKGROUND

The molecular orbital theory of the g- and A-tensors of square-planar (tetragonal) copper complexes was written by Maki and McGarvey (13) and developed in a more complete form by Kivelson and Neiman (11). The expressions given by these latter authors for $g_\parallel$ and $A_\parallel$ are:

$$g_\parallel - 2.0023 = -8\lambda_o \alpha B_1/\Delta E_{xy}\left[\alpha B_1 - \alpha' B_1 S - \alpha'(1-B_1^2)^{\frac{1}{2}}T(n)/2\right] \tag{1}$$

$$A_\parallel = P\left[-\alpha^2(\tfrac{4}{7}+k_o)+(g_\parallel-2)+\tfrac{3}{7}(g_\perp-2)\right.$$

$$-8\lambda_o \alpha B_1/\Delta E_{xy}(\alpha' B_1 S + \alpha'(1-B_1^2)^{\frac{1}{2}}T(n)/2)$$

$$\left.-\tfrac{6}{7}\lambda_o \alpha B/\Delta E_{xz}(\alpha' BS + \alpha'(1-B^2)^{\frac{1}{2}}T(n)/2^{\frac{1}{2}})\right] \tag{2}$$

Here α, α', B, B_1 are bonding parameters in the MO formulation; the quantities k_o, λ_o, and P are semi-empirical constants associated with the free copper ion; $T(n)$ and S characterize the covalent character of the bond; and ΔE_{xy}, ΔE_{xz} are optical transitions. All-in-all, it has not been easy to relate experimental EPR observables to the unknowns in the problem, although there is a large literature in which this has been the goal.

For our present purposes, it is sufficient to remark that $g_\parallel$ and $A_\parallel$ are expected on theoretical grounds to be correlated. The quantity $(g_\parallel-2)$ appears in the expression for $A_\parallel$, as do $\alpha, \alpha', B_1, S, T(n)$ and ΔE_{xy}. When there is a change in any of these molecular parameters, each of the spectroscopic observables must change. If there is a distribution of g-values arising from g-strain, there must be a concomitant distribution of A-values. On should therefore speak of "g-A" strain.

One can simplify the above equations by retaining only the leading terms

$$(g_\parallel - 2.0023) \simeq -8\lambda_o a B_1 /\Delta E_{xy} (aB_1 - a'B_1 S) \tag{3}$$

$$A_\parallel \simeq P \left[-a^2(\tfrac{4}{7}+k_o) + (g_\parallel - 2) - 8\lambda_o a a' B_1^2 S/\Delta E_{xy} \right.$$

$$\left. -\tfrac{6}{7}\lambda_o a a' B^2 S/\Delta E_{xz} \right] \tag{4}$$

Some further simplification is possible using the normalization equation

$$a^2 + (a')^2 - 2aa'S = 1 \tag{5}$$

and by making the assumption that $\Delta E_{xz} \simeq \Delta E_{xy}$.

Let us assume that changes occur in the molecular parameters such that changes in $\Delta g_\parallel$ and $\Delta A_\parallel$ occur.

$$\Delta g_\parallel \simeq \partial g_\parallel /\partial a \cdot \Delta a + \partial g_\parallel /\partial B_1 \cdot \Delta B_1 + \partial g_\parallel /\partial \Delta E_{xy} \cdot \Delta E_{xy} \tag{6}$$

$$\Delta A_\parallel \simeq \partial A_\parallel /\partial u \cdot \Delta a + \partial A_\parallel /\partial B_1 \cdot \Delta B_1 + \partial A_\parallel /\partial B \cdot \Delta B + \partial A_\parallel /\partial \Delta E_{xz} \cdot \Delta E_{xy} \tag{7}$$

Values of a, B_1, B, ΔE_{xy} are available from the literature (12,14). Inserting typical values in order to determine the leading terms.

$$\Delta g_\parallel = 0.99\Delta a + 0.99\Delta B_1 + 0.4\Delta (\Delta E_{xy})/\Delta E_{xy} \tag{8}$$

$$\Delta A_\parallel = P\left[-0.83\Delta a + 0.85\Delta B_1 + 0.38\Delta (\Delta E_{xy})/\Delta E_{xy}\right] \tag{9}$$

In this approximation, all changes in $g_\parallel$ and $A_\parallel$ can be associated with three parameters: the in-plane σ and π bonding parameters α and β_1 and the ground to excited state optical transition frequency ΔE_{xy}. The available experimental evidence (11, 13) for α, β_1 and ΔE_{xy} does not permit any further simplification. We are left with three unknowns and two equations. The physical concepts are that α, β_1, give information on the covalency of the in-plane bonds and ΔE_{xy} is a measure of the distortion from octahedral symmetry (15).

If we define a new quantity ΔQ

$$\Delta Q = \Delta \beta_1 + 0.4\Delta(\Delta E_{xy})/\Delta E_{xy}, \tag{10}$$

then

$$\Delta g_{\|} \simeq 0.99\Delta\alpha + 0.99\ \Delta Q \tag{11}$$

$$\Delta A_{\|} \simeq P\left[-0.83\Delta\alpha + 0.85\ \Delta Q\right]. \tag{12}$$

Equations (11) and (12) express clearly the correlation that is expected between changes in $g_{\|}$ and changes in $A_{\|}$.

Assume a distribution of $\Delta g_{\|}$ values $\sigma g_{\|}$ and a distribution of $\Delta A_{\|}$ values $\sigma A_{\|}$. Then Froncisz and Hyde (1) show that the width of each hyperfine turning point, using field modulation is

$$\Delta W_{\frac{1}{2}} = \left[(W_{\frac{1}{2}}^{R})^2 + (M_I\sigma A_{\|})^2 + (h\nu\sigma g_{\|}/g_{\|}^2 B_o)^2 + (\sigma g_{\|}\sigma A_{\|})2M_I h\nu/g_{\|}^2 B_o\right]^{\frac{1}{2}} \tag{13}$$

Since M_I can change sign, for both $M_I = -1/2$, $M_I = -3/2$, there exists a frequency such that there is perfect cancellation and the observed width is the residual width, $W_{\frac{1}{2}}^{R}$. If for some reason the correlation is not perfect, an empirical correlation coefficient ε that varies from 0 to 1 can be introduced as a factor in front of the last term of Eq.(13).

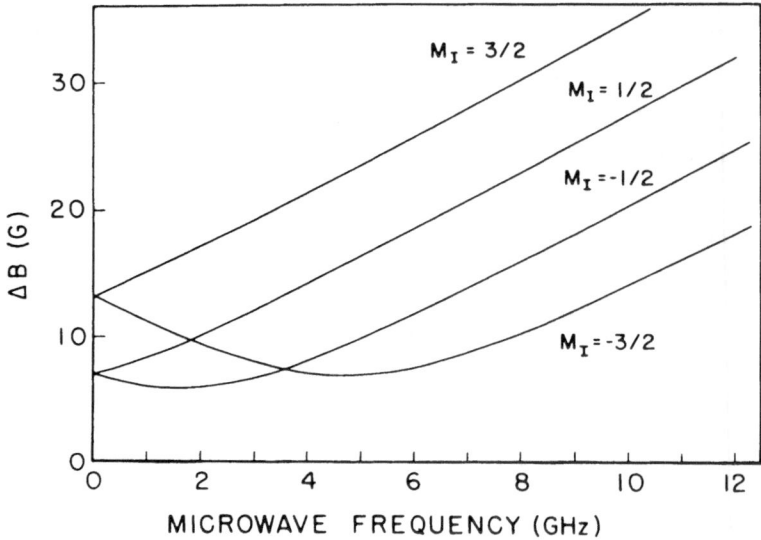

Figure 1 – Widths of the $g_\parallel$ turning points as calculated by the theory of Froncisz and Hyde, using the data of Cu(catechol)$_2$ (Ref.(1)) to fix the adjustable parameters in the theory

Figure 1 shows a plot of Eq.(13) as a function of microwave frequency. Minima, where best resolution is expected, occur at 5 GHz for $M_I = -3/2$ and at 2 GHz for $M_I = -1/2$. If $\varepsilon = 1$, the depths of the minima are the same for both quantum numbers; however if the correlation is imperfect and $\varepsilon < 1$, the $M_I = -1/2$ line exhibits a deeper minimum than does the $M_I = -3/2$ line.

By fitting experimental data to Eq.(13), four unknowns, the residual width, $\sigma g_\parallel$, $\sigma A_\parallel$ and ε can be obtained. If desired, $\sigma\alpha$ and σQ can be calculated. One requires four linewidths, which might correspond in principle to the four quantum numbers at one frequency, or one quantum number at four frequencies, or any other combination. In practice at least two frequencies are needed because of overlap of $g_{\parallel}$ and $g_{\perp}$ features.

III. SAMPLE PREPARATION

There is an exception to the general behavior predicted by the theoretical development of the previous section. When copper replaces iron in the porphyrin ring, all $g_{\parallel}$ hyperfine lines exhibit the same width at X-band and there is essentially no further improvement on going to S-band. See for example figure 2. There is neither g- nor A-strain.

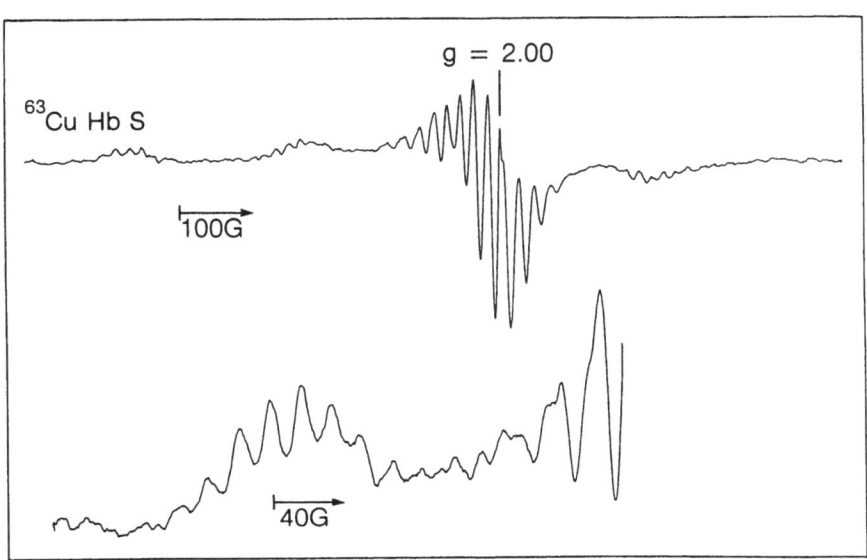

Figure 2 - ESR spectrum of sickle cell hemoglobin for which ^{63}Cu is substituted for iron in the porphyrin ring. Lower spectrum is an expansion of $M_I=-\frac{1}{2}$ line in $g_{\parallel}$ region. Spectrometer conditions: microwave frequency 3.435 GHz, T=-150°C, time constant 1 s, mod-amplitude 5G, microwave power 3dB. Spectra obtained at ESR Center for P.T.Manoharan,K.Alston and J.M.Rifkind with permission

However for all other copper containing proteins and for all copper complexes in frozen solution, the predicted behavior is observed. This exception, fig. 2, underscores the importance of good short range order. If the g- and A-strain can be eliminated altogether, than that is very good; but if the correlation coefficient, ε , is 1, information of equivalent quality can be obtained by selection of the appropriate microwave frequency. It is not surprising that sample preparation is extremely important if ε is to approach 1. Outlined in this section are the recipes and tricks for making good samples as practiced largely by W.E. Antholine. The focus is on frozen solutions of copper complexes of relatively low molecular weight. With care in sample preparation, well resolved superhyperfine lines on the $M_I=-\frac{1}{2}$ copper $g_\parallel$ line <u>always</u> can be obtained.

The following is a check list on sample preparation.

(1) A single copper isotope, either ^{63}Cu or ^{65}Cu must be used.

(2) Some improvement in resolution can be obtained by use of deuterated solvents.

(3) Aggregation or formation of dimers and oligomers of the copper complexes must be avoided.

(4) Strategies are required to avoid having complexes in the sample with mixed ligation.

(5) The signal-to-noise ratio (i.e., the concentration) should be as high as possible, since low field modulation amplitudes are required for best resolution.

(6) Solvents known to form good glasses are preferred.

The following is a check list of possible but as yet unknown effects in sample preparation:

(1) The rate of freezing.

(2) The quality of the glasses formed (i.e. opaque and polycry-

stalline or clear and glass-like).

(3) The diameter of the sample tube (related to the rate of free-zing and heat transfer).

(4) The material used for the sample tube.

(5) Dissolved oxygen (but effects are believed to be minimal).

(6) The nature of the axial ligands.

Initial patience is required. Much time and effort can be saved if a few additional samples are prepared for each change in the preparation.

At relatively low copper concentrations where the problems of aggregation are reduced, the ratio of ligand concentration to cupric ion concentration should be varied from 1:1 to 100:1. The tendency to form dimers and other adducts is screened. For example, addition of excess ligand increased the formation of monomers relative to dimers of copper carnosine (3). The pH should be varied through several orders of magnitude when using water as solvent in order to examine the effects of ligand pK values and whether -OH can displace axial or equatorial ligands. This mapping or characterization of the system is required in order to make a homogeneous sample: that is, the sample should not be a mixture of complexes differing somewhat from each other such that there is overlap in the $g_{\parallel}$, $M_I = -\frac{1}{2}$ region.

Concentrations in the range of 1 to 5mM are necessary to detect the $g_{\parallel}$, $M_I = -\frac{1}{2}$ line using only a few gauss of field modulation in our S-band spectrometer (which has a loop-gap resonator (16)) at 77 K. Aggregation can be diminished by the addition of perchlorate salt (about 2M) (17). Sodium perchlorate as well as sucrose has been used to prevent aggregation of copper carnosine dimers (3). Addition of a 12 fold excess of histidine prevented aggregation of cupric-histidine complexes (this laboratory, unpublished), suggesting another approach to the aggregation problem.

In some situations, the solvent has been mixed with DMSO up to 50%. Somewhat better glasses are formed, the tendency for aggregation is reduced, and the DMSO may coordinate in the axial position in a manner that yields more uniform samples. In studies of the tridentate complex 2-formylpyridine monothio-semicarbazone, DMSO completed the in-plane coordination, reducing the tendency for aggregation (18).

IV. SIMULATIONS

Figure 3 shows simulations of ESR spectra for the $M_I = -\frac{1}{2}$ line for 1,2,3 and 4 equivalent nitrogens assuming typical values for the input parameters to the program. It is apparent that 1 and 2 nitrogen ligation is easily and reliably determined.

Distinguishing between 3 and 4 is a more difficult problem that has been and continues to be a particular goal as we refine our methodological approaches.

The ratios of intensities of the superhyperfine lines in a stick diagram are:

Line Number	4	3	2	1	0	1	2	3	4
3N:		1	3	6	7	6	3	1	
4N:	1	4	10	16	19	16	10	4	1

The problem is that the ratio of 6:7 is almost indistinguishable from 16:19, so the central 3 lines look almost the same. Of course if the signal-to-noise ratio is good enough, the outside line for 4N is definitive. Long signal acquisition times and signal averaging are highly desirable. The best ratios are, however, the intensities of the second and third lines compared to that of the center. One has from the stick diagram:

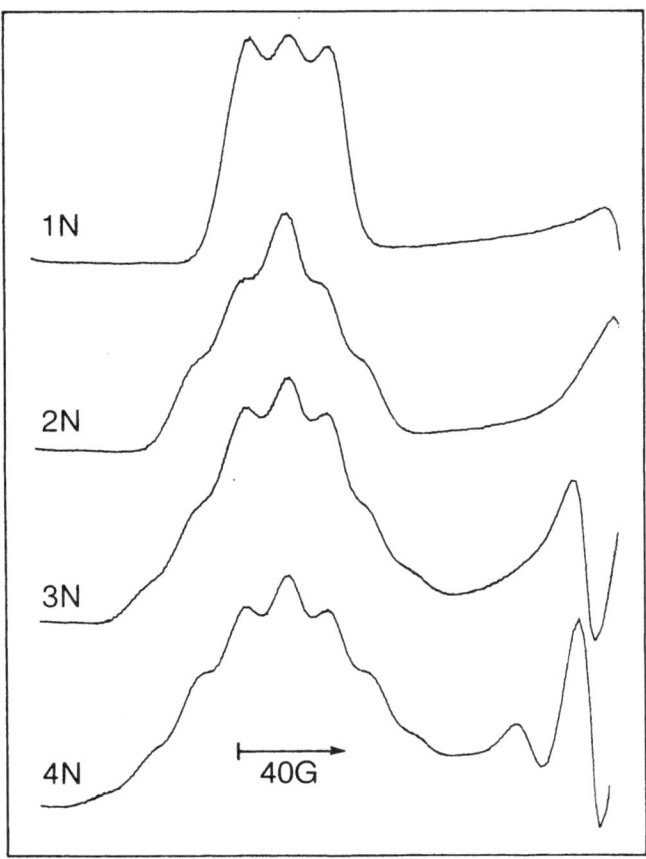

Figure 3 — Computer simulation of ESR spectra for $M_I=-\frac{1}{2}$ line in $g_{\parallel}$ region assuming 1,2,3 and 4 equivalent nitrogen donor atoms. ESR parameters: $g_{\parallel}=2.22$, $g_{\perp}=2.02$, $A_{\parallel}=181\times10^{-4}$ cm^{-1}, $A_{\perp}=20\times10^{-4}$ cm^{-1}, $A_N=16\times10^{-4}$ cm^{-1}, linewidth 6.5 G, microwave freq. 3.7 GHz

	1:0	2:0	3:0
3N	.86	.43	.14
4N	.84	.53	.21

Our procedure in comparing simulated and experimental spectra is to adjust the intensities and input parameters for the best match to lines 0
and 1, and to overlay simulations with 3 and 4 nitrogens onto the experimental spectrum. We then look for best fit at the second and third
lines.

V. EXPERIMENTAL RESULTS

A) ONE AND TWO NITROGENS

Figure 4 shows on the left side from the work of Froncisz and Aisen (19)
the $M_I=-\frac{1}{2}$ line of cupric ion bound to transferrin (one nitrogen), together with simulations assuming residual linewidths of 6.5 and 5.5 G respectively. On the right side is the spectrum of a copper complex with
two cis nitrogens and two sulfurs, together with simulations again with
6.5 and 5.5 G widths.

This figure illustrates how very important it is to achieve the best
possible resolution. From Fig.1, it is apparent that the minimum in
linewidth is broad and that any microwave frequency between 2 and 3.5
GHz is satisfactory. But any frequency that increases the residual width
by as much as 1/3 of a gauss will significantly lower the resolution.
Good techniques in sample preparation also are required.

Referring to the right hand spectra, we have come to emphasize the importance of the low field half of the $M_I = -\frac{1}{2}$ hyperfine line relative to the high field half. In this example there is some residual overlap with $g_\perp$ features. The 1:2:3:2:1 intensity ratios are evident.

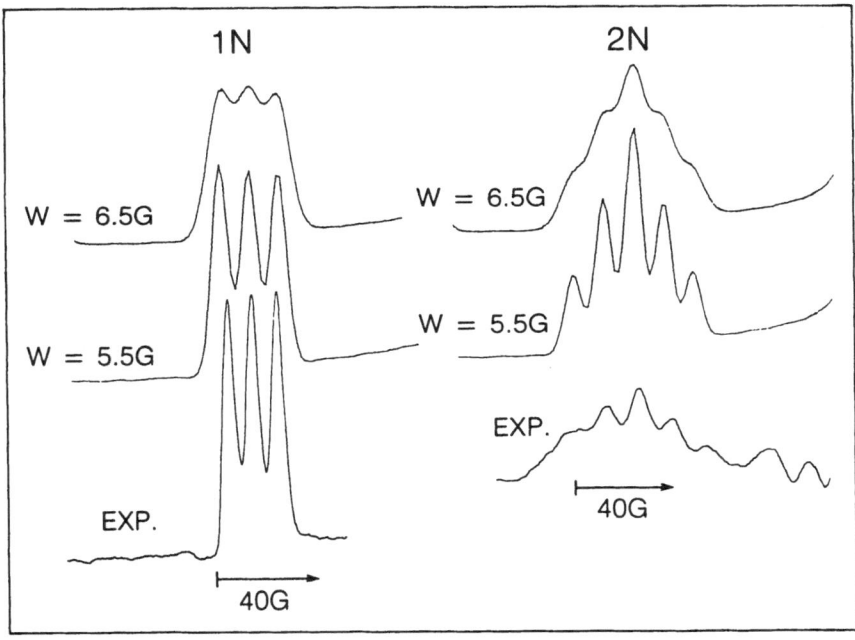

Figure 4 – LEFT HAND SIDE: Computer simulated spectra assuming one nitrogen donor atom and linewidths of 6.5 and 5.5 G respectively (top two spectra) and experimental spectrum for cupric ion bound to transferrin.
RIGHT HAND SIDE: Computer simulated spectra assuming two equivalent nitrogen donor atoms and linewidths of 6.5 and 5.5 G (top two spectra) and experimental spectrum of 3-ethoxy-2-oxobutyraldehyde bis (N^4, N^4-dimethylthiosemicarbazonato) copper(II), $CuKTSM_2$, i.e.

Rather generally the resolution of the superhyperfine lines tends to decrease as the number of nitrogen ligands increases. This suggests a systematic trend in the correlation coefficient ε.

B) THREE AND FOUR NITROGENS: HISTIDINE

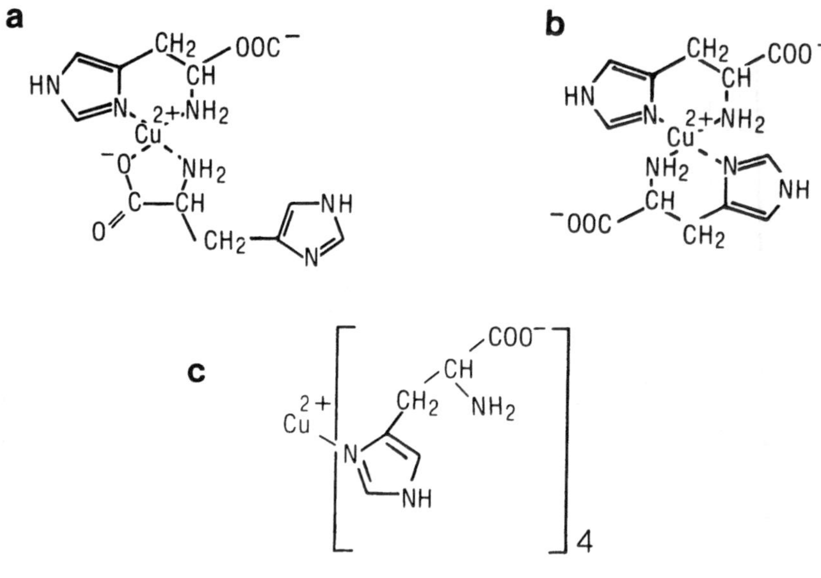

Figure 5 - Tentative structures for Cu(L-His)$_n$ complexes in frozen solution

The difficulties in distinguishing three and four nitrogens are illustrated in this section by a specific example from work in progress. Figure 5 illustrates tentative structures for Cu(L-His)$_2$ and Cu(L-His)$_4$ in aqueous solution at pH 7.3. Histidine is tridentate; combinations involving two nitrogens and two oxygens can immediately be seen to be incompatible with the spectra. As will be seen, compound c, four histidines, cannot be eliminated on the basis of existing information, although it seems somewhat improbable on chemical grounds. The controversy in the literature has been between structures a and b. The mixed histamine-like-glycine-like complex, a, was favored by Rotilio and Calabrese (20) and supported by recent electron spin echo studies (21). An x-ray analysis favored structure b (22). EPR data (23) suggested that two compounds could exist at physiological pH and temperature.

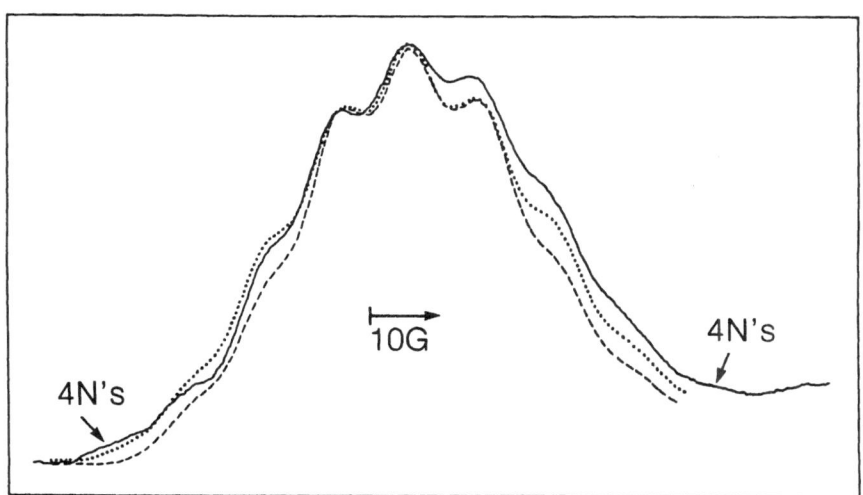

Figure 6 - The M_I=-½ line in the $g_\parallel$ region for Cu(L-His)$_2$ in D$_2$O at liquid nitrogen temperature, pH 6.8, signal averaged from 50 scans. The dotted line is a computer simulations assuming four equivalent nitrogens and the dashed line is a simulation assuming three equivalent nitrogen donor atoms

Figure 6 shows the $M_I=-\tfrac{1}{2}$ line in a standard display with simulated spectra for three and four nitrogens superimposed and forced to fit at the center of the line. The fit is substantially better for 4 nitrogen nuclei. Thus structures b and c would be favored.

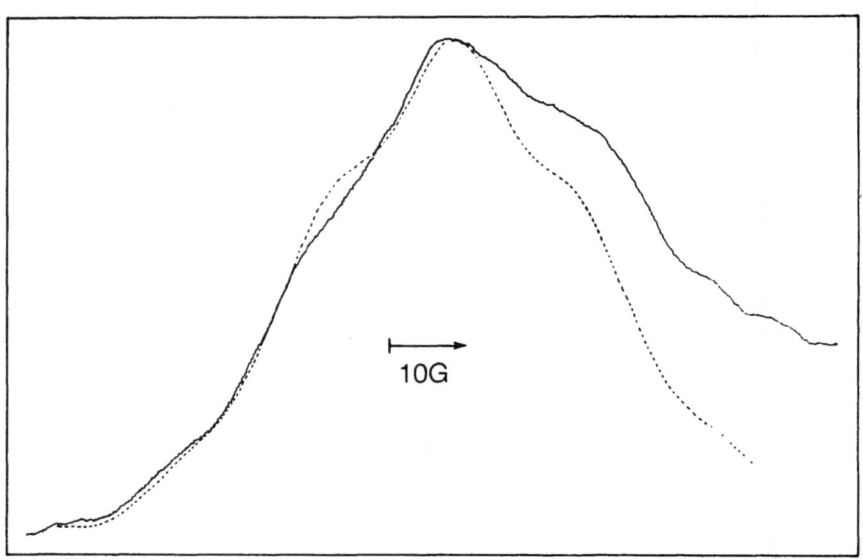

Figure 7 – ESR spectrum for $M_I=-\tfrac{1}{2}$ line in $g_{\parallel}$ region for ^{63}Cu in the presence of excess histidine for which ^{15}N was substituted for ^{14}N in the imidazole ring. Dashed line (---) simulated spectrum

In order further to confirm the assignment, a histidine was obtained with the imidazole ring nitrogens substituted by ^{15}N but not the glycine nitrogen. The $M_I=-\tfrac{1}{2}$ line is shown in figure 7. Because there is an odd number of superhyperfine lines evident in fig.7, the number of ^{15}N nuclei must be even, either 2 or 4. Therefore compound a is not possible, positively confirming the assignment of fig.5: namely that the coordination is to four nitrogens.

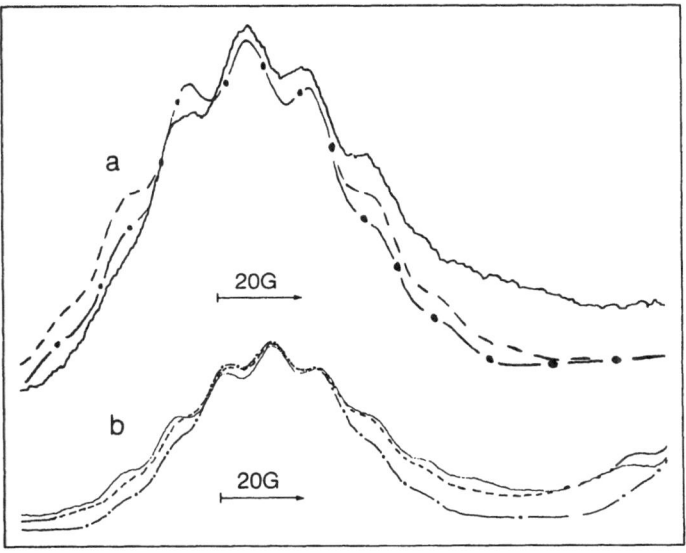

Figure 8 – TOP SPECTRA: ESR spectrum for $M_I = -\frac{1}{2}$ line in $g_{\parallel}$ region and simulated spectra for Cu^{2+} bound to ribulose-1,5 bisphosphate carboxylase. Spectra obtained at ESR Center for Rolf Branden et al.,with permission.
BOTTOM SPECTRA: ESR spectrum for $M_I = -\frac{1}{2}$ line in $g_{\parallel}$ region and simulated spectra for Cu^{2+} bound to serum albumin obtained at ESR Center for Gopa Rakhit et al.,with permission (9). Dashed lines (---) assume four equivalent nitrogen donor atoms; dot dash (.-.-) assume three nitrogen donor atoms

Consider, however, whether or not compounds b and c can be distinguished: namely $2^{15}N, 2^{14}N$ or $4^{15}N$. The simulated spectrum for four ^{15}N nuclei, fig.7, is fairly good, but the simulated spectrum (not shown) for $2^{15}N, 2^{14}N$ can be made to fit fairly well depending on the choice of the residual linewidth. Distinguishing between structures b and c can be done in an experimentally definitive manner by using fully ^{15}N substituted histidine in comparison with partially substituted histidine. The compound is available and the work in progress.

C) THREE AND FOUR NITROGENS: SERUM ALBUMIN AND CARBOXYLASE

Following the standard procedure, figure 8 shows $M_I = -\frac{1}{2}$ for Cu-Serum Albumin (9) and Cu-Carboxylase, with superimposed simulated spectra for three and four nitrogens with central lines forced to coincide. Agreement for three nitrogens for carboxylase and for four nitrogens for serum albumin is fairly good, and the reverse assignment, four nitrogens for carboxylase and three for serum albumin is poor. There are puzzles as yet unresolved, however. The fit for serum albumin is better on the high field side of the line; for carboxylase the fit is better on the low field side. For carboxylase the resolution is poorer than the simulation for the outer superhyperfine lines when agreement is forced in the central region of the spectrum. For serum albumin the reverse is true: outer line resolution is good.

D) SUMMARY OF RESULTS

In Table I the nitrogen hyperfine couplings are listed for a number of copper complexes with three or four nitrogens. These are the couplings that gave best agreement between experimental and simulated spectra in the $g_{\parallel}, M_I = -\frac{1}{2}$ region. The range is from 10 to 12 G, confirming the assumption in the present paper that nitrogen hyperfine couplings tend to vary rather little in square planar copper complexes.

TABLE I

Nitrogen hyperfine coupling (A^N) and linewidth

in $g_{\parallel}$ region at low frequency

Copper complexes	Coordination	A^N(G)	M_I =	$\Delta H_{\frac{1}{2}}$(G)*		ref.
				-3/2	-1/2	
Transferrin	1N	10		28(28)	24(29)	(19)
CuKTSM$_2$	2N2S	12		34	33	(8)
CuL$^+$+GSH	2N2S	12		42(45)	34(50)	(6)
Laccase	3N	12		63(47)	58(60)	unpublished, with permiss.
Carboxylase	3N	10		53	51	unpublished, with permiss.
Serum Albumin	4N	11		51(54)	53(60)	(9)
metHb	4N	12		52(51)	57(54)	(5)
excess Carnosine	4N	11		67	58	(3)
excess Histidine ^{14}N	4N	10		57(53)	50(55)	unpublished
^{15}N in imidazole ring	4N	16		50(58)	50(62)	

*Linewidth in parenthesis for X-band. $\Delta H_{\frac{1}{2}}$ is full width at half height of the envelope.

The table also shows the envelope widths of the $M_I=-3/2$ and $M_I=-1/2$ lines at both X and S-band. It has occurred to us that one might distinguish between three and four nitrogens by calculating the second moment of the line, but the table does not encourage this idea. There is enough variation in nitrogen hyperfine couplings and uncertainty in baseline position that the second moment would not permit a precise choice between three or four nitrogens. Linewidths or second moment analysis will distinguish, as evident in the table, between 2 and 3 nitrogens.

The narrower envelope of the -3/2 line compared to the -1/2 line at X-band is consistent with fig.1 If one is limited to X-band spectroscopy, the -3/2 line will exhibit best resolution.

CONCLUSIONS

The $M_I=-\frac{1}{2}$ line in the $g_{\parallel}$ range of the EPR spectra of copper complexes is convincingly diagnostic for assignment to 1 or 2 nitrogen ligands and to 3 or 4 if ^{15}N substitution is possible. Assignment to 3 or 4 in the absence of ^{15}N substitution may or may not be convincing depending on the quality of the experimental spectrum and the goodness-of-fit to the simulated spectrum.

One wonders if improved methodology would improve one's ability to distinguish between 3 or 4 ^{14}N ligands. One excellent possibility would be to optimize simultaneously the fit to the -3/2 line as measured at 5 GHz and to the -1/2 line as measured at 2 GHz. A microwave bridge operating in the octave bandwidth between 4 and 8 GHz is under construction in our laboratory for this purpose. It is also possible that despite the predicted poorer resolution for M_I-3/2, the baseline and lineshape problems would be reduced since there would be no underlying contributions from spectral fragments associated with other copper nuclear quantum numbers.

Heavy weight has been placed in this paper on the importance of sample preparation. The discipline of optimization outlined here using resolution of $M_I=-\frac{1}{2}$ line as the experimental observable might well be an appropriate initial step for workers primarily interested in other experimental modalities including ENDOR and electron spin echo envelope modulation. We suspect that various types of heterogeneity are very common occurrences in samples containing copper complexes.

ACKNOWLEDGEMENT

This work was supported by Grants GM-27665 and RR-01008 from National Institutes of Health.

REFERENCES

1) W.Froncisz and J.S.Hyde, J.Chem.Phys. **73**, 1 (1980).

2) J.S.Hyde and W.Froncisz, Ann.Rev.Biophys.Bioeng. **11**, 391 (1982).

3) C.E.Brown,W.E.Antholine and W.Froncisz, J.Chem.Soc., Dalton Trans. 590 (1980).

4) W.E.Antholine and F.Taketa, J.Inorg.Biochem. **16**, 145 (1982).

5) F.Taketa and W.E.Antholine, J.Inorg.Biochem. **17**, 109 (1982).

6) W.E.Antholine and F.Taketa, J.Inorg.Biochem. **29**, 69 (1984).

7) W.E.Antholine,J.S.Hyde,R.C.Sealy and D.H.Petering, J.Biol.Chem. **259**, 4437 (1984).

8) W.E.Antholine,R.Basosi,J.S.Hyde,S.Lyman and D.H.Petering, Inorg. Chem. **23**, 3543 (1984).

9) G.Rakhit,W.E.Antholine,W.Froncisz,J.S.Hyde,J.R.Pilbrow,G.R.Sinclair and B.Sarkar, J.Inorg.Biochem. in press.

384

10) G.H.Rist and J.S.Hyde, J.Chem.Phys **52**, 4633 (1970).

11) D.Kivelson and R.Neiman, J.Chem.Phys. **35**, 149 (1961).

12) I.V.Ovchinnikov and V.N.Konstantinov, J.Magn.Reson. **32**, 179 (1978).

13) A.H.Maki and B.M.McGarvey, J.Chem.Phys. **29**, 31 (1958).

14) O.M.Petrukhin, I.N.Marov, V.V.Zhukov, Yu.N.Dubrov and A.N.Ermakov, Russ.J.Inorg.Chem. **17**, 973 (1972).

15) T.Vanngard, "Copper proteins" in "Biological Applications of Electron Spin Resonance" (H.M.Swartz,J.R.Bolton and D.C.Borg, eds.) Wiley-Interscience, New York (1972).

16) W.Froncisz and J.S.Hyde, J.Magn.Reson. **47**, 515 (1982).

17) K.-E.Falk,E.Ivanova,B.Roos and T.Vanngard, Inorg.Chem. **9**, 556 (1970).

18) A.Saryan,K.Mailer,C.Krishnamurti,W.E.Antholine and D.H.Petering, Biochem.Pharmacol. **30**, 1595 (1981).

19) W.Froncisz and P.Aisen, Biochim.Biophys.Acta **700**, 55 (1982).

20) G.Rotilio and L.Calabrese, Arch.Biochem.Biophys. **143**, 218 (1971).

21) J.H.Freedman,J.L.Davis,W.B.Mims and J.Peisach, Inorg.Chim.Acta **79**, 218 (1983).

22) W.Camerman,J.K.Fawcett,T.P.A.Kruck,B.Sarkar and A.Camerman, J.Amer. Chem.Soc. **100**, 2690 (1978).

23) B.A.Goodman,D.B.McPhail and H.K.J.Powell, J.Chem.Soc.,Dalton Trans. 822 (1981).

PBB, Vol. 2
Advanced Magnetic Resonance Techniques
in Systems of High Molecular Complexity
© 1986 Birkhäuser Boston, Inc.

AN EPR STUDY OF THE PHOTOEXCITED TRIPLET STATE

OF TETRAHEDRAL d° TRANSITION METAL ANIONS

W.Barendswaard,C.J.M.Coremans,W.A.J.A.van der Poel,

J.van Tol and J.H.van der Waals

Huygens Laboratory, P.O. Box 9504,

2300 RA LEIDEN, The Netherlands

The excited states of tetrahedral anions which in their groundstate have a d° transition metal at the centre - such as VO_4^{3-}, CrO_4^{2-}, MnO_4^{-} and similar compounds with 4d and 5d metals - have since long been the object of study. In an MO description the first excitation corresponds to the "charge-transfer" $t_1 \rightarrow 2e$, in which the t_1 MO's are linear combinations of 2p AO's on the oxygens, and the 2e MO's involve the $d_{x^2-y^2}$ and d_{z^2} AO's on the metal (1). Four electronic states result, the energies of which have been predicted to lie in the order $^1T_2 > {}^1T_1 > {}^3T_2 \gtrsim {}^3T_1$ (2). In this picture the long-wavelength absorption band of such an ion is the crystal-field-induced $^1T_1 \leftarrow {}^1A_1$ transition while the second, much stronger band is the dipole-allowed $^1T_2 \leftarrow {}^1A_1$ transition.

So far the spin triplets $^3T_1$, $^3T_2$ have remained elusive. Although Blasse made it plausible (3) that the luminescence of the tetroxo-ions originates from them, no assignments have been made. The problem is that, whereas the CrO_4^{2-} and MnO_4^{-} ions (which exhibit well-resolved bands in their absorption spectra) do not luminesce, the other in general strongly luminescent compounds (e.g. VO_4^{3-}, MoO_4^{2-}) have featureless spectra.

An intermediate position is taken up by the bichromate ion, $Cr_2O_7^{2-}$, which consists of two tetrahedra linked by a common oxygen. The $K_2Cr_2O_7$

crystal has recently been shown to become luminescent at low temperature, while it also exhibits nicely structured absorption and emission
spectra (4). In a comparative study of the absorption spectra of $K_2Cr_2O_7$
and $KCrO_3Cl$ Ballhausen et al. (5) came to the conclusion that when
$Cr_2O_7^{2-}$ is excited to its first singlet state the excitation is trapped
on one half of the anion. As we shall presently see, a similar trapping
occurs in the metastable triplet state. The lower excited states of the
$Cr_2O_7^{2-}$ ion therefore do not reflect the approximate C_{2v} symmetry which
this complex has in the ground state, but rather resemble those of a
(perturbed) trigonal pyramid.

In our laboratory we have set up a programme to investigate the lower
triplet states of the d^o tetroxy-anions by optical spectroscopy and
electron paramagnetic resonance. Here we report the results obtained
thus far with EPR on three different crystals: $K_2Cr_2O_7$, $CaMoO_4$ and YVO_4.

$K_2Cr_2O_7$

This was the first system to be investigated, and by a combination of
optical and magnetic resonance techniques Van der Poel et al.(6) managed
to provide the answers to two outstanding questions. They arrived at an
assignment of the long-wavelength part of the excitation spectrum -
which had eluded previous attempts at a consistent interpretation - and
identified the triplet state from which the emission originates. The key
to this success was provided by the discovery that "tickling" of a
luminescent crystal with microwaves led to a change in the intensity of
the emission at two specific frequencies, 4.63 and 7.37 MHz, but neither
the sum nor difference frequency had any effect.

It thus became apparent that the two frequencies do not connect the spin
sublevels of a single luminescent triplet state, but that they must oc-

cur in metastable triplet states of two distinghuishable species. Subsequently these species were identified with the two inequivalent sites A and B that the $Cr_2O_7^{2-}$ ions occupy in the triclinic $K_2Cr_2O_7$ crystal. (see figure 1. At the time no oscillators were available to us that could go beyond 12 GHz, and zerofield transitions other than those between the two upper spin levels could not be probed). The complexity of the optical spectra is caused by the circumstance that the $S_1 \leftarrow S_0$ and $T_0 \leftarrow S_0$ transitions of the A site both lie below the $T_0 \leftarrow S_0$ transition of the B site. On uv excitation of the crystal one generates the partly overlapping emission spectra of both sites, but on excitation in the $S_1 \leftarrow S_0$ 0-0 band of the A site at 550.3 nm the emission spectrum becomes much simpler because then the luminescence originates from A only. The remarkable site-dependence of the optical spectra (and also of the triplet zero-field splitting, see below) is attributed to the proximity of

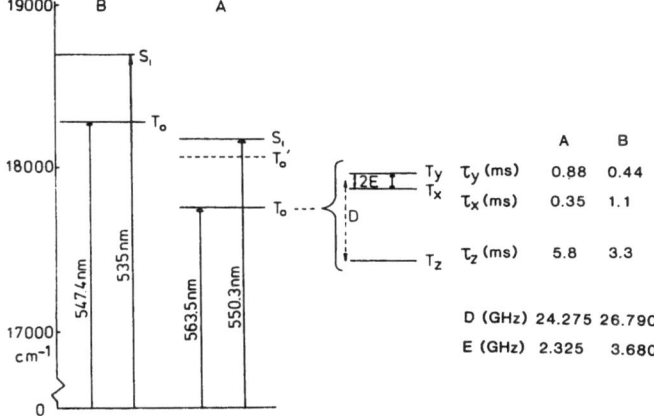

Figure 1 - Lower excited states of the $Cr_2O_7^{2-}$ ion in the two sites (A and B) in the $K_2Cr_2O_7$ crystal according to Van der Poel et al.(6). At the right the zero-field splitting of the metastable triplet state T_0 is indicated with the lifetimes of the sublevels and values of the splitting parameters determined in EPR experiments (7,8) (because of inhomogeneous broadening the parameter values vary slightly with the wavelength of excitation)

a number of orbital states that stem from the $^3T_1$, $^3T_2$ multiplets of the tetrahedral ion and thus lead to a high polarizability.

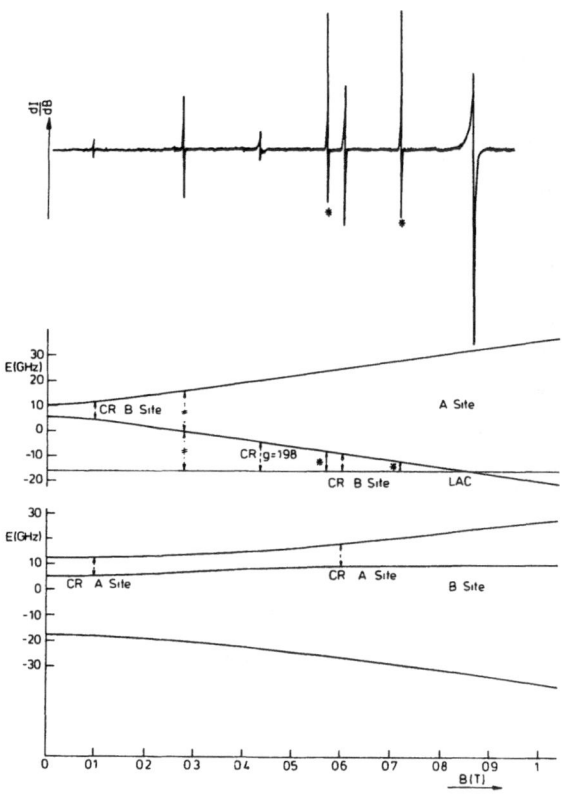

Figure 2 – Top: Optically detected EPR spectrum of a luminescent $K_2Cr_2O_7$ crystal at 1.2 K in a field $\underline{B}$ parallel to the $\underline{z}$ fine structure axis of the A site; magnetic field modulation at 63 Hz and a microwave source with a nominal frequency of 4.1 GHz were used.
Bottom: Zeeman diagrams of the $Cr_2O_7^{2-}$ ions in the two sites for the above orientation of $\underline{B}$.
The signals marked with a * are genuine EPR transitions at 4.1 and 8.2 GHz. The other signals are caused by level anticrossing (LAC) or cross relaxation (CR) as indicated in the diagram.

EPR experiments at 1.2 K have been carried out by Van der Poel et al.(7) in an external field and further experiments in zero field by Barendwaard et al.(8), all with optical detection. Figure 2 gives an example of an EPR spectrum obtained with the field $\underline{B}$ aligned parallel to a fine structure axis of the A site. Field modulation and lock-in detection of the photo-multiplier output were used. In order to have good optical access the crystal was placed in a helix rather than a cavity, and the microwaves were provided by a HP sweep oscillator set at 4.1 GHz followed by a Varian travelling wave tube amplifier. The spectrum, which represents a nice example of the various types of signal one may encounter in such an experiment, may be understood with reference to the Zeeman diagrams in the lower part of the figure. The two signals marked with an asterisk are the only real EPR signals: the one on the right is induced by microwave quanta at the nominal frequency of 4.1 GHz, the one on the left by the second harmonic at 8.2 GHz. The other signals – which remain even after the microwave source has been switched off – originate from cross-relaxation between two transitions, or level (anti-) crossing of two Zeeman states (9).

From the EPR spectra as a function of the orientation of the crystal in the field we determined the principal axes directions of the zero-field splitting tensors of the two sites. An, at first unexpected, result was that the directions of these spin axes do not bear out the (approximate) C_{2v} symmetry of the $Cr_2O_7^{2-}$ ion in its groundstate. For each of the two sites the $\underline{z}$ axis (associated with the spin sublevel of lowest energy) points approximately from one of the Cr atoms to the bridge oxygen (for details see fig.2 of ref.(7)). With the excitation trapped on one part of the ion this indeed is what one would expect. If the excited segment were isolated from its environment it would have the symmetry of trigonal pyramid (C_{3v}), with two of the three spin sublevels T_x and T_y

degenerate (i.e. E=O in the usual nomenclature of zero-field splittings). Because of the coupling to the neighbouring segment and interaction with the crystal field, the actual situation has no real degeneracy but E D.

The zero-field splitting patterns of the two sites are as sketched at the right of fig. 1. By making use of microwave sources covering a wider frequency range it has recently been possible to study five of the EPR transitions in zero field and to determine the lifetimes of the individual sublevels from so-called MIDP transients (8,10). Qualitatively speaking, the results are in good agreement with the idea that the excitation is localized on one side of the anion. The triple degeneracy of the t_1 MO of the tetrahedral CrO_4^{2-} ion here is lifted, with the excitation now occurring from the component which belongs to the irreducible representation A_2 of the C_{3v} pointgroup. The hole in the closed shell thereby avoids the bridge oxygen. It can be shown (7) that in this situation the spin-orbit coupling to nearby singlet states indeed is such that two of the spin sublevels acquire radiative character, whereas the third one with the spin axis along the C_3 axis in a first approximation does not.

$CaMoO_4$

We are presently engaged in a study of its luminescent centres at 1.3 K with the aid of a 75 GHz EPR spectrometer with optical detection built by Van Kesteren (11). This tetragonal crystal has the Scheelite structure in which the MoO_4^{2-} tetrahedra (with D_{2d} symmetry in the groundstate) occupy sites of symmetry S_4. If the S_4 symmetry is preserved one expects all MoO_4^{2-} ions in the crystal to remain magnetically equivalent on excitation. However, the EPR experiments have revealed a quite different state of affairs. For a general orientation of the magnetic field $\underline{B}$

relative to the crystal <u>four</u> signals appear, which coalesce when <u>B</u> is brought along the tetragonal <u>c</u> axis.

The MoO_4^{2-} units, apparently, become distorted on excitation whereby the equivalence of the four oxygen ligands that exists in the point group S_4 is lost. All signals can be described by a single spin hamiltonian for an orbitally nondegenerate triplet state with four different, symmetry related, orientations of the fine structure axes. The <u>z</u> axes (associated with the triplet sublevel of lowest energy) do not coincide with the crystal <u>c</u> axis but subtend angles of 33° with it. What one observes must be the manifestation of a static Jahn-Teller effect. On excitation the MoO_4^{2-} ion lowers its energy through distortion, and at the temperature of the experiment (1.2 K) each of the ions is "locked" in that distorted form amongst four equivalent ones which is favoured by random strain in the crystal (12). The zero-field splitting is appreciable with three almost equidistant sublevels; in terms of the conventional parameters: $D = 96.2$ and $E = 30.8$ GHz.

The analysis of the data presently available – which requires corroboration by further EPR experiments at a different microwave frequency – has yielded the fine-structure axis directions shown in figure 3 (an arbitrary set out of the four has been chosen, its three partners follow via the operation S_4). As for the metastable triplet state of the excited fragment of the $Cr_2O_7^{2-}$ ion, one of the sublevels (T_y) proves to be markedly less radiative than the other two; the axis associated with this sublevel happens to be almost exactly perpendicular to one of the symmetry planes of the MoO_4^{2-} tetrahedron in its ground state (the hatched plane σ_d in fig.3). We think the following, speculative argument may provide a qualitative explanation for these results.

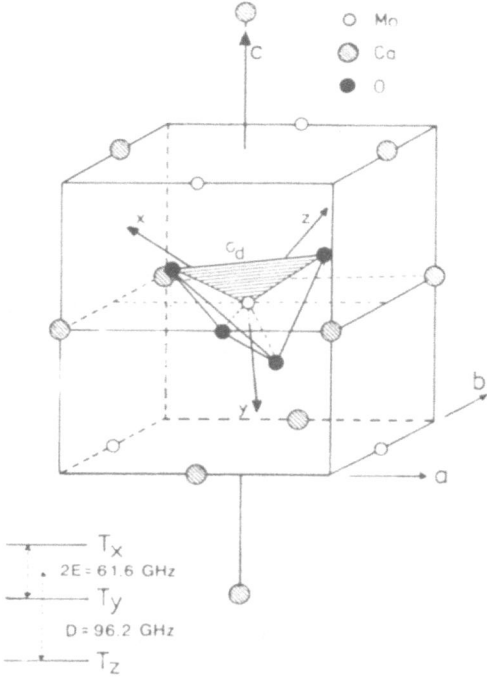

Figure 3 – Directions of the fine structure axes, x, y, z, of the metastable triplet state of the MoO_4^{2-} ion in a $CaMoO_4$ crystal determined in EPR experiments at 75 GHz. The picture
shows a MoO_4^{2-} ion at the centre of the unit cell in relation to the surrounding metal positions; the oxygens surrounding the other Mo atoms have been omitted. (The rectangular box drawn extends from -2/4 to +2/4 along the c axis
and thus encompasses only one half of the unit cell). At
the lower left the assignment of the fine structure axes to
the spin levels in zero field is given. The y axis is very
nearly perpendicular to the plane σ_d

From ab-initio calculations on the ground state of the CrO_4^- ion (13) it
is expected that this singly charged ion with its t_1^5 electron configuration will be unstable in a conformation of T_d symmetry - in agreement with the Jahn-Teller theorem. The system should have a tendency to
distort to a conformation of C_{3v} symmetry with the "hole" in the t_1^6
closed shell localized on the oxygens forming the base of the pyramid.

It is then very likely that our doubly charged MoO_4^{2-} ion will show a similar instability when it is excited and a hole is thereby likewise created in the t_1^6 closed shell. But if the MoO_4^{2-} tetrahedron, which has a D_{2d} symmetry in its ground state, indeed distorts in such a manner, the one plane σ_d will remain as the only element of symmetry. The orbital and spin states can then be classified according to whether they are symmetric (A') or antisymmetric (A") with respect to reflection in this plane. More in particular, two of the fine structure axes (here x,z) have to lie in the plane, and the third (y) should be perpendicular to it. As a result of the distortion the original $^3T_1$ multiplet will split into $^3A'+2^3A"$. If the $^3A'$ multiplet should be lowest in energy then one can show, by arguments very similar to those used in explaining the radiative decay of $K_2Cr_2O_7$ (7), that the T_y component of this multiplet cannot effectively acquire radiative character from nearby excited singlet states since the orbital and spin parts of T_y both are of species A'.

YVO_4

We have just begun the investigation by magnetic resonance techniques of the luminescent centres in the tetragonal YVO_4 crystal, where the site symmetry for the anions likewise is S_4. The results of experiments with variable frequency microwave sources in zero field and those of conventional EPR experiments in an applied field all can be interpreted by a spin hamiltonian for an orbitally nondegenerate triplet state. The zero-field splitting is given by $|D|$ = 20.25, $|E|$ = 7.15 GHz (8). We are convinced that this pattern with three nearly equally spaced levels again results from a lowering of the symmetry of the VO_4^{3-} ions on excitation, in accordance with the Jahn-Teller theorem. If the crystal field were primarily responsible for the lifting of the orbital degeneracy of the $^3T_1$ state of the tetrahedral ion with a ^3A assignment for the luminescent triplet state in S_4, then the parameter E should vanish by symmetry.

We think two major conclusions may be drawn from the magnetic resonance experiments done thus far.
a) The luminescence of the d° transitions metal oxy-anions stems from a metastable triplet state.
b) For the MoO_4^{2-} and VO_4^{3-} ions in two crystalline environments with S_4 site symmetry, this high symmetry is not reflected in the EPR results.

Apparently there is an intrinsic lowering of symmetry on excitation which is not primarily determined by the crystal field. (The CrO_3Cl^- ion is known to exhibit a somewhat analogous behaviour. From Stark experiments on the $KCrO_3Cl$ crystal Høg et al. (14) have concluded that this ion, which has a C_{3v} groundstate symmetry, distorts on excitation to its first excited singlet state ('E in C_{3v}).)

Of course, more work needs to be done and two types of experiment would seem particularly desirable. In the $CaMoO_4$ and YVO_4 crystals used in our investigations it is not known whether the emission originates from the anions in general, or from anions near crystal defects or impurities that act as traps. Studies on mixed crystals in which the anion of interest is diluted in an inert host, therefore, are planned. Second, one would like to obtain an idea of how much higher in energy the other 15 spin-orbital components of the $^3T_1$, $^3T_2$ multiplets lie above the metastable triplet observed in the present experiments. If some of these components would lie at a distance of 10-100 cm^{-1} their presence might be revealed in temperature dependent EPR studies in combination with luminescence decay experiments (3).

ACKNOWLEDGEMENT

An important part of the present work owes its success to instrumentation developed by J.Disselhorst,H.W.van Kesteren, J.Schmidt and W. Th.Wenckebach. We further thank M.Noort for preparative help and G. Blasse for his continuous advice. These investigations were supported by the Netherlands Foundation for Chemical Research (S.O.N.) with financial aid from the Netherlands Organization for the Advancement of Pure Research (Z.W.O.).

REFERENCES

1) C.J.Ballhausen and A.D.Liehr, J.Mol.Spectr. **2**, 342 (1958); **4**, 190 (1960).

2) C.J.Ballhausen, Theoret.Chim.Acta, **1**, 285 (1963).

3) G.Blasse, Structure and Bonding **42**, 1 (1980), with further refs.

4) A.Freiberg and L.A.Rebane, J.Luminescence **18/19**, 702 (1979).

5) C.L.Ballhausen,J.P.Dahl and I.Trabjerg, Colloques Int.Cent.Natl. Rech.Scient. **191**, 69 (1970); V.Miskowski,H.B.Gray and C.J. Ballhausen, Mol. Phys. **28**, 729 (1974).

6) W.A.J.A.van der Poel,M.Noort,J.Herbich,C.J.M.Coremans and J.H. van der Waals, Chem.Phys.Letters, **103**, 245 (1984)

7) W.A.J.A.van der Poel,J.Herbich and J.H. van der Waals, Chem.Phys. Letters, **103**, 253 (1984)

8) W.Barendswaard and J.H.van der Waals, Chem.Phys.Letters, to be published.

9) W.S.Veeman and J.H.van der Waals, Chem.Phys.Letters **7**, 65 (1970)

10) J.Schmidt,W.S.Veeman and J.H.van der Waals, Chem.Phys.Letters **4**, 341 (1969).

11) H.W.van Kesteren, Thesis, University of Leiden, 1985.

12) N.S.Ham, in "Electron Paramagnetic Resonance", (S. Geschwind ed.),

Plenum Press, Oxford, 1972, chap.1.

13) H.B.Broer-Braam, Thesis, University of Groningen, 1981.

14) J.H.Høg,C.J.Ballhausen and E.I.Solomon, Mol. Phys. **32**, 807 (1976).

PBB, Vol. 2
Advanced Magnetic Resonance Techniques
in Systems of High Molecular Complexity
© 1986 Birkhäuser Boston, Inc.

EPR, ENDOR AND DF-ODMR STUDIES OF EXCITED TRIPLETS

IN WEAK CHARGE-TRANSFER MOLECULAR CRYSTALS

Luigi Pasimeni

Department of Physical Chemistry, University of Padova

Via Loredan, 2 - 35131 Padova (Italy)

INTRODUCTION

In recent years considerable efforts and progresses have been made in
the study of molecular crystals of weak charge-transfer (CT) complexes.
These materials are built up of stacks of alternating donor and acceptor
molecules with the molecular planes mutually parallel.

One of the most important tools for the study of their lowest excited
triplet states has become EPR-spectroscopy and related methodologies
like ENDOR and ODMR that measure the dipolar interaction of the triplet
electrons characterized by the ZFS parameters as well as the hyperfine
interactions between electrons and magnetic nuclei.

For measurements performed near room temperature, mobile excitations are
suitably described as excitons that move through the crystal by an
incoherent motion. The exciton hops from one site to another of the
crystal until either decays at a site of the host lattice by
annihilation with emission of a quantum of light or is captured by an

impurity and decays there.

Work presented here has been done under conditions of continuous illumination and of continuous microwaves irradiation. We report on results obtained in our Laboratory by performing EPR, ENDOR and ODMR measurements. Experimentally we selected the series of TCNB charge-transfer complexes with several donor molecules like Anthracene (A), Biphenyl (B), Naphthalene (N) and Pyrene (Py). Deep traps are obtained by doping the crystals with small amounts of acceptor molecules stronger than TCNB like TCNQ.

Results are presented in three separate sections. In the first one we report our recent EPR results on the angular dependence of spin polarization (SP) of triplet excitons as a function of the orientation of magnetic field with respect to the crystal. Then we consider the process of the SP transfer into crystal sites occupied by impurities that capture the mobile excitations.

In the second section we analyse ENDOR data of the Pyrene-TCNB triplet trap in the Naphthalene-TCNB host crystal with the aim of elucidating the molecular motion of the guest donor.

Finally, we give an account of our recent results on the delayed fluorescence (DF)-ODMR spectra of the B-TCNB crystal doped with TCNQ which have served to put out the presence of homo- and heterofusion processes in the annihilation of their triplet states.

1 - EPR

1.1 - Spin polarization of A-TCNB excitons.

Recently (1) we have found that in the A-TCNB crystal the orientational dependence of spin polarization reflects the position of the acceptor molecule in the crystal lattice, whereas principal directions of the ZFS tensor are given by those of the donor molecule. A similar behavior has been noted in the spin polarized triplet excitons of the Ph-PMDA crystals (2).

This feature of the SP angular dependence relied so far on the comparison of the calculated values of the EPR intensity I_+ for the 0+ transition with the experimental signal phases. Here we give a more direct evidence of the involvement of the TCNB molecule in the ISC of the A-TCNB complex.

The crystal of this complex undergoes, on cooling, a phase transition at about T=204 K. At low temperature the crystal belongs to the monoclinic crystallographic system, space group $P2_1/a$, with two complexes per unit cell. They differ for the orientation of the long in-plane axes of the anthracene molecules which are tilted by $\pm8.6°$ from the direction of the long in-plane symmetry axis of the acceptor TCNB by rotating around the axis normal to the molecular planes (3,4). In Fig. 1 we show the angular dependence of the resonance fields and phases of the EPR signals with the magnetic field rotated in the planes of the TCNB symmetry axes. With respect to this frame there is no difference at high and at low temperature in the orientations indicated by the arrows in Fig. 1 where the EPR signals of the sites change their phases clearly showing that spin polarization of the two A-TCNB complexes in the unit cell follows identical angular dependence. In fact, this is expected if the TCNB molecule is responsible for the ISC process.

1.2 - Spin polarization of TCNQ triplet traps

For excitons the intensity of the EPR line

$$I_{ij} \propto (N_i - N_j) \qquad\qquad (1)$$

proportional to the difference of Zeeman populations of the two spin

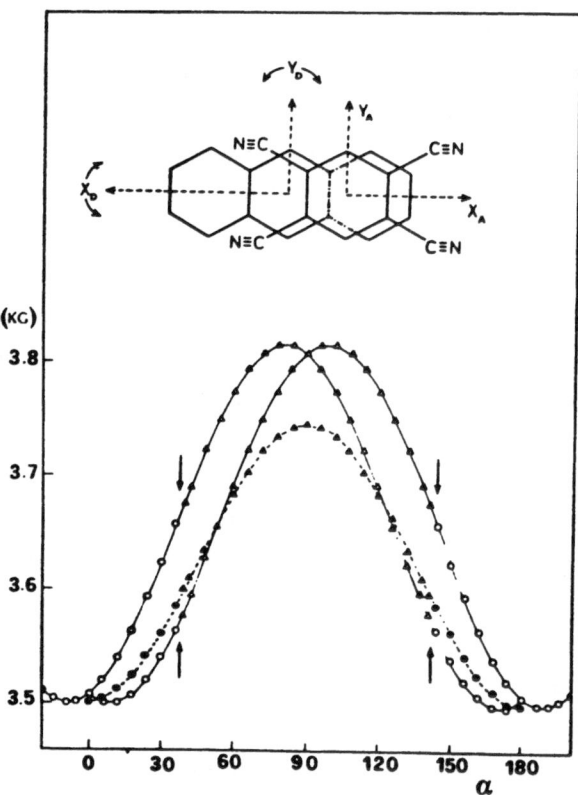

Figure 1 – A-TCNB. Resonant fields and phases of the 0+ EPR line at 250 K ($\bullet$,$\blacktriangle$) and at 120 K (O,$\triangle$) with B_o rotated in the $X_A Y_A$ symmetry plane of TCNB ($\alpha=0$, $B_o||X_A$). Libration of the donor molecule about the Z_D axis by $\pm 8.6°$ is indicated (3). (O,$\bullet$)=Emission; ($\triangle$,$\blacktriangle$)=Absorption

states joined by microwaves depends on the amount of spin polarization given by $(N_i - N_j)/(N_i + N_j)$ and on the absolute concentration of the triplet species. Polarization is controlled by the interplay of spin-lattice relaxation time among the triplet sublevels and of the triplet life-time while the triplet concentration is governed only by the decay rate provided that the generation conditions of the triplet state are maintained constant. The intensity I_+ of the 0+ EPR line as a function of the ratio W_1/K for a given orientation of B_0 reveals the presence of a maximum at about $W_1 = K$. When $W_1 \ll K$, the spin-lattice relaxation time is long enough to preserve a large amount of spin polarization, but the shortening of the exciton life-time with respect to the case $W_1 = K$ causes a decrease of the EPR signal. In the other limit $W_1 \gg K$ the efficient spin-lattice relaxation destroys SP and now it becomes responsible for the diminished value of I_+.

The spin level populations n_i of the traps and hence the intensity of their EPR lines at different orientations of B_0 can be calculated by knowing the steady-state populations N_i of the exciton Zeeman levels projected along those of the trap. The values of n_i are obtained by solving the set of kinetic equations written in vectorial form

$$\dot{\underline{n}} = -R \underline{n} + \underline{N}^p \tag{2}$$

where matrix R is the relaxation-decay matrix (1). In Fig. 2 we report the calculated values of I_+ for the A-TCNQ trap in the A-TCNB crystal for the two values of log $W_1/K=-1$ and log $W_1/K=1$ corresponding to the TCNQ doped ($\approx 10^2$ ppm) and undoped crystal, respectively.

2 − ENDOR

Crystals of weak CT complexes in some cases show the feature of large
degrees of freedom for in-plane librational motions of donors. Freezing-
-in of these librations is often accompanied by phase transitions.

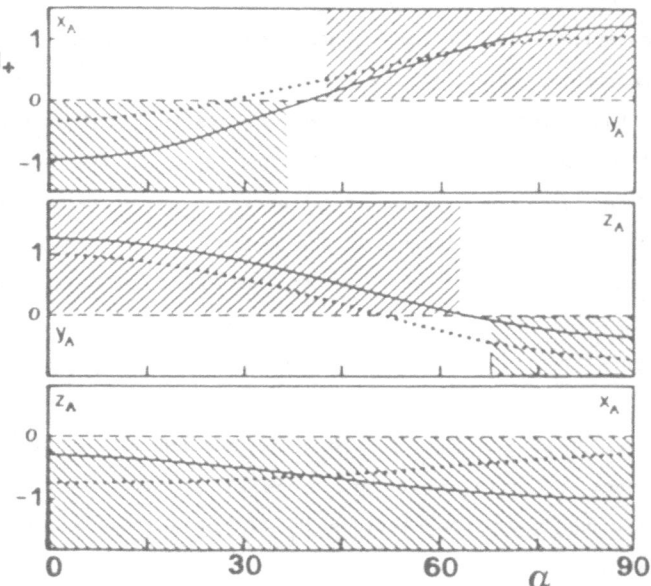

Figure 2 - A-TCNQ in the A-TCNB single crystal. Calculated angular dependence of I_+ with B_o rotated in the principal planes of
the trap ZFS tensor for the values of log $W_1/K=-1$ (solid
line) and of log $W_1/K=1$ (dotted line). Shaded areas show
the experimental sign of I_+. Populating rate constants of
the exciton $P_x:P_y:P_z=1:0.32:0.16$ (1) and decay rate constants of the trap $K_x:K_y:K_z=0.30:1:0.05$

In this context the CT complex of the donor Naphthalene (N) with the
electron acceptor 1,2,4,5-Tetracyanobenzene (TCNB) has received a great
deal of attention. Its room temperature crystal structure (5) consists
of TCNB molecules that occupy a fixed position in the crystal lattice,
while the orientation of the N molecules is disordered adopting two
alternative positions, with equal probability, that differ by a rotation

of $\pm18°$ around an axis perpendicular to the molecular plane. Raman studies (6) indicate that N-TCNB undergoes a phase transition at 63 K but no X-ray crystal structure is known of the low temperature phase.

X-traps in their first excited triplet state were also investigated by EPR and ENDOR at low temperature and an indication of their structure in the low temperature phase was obtained (7). It was found that the long in-plane axis of N forms an angle of $\pm16°$ with respect to the $\mathbf{a'}$ ($\mathbf{a'=bxc}$) crystallographic axis. Interestingly, also the normal to the N molecular plane is tilted with respect to the stacking axis c.

The host crystal N-TCNB doped with the guest donor Pyrene (Py) has been studied by EPR (8). Measurements were performed below and above the phase transition temperature T_c=63 K. The orientation of the symmetry axes of the guest donor with respect to the crystallographic axes at 30 K was determined.

We have performed ENDOR measurements at 130 K from which we obtain the components of the proton hyperfine tensors of the Py-TCNB complex. This data allowed us to determine spin distribution in the complex and hence the degree of spin delocalization from the donor to the acceptor molecules.Once spin distribution is known, hyperfine tensor components can be calculated and compared to the experimental ones. The dipolar contribution to the hyperfine tensors observed experimentally results averaged by motion of the CT complex and therefore a comparison between calculated and experimental values can give valuable information on the extent of motion.

The symmetry properties of the proton hyperfine dipolar tensors reported in TABLE I can help to find a correct dynamical model for reproducing the experimental hyperfine tensors of Pyrene. We have found that, in the a', b, c system the γ proton tensor is diagonal and it must be noticed that a jumping process only between two sites cannot explain this feature. In fact, in a monoclinic system like the N-TCNB crystal, the average of the hyperfine tensors diagonal in the two different crystal sites has the binary axis b as principal axis while the other two ones are generally rotated with respect to a' and c.

Table I

Experimental and calculated hyperfine dipolar tensors (in MHz) of Pyrene protons. A^{calc} is calculated with pyrene in a fixed position, its symmetry axes coinciding with the a',b,c crystallographic axes. A_{av}^{calc} denotes dipolar tensor after averaging by motion

Molecule	Proton	Tensor component	A^{Calc}	A^{Exp}	A_{av}^{Calc}
	α	A_{xx}/h	-2.79	-2.79	-2.76
		A_{xy}/h	∓6.22	∓5.99	∓5.89
		A_{yy}/h	3.43	3.09	3.05
		A_{zz}/h	-0.64	-0.30	-0.29
	β	A_{xx}/h	-0.75	-0.70	-0.73
		A_{xy}/h	∓2.61	∓2.23	∓2.47
		A_{yy}/h	2.07	1.71	1.76
		A_{zz}/h	-1.32	-1.01	-1.03

We have found that a four sites dynamical model enables to reproduce the
α and β proton hyperfine dipolar tensors of Pyrene observed
experimentally. It consists of a rapid libration around the normal plane
with amplitude of about $\pm 5°$ together with a rotation around the long
in-plane axis of Pyrene of about $\pm 16°$. The presence of the two motions
was inferred from the EPR data of the X-trap of the N-TCNB crystal in
the low-temperature phase (9). Averaged values of the calculated
hyperfine dipolar tensors of α and β protons are reported in TABLE I.

Therefore, the proton hyperfine tensors of Pyrene and TCNB in the CT
complex, determined by ENDOR measurements enabled us to obtain spin
distribution and CT character of Py-TCNB in the high temperature phase.
Experimental hyperfine dipolar tensors of Pyrene are reproduced by
allowing for donor librational motion in its molecular plane with an
amplitude by far less large than that of 40° expected from the
low-temperature orientation of Pyrene (8).

3 - DF-ODMR SPECTRA

Besides magnetic resonance experiments of excited triplets performed by
magnetic detection, resonant transitions between triplet exciton spin
levels can be detected through their change in the delayed fluorescence
(DF) yield occurring after triplet-triplet annihilation of triplet
exciton pairs. The method has been applied in the absence (10) and in
the presence of a static magnetic field (11). We will focus on the
latter type of experiments reporting on the features of the DF-ODMR
spectra of the B-TCNB crystals doped with TCNQ.

3.1 - Triplet-Triplet annihilation between unlike species.

The annihilation process is called heterogeneous (heterofusion) when the two constituents A and B of the reactive pair belong to different species. Their difference can arise either from the different values of the ZFS parameters or from the unlike orientation of the principal directions of the fine structure tensor, or from both causes. The latter case usually occurs when the pair is formed by a triplet exciton and a deep trap.

Following the derivation given for homofusion (12) we have calculated the variation $\Delta \gamma_{i\pm}$ in the delayed fluorescence yield by distinguishing case 1: microwaves acting on the free triplet sublevels (either exciton or trap), from case 2: microwaves acting inside the heterogeneous pair.

In case 1 we derived the relationship (13)

$$\Delta \gamma_{1\pm}^{A,B} = - \Gamma_1 \; (R_\pm^{A,B} - 1) \; (R_\pm^{B,A} - 1) \tag{3}$$

valid for low microwave fields. The coefficient Γ_1 contains kinetic parameters while the quantities $R_\pm$ represent the ratios $R_\pm = n_\pm / n_o$.

In case 2, when the sublevels of the heterogeneous pair are involved, the major difference with respect to the case of the homofusion stems from the fact that now we must consider three distinct pair states, 00, +- and -+, that have singlet character, instead of two, degeneracy of the +- and -+ states being lost. Our derivation gives the expression (13).

$$\Delta \gamma_{2\pm}^{A,B} = \Gamma_2 \; (R_\pm^{A,B} - \varepsilon_1) + R^{A,B} (1 - R_\pm^{A,B} \varepsilon_1) \tag{4}$$

where kinetic parameters are included in Γ_2 and ε_1 is related to the branching ratios of species A and B.

3.2 B-TCNB crystal doped with TCNQ.

The total emission spectrum of the single crystal of B-TCNB containing TCNQ molecules has been measured at 300 K and at 77 K. In this temperature range emission is independent of the wavelength of incident light for excitation wavenumbers larger than 24000 cm^{-1}. It consists of the superposition of the emission spectrum of the B-TCNB CT complex with two broad bands centered at 21740 cm^{-1} and at 20620 cm^{-1} and of a lower energy band with maximum at 14810 cm^{-1} lacking in the emission of the pure B-TCNB crystals and attributed to emission of the B-TCNQ complex.

ODMR spectra have been detected by collecting the light that passed through a narrow band optical filter centered either at ν_T = 15200 cm^{-1} or at ν_E =20000 cm^{-1}. The former case corresponds to light emitted at about the maximum of the emission band of the B-TCNQ complex. The large separation of the maxima in the emission spectra of B-TCNB and B-TCNQ ensures that light emitted at ν_T is due to the B-TCNQ trap only while that collected at ν_E is emitted by the B-TCNB crystal host.

When we collect light at ν_E, we observe a positive ODMR signal for each orientation of the static magnetic field rotated in the fine structure principal planes of B-TCNB, irrespective of the fact that the microwave field acts on the triplet sublevels of the exciton or of the trap. A similar behavior is displayed by the ODMR signal when light is collected at ν_T and exciton sublevels are perturbed by microwaves. ODMR signal due to transitions between the trap levels while detected at ν_T was too weak to be analyzed in detail.

The features of the ODMR spectra have been interpreted in terms of: (i) Triplet-triplet annihilation between the excitons of the host crystal and (ii) Triplet-triplet annihilation between exciton and trap.

We distinguish:

1) – **The light emission at ν_E derives from both i) and ii) mechanisms.**
   The experimentally determined angular dependence of the ODMR signal
   at ν_E cannot give information about the second contribution coming
   out from the ii) process since it cannot be separated from the
   first one by this experiment.

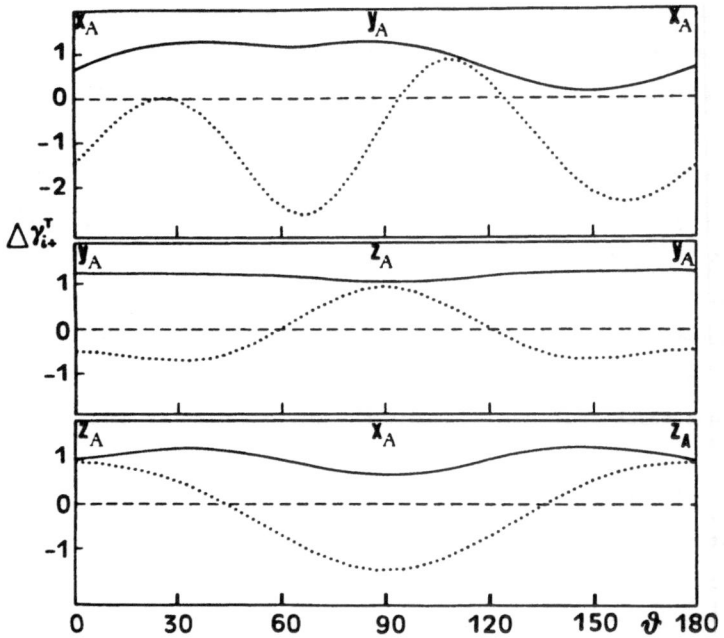

Figure 3 – Calculated curves of $\Delta\gamma_{1+}^{T}$ (dotted line) and of $\Delta\gamma_{2+}^{T}$
(solid line) according to eqs.(3) and (4) for the 0+ tran-
sition of the trap. Evaluations are made for $\varepsilon_1 = 0.1$ and
plots are given in Γ_1 and Γ_2 units, respectively

2) – **Light collected at ν_E or ν_T and microwafe field acting on the spin
   levels of the trap or of the exciton, respectively.**
   We have calculated the value of $\Delta\gamma_{i+}^{T}$ by distinguishing case a):
   microwave acting on the free triplet sublevels either of exciton or

trap, from case b): microwave field perturbing levels inside the heterogeneous pair.

So we have calculated the values of $\Delta\gamma^T_{1+}$ and $\Delta\gamma^T_{2+}$ relative to the 0+ transition between trap sublevels for B_0 rotated in the ZFS principal planes of the trap.Symbols that appear in eqs (3) and (4) denote A=trap and B=exciton and the curves are plotted in Fig. 3. An analogous calculation was also performed by changing the role of the exciton with that of the trap in eqs (3) and (4). Again we obtain a change in sign of $\Delta\gamma^T_{1+}$ for certain orientations of B_0 whereas $\Delta\gamma^T_{2+}$ remains always positive. Curves reported in Fig. 3 allow to assign the angular dependence of the DF-ODMR signal to the effect of the microwave transitions between the spin levels inside the triplet pair in line with our previous results obtained for the B-TCNB crystal (12).

ACKNOWLEDGEMENTS

The sections in this article are based on extensive work performed in collaboration with Prof. C. Corvaja, Dr. G. Agostini, Prof. G. Giacometti and Dr. A.L. Maniero. I would like to take this opportunity to thank them.
The work was supported in part by the Italian National Research Council (CNR) through its "Centro di Studio sugli Stati Molecolari Radicalici ed Eccitati" and in part by "Ministero della Pubblica Istruzione".

REFERENCES

1) L.Pasimeni,G. Guella and C.Corvaja, Chem.Phys.Letters, **84**, 347 (1981).

2) C.P.Keijzers,J.Duran and D.Haarer, J.Chem.Phys., **68**, 3563 (1978).

3) H.Tsuchiya,F.Marumo and Y.Saito, Acta Crystallogr., **B28**, 1935.

4) J.J.Stezowski, J.Phys.Chem., **83**, 550 (1979).

5) S.Kumakura,F.Iwasaki and Y.Saito, Bull.Chem.Soc. Japan, **40**, 1826 (1967).

6) R.M.McFarlane and S.Ushioda, J.Chem.Phys., **67**, 3214 (1977).

7) A. Grupp,H.C.Wolf and D.Schmid, Chem.Phys.Letters, **85**, 330 (1982).

8) H.Möhwald and E.Erdle, Phys.Stat.Sol. (b), **103**, 757 (1981).

9) E.Erdle and H.Möhwald, Chem.Phys., **36**, 283 (1979).

10) J.U.von Schütz,W.Steudle,H.C.Wolf and V.Yakhot, Chem.Phys., **46**, 53 (1980).

11) E.L.Frankevich,A.I.Pristupa and V.I.Lesin, Chem.Phys. Letters, **54**, 99 (1978).

12) G.Agostini,C.Corvaja,G.Giacometti and L.Pasimeni, Chem.Phys., **77**, 233 (1983).

13) L.Pasimeni,C.Corvaja,G.Agostini and G.Giacometti, Z.Naturforsch., **39A**, 427 (1984).

PBB, Vol. 2
Advanced Magnetic Resonance Techniques
in Systems of High Molecular Complexity
© 1986 Birkhäuser Boston, Inc.

ESR AND ESEM STUDY OF SPIN PROBE AQUEOUS SOLUTIONS

ADSORBED ON SILICA GEL

M.Romanelli[a],G.Martini[a] and L.Kevan[b]

[a]Dipartimento di Chimica, Università di Firenze,

50121 Firenze, Italy.

[b]Department of Chemistry, University of Houston,

Houston, TX 77004, USA.

It is well known that water adsorbed on solid surfaces exhibits peculiar

proporties (1-4). The investigation of the molecular interactions chara-

cterizing these systems assumes a considerable interest in colloidal,

catalytic and biological fields. Silica and silicates are the most used

porous supports for water adsorption studies (1,2).

Continuous wave electron spin resonance has been extensively used to

study radical or transition metal ion solutions adsorbed on various po-

rous supports (5-10). Nitroxides were the most useful spin probes in

this kind of investigations (11-13).

In the present study both continuous wave and pulsed ESR techniques were

used to obtain information on water solutions of a negative radical (2,

2,5,5-tetramethylpyrrolidine-1-oxyl-3-carboxylate, TEMPYDO⁻) at diffe-

rent concentrations on silica gels with different pore sizes.

By combining the data from the two techniques some further details were

got about:

　　a) the dynamics of the spin probe in the adsorbed water in a large range of temperature;

b) the dependence of the rheological properties of the adsorbed water on the pore size;

c) the structural changes to which adsorbed water undergoes at different distances from the surface;

d) the effect of the protons of water and of the nitroxide ring on the modulation of the spin echo signal.

EXPERIMENTAL

The spin probe 2,2,5,5-tetramethyl pyrrolidine 1-oxyl-3-carboxylate (TEMPYDO$^-$) was purchased as a sodium salt from Molecular Probe, Eugene, OR. Solutions at concentration 1×10^{-2} and 2.5×10^{-4} M of this radical were prepared in twice distilled water. Careful outgassing was carried out to remove dissolved oxygen. S4,S10, and S20 symbols represented the three silica gels (Merck adsorbent for chromatography) used as supports for the adsorption experiments: the pore diameters were 4, 10, and 20 nm respectively. The adsorption of the nitroxide solutions and the sample manipulation were as described in previous papers (11-13).

The ESR spectra were registered with a Bruker ESR spectrometer model 200D operating in the X-band. Temperature variation was achieved with a Bruker ESR ST100/700 variable temperature assembly.

The ESE two pulse spectra were registered with a home-built spectrometer at the University of Houston. The field was set at 3295 G at 4K, the repetition rate ranged between 55 and 25 ms, the pulses were 50 ns wide with ∼100 W power.

RESULTS AND DISCUSSION

CONTINUOUS WAVE ESR

Water solutions of TEMPYDO$^-$ after adsorption on silica gels gave rise at 298 K to the usual three line spectrum due to the radical in almost free motion, i.e. with correlation time $\tau < 10^{-9}$ s (FM spectrum), independently of the concentration and of the pore size. As observed with other nitroxides in the same systems (11-13), the broader the line resulted the narrower the pores of the support were.

Overimposed to the free motion spectrum a slow motion signal ($\tau > 10^{-9}$ s; SM signal) had an appreciable intensity in the ESR spectrum from S4 sample. The SM spectrum appeared on S20 and S10 samples only below 298 K and, as usual, its intensity increased with decreasing temperature. The spin probe in this case too occupied zones in the pores that possessed different degrees of mobility and its exchange between the two motion domains occurred at a slow rate as compared with the ESR time scale. The linewidth dependence on the concentration is analyzed elsewhere (14,15) and it results from the superposition of two different spin-spin intera-ctions: a) the dipole-dipole effect that contributes at low concen-tration, in narrower pores and at low temperature; b) the Heinsenberg spin exchange effect that contributes only in larger pores at high concentrations and high temperatures. Moreover, the correlation times for the motion from room temperature upwards are also calculated from the linewidth and they are indicative of an anisotropic motion of TEMPYDO$^-$ molecules in the solution adsorbed on the porous support (15).

Figure 1 shows the ESR spectra of a 2.5×10^{-4} M TEMPYDO$^-$ solution adsorbed on S4 samples registered at different temperature. At 270 K most of the solution behaved as a liquid and the SM spectrum contribution in the overall signal was relatively small. Its intensity increased with decre-asing temperature and at 243 K both types of spectra till contributed.

At 210 K no appreciable contribution from the free motion signal appeared while at 160 K the probe mobility closely resembled that one observed in a glassy matrix. Results in line with those obtained with S4 samples were also got from S10 and S20 samples, the largest differences being in the relative SM/FM intensity ratios and in the temperature of the FM spectrum disappearance.

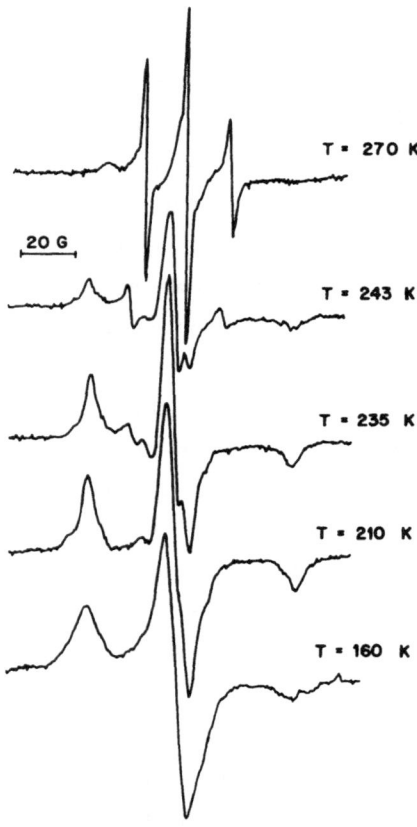

Figure 1 – ESR spectra in the range 270-160 K of a 2.5x10^{-4}M water solution of TEMPYDO$^-$ adsorbed on S4

The analysis of the c.w. ESR spectra was mainly based on the slow-motion ESR spectra as developed by Freed and coworkers (16) and computer simulation was necessary in this kind of study. The Freed program was adapted to compute spectra resulting from the combination of signals due to spin systems in different motion conditions. The brownian reorientational model was used, and two different diffusional constants were chosen to represent the anisotropic motion of the radical in the water adsorbed into the pores. This agreed with the anisotropy of the motion previously observed on these supports (15).

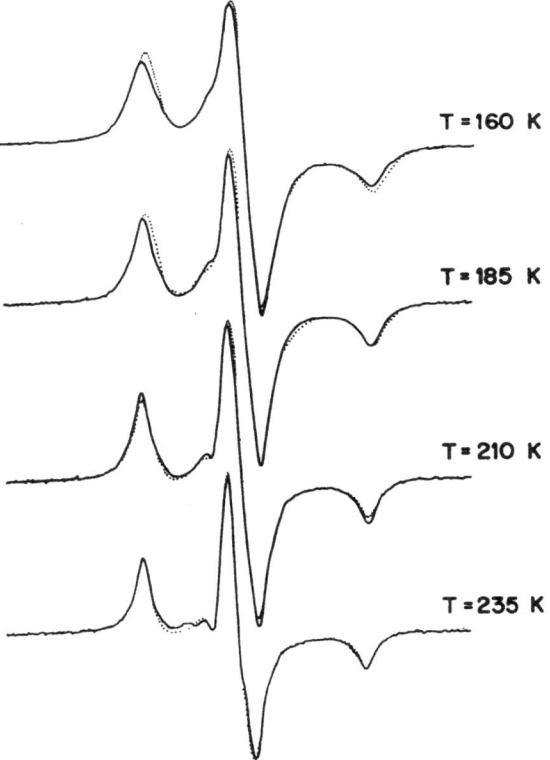

Figure 2 - ESR spectra (full lines) of a 1×10^{-2}M aqueous solution of TEMPYDO⁻ adsorbed on S20 samples. Dotted lines: computed spectra

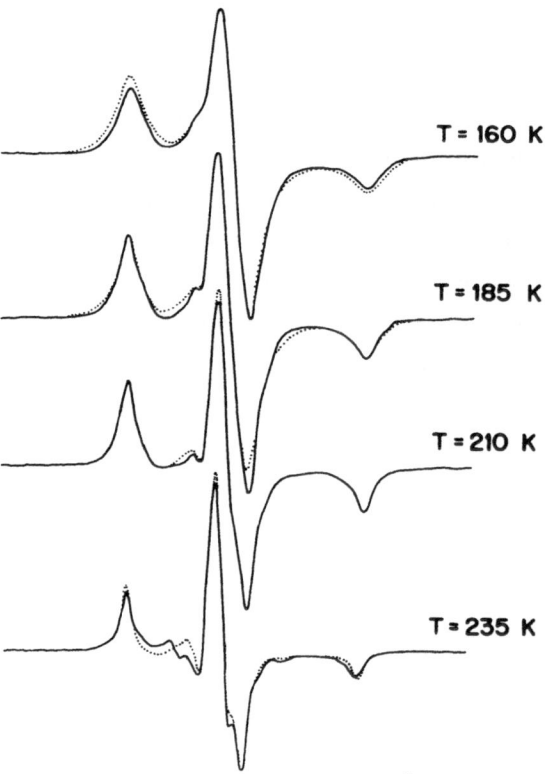

Figure 3 - ESR spectra (full lines) of a 1×10^{-2} M aqueous solution of
TEMPYDO‾ adsorbed on S4 samples. Dotted lines: computed
spectra

Figures 2-3 show the comparison between experimental and computed
spectra in slow motion conditions. The fit was quite good for all cases
up to 210 K. At 235 K the assumption of simple slow motion domain was
not sufficient. The disagreement was greater in S4 than in S20 samples
at the same nitroxide concentration.

To get a better fit when superposition of FM and SM spectra occurred, it
was necessary to add to the calculated SM component a second signal in
more rapid motion condition. This procedure is shown in figure 4. The

fit resulted thus very good. The amount of the species responsible for the FM and SM spectra may be estimated from the relative weights attributed to the components in the spectral simulations. The following relative weights were used, independently of the radical concentration:

	slow motion	fast motion
S4	95%	5%
S20	99%	1%

Some of the correlation times used for the computation of the spectra are reported in table I.

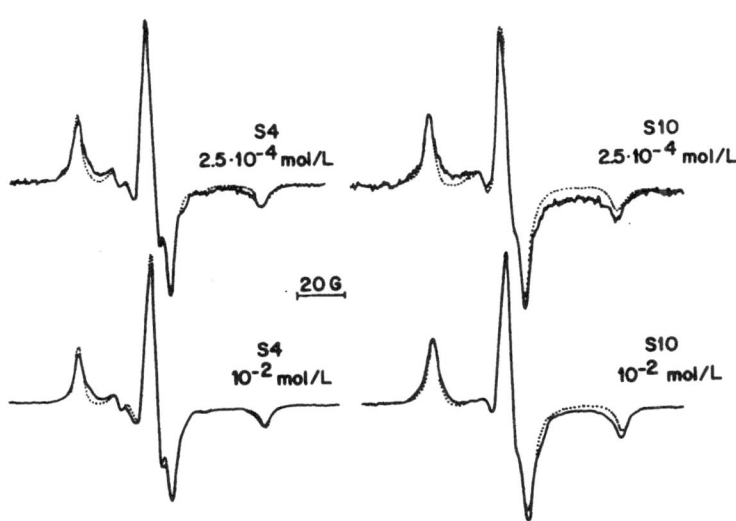

Figure 4 - Experimental (full lines) and computed (dotted lines) spectra of 2.5×10^{-4} M and 1×10^{-2} M aqueous solutions of TEMPYDO⁻ adsorbed on S4 and S10. Registration temperature: 235K

TWO-PULSE ESE

Both structural and dynamical information can be obtained from electron spin echo signals. The theory of the ESE effect predicts that the echo amplitude $V(\tau)$ can be expressed as (17);

$$V(\tau)_{exp} = V(\tau)_{mod} \cdot V(\tau)_{dec}$$

where $V(\tau)_{mod}$ includes effects due to the nuclear modulation of the electron echo and $V(\tau)_{dec}$ represents decay effects due to the loss of phase memory of the spin system; τ is the time interval between the two pulses. We assume that the mechanisms which determine $V(\tau)_{mod}$ and $V(\tau)_{dec}$ are independent. $V(\tau)_{mod}$ depends on the identity, the number and the distance of nuclei surrounding the paramagnetic center; in addition $V(\tau)_{mod}$ is also affected by the contact hyperfine constants of the nuclei in/and near the paramagnetic probe.

$V(\tau)_{dec}$ is due to the dephasing of the Larmor frequencies arising from the modulation of the spin-spin interactions as a consequence of T_1 or T_2 relaxation mechanisms.

To simulate $V(\tau)_{mod}$ the following approximations were introduced (18):

   a) two types of proton can be distinguished: N_i protons closer to the paramagnetic center and the remaining protons at higher distance (the latter protons are called matrix protons);

   b) the N_i protons are spherically distributed in shells with different radii r_i whose maximum value is denoted as r_{ext}; the matrix protons are distributed at distance from $\underline{r} > \underline{r}_{ext}$ to $\underline{r}_{\infty}$.

The overall ESEM amplitude was calculated from (17):

$$V(\tau)_{\text{mod}} = \Pi \left\{ <V(\tau,\underline{r}_i)>^{N_i}_{\text{mod}} \right\} \cdot V(\tau,\underline{r}>\underline{r}_{\text{ext}})_{\text{mod}}$$

where:

$$<V(\tau,\underline{r})>^{N}_{\text{mod}} = (\Pi)<V(\tau,\underline{r})>^{N}$$

and:

$$<V(\tau,\underline{r})> = \int_{0}^{\pi} V(\tau,\underline{r},\vartheta)\sin\vartheta d\vartheta / \int_{0}^{\pi} \sin\vartheta d\vartheta$$

$$V(\tau,\underline{r},\vartheta) = (1-k/2)+(1/k)\left\{ \cos\omega_{\alpha}\tau+\cos\omega_{\beta}\tau -\tfrac{1}{2}\cos(\omega_{\alpha}-\omega_{\beta})\tau \right\}$$

$$V(\tau,\underline{r}>\underline{r}_{\text{ext}})_{\text{mod}} = 1 - (4\pi/5)(g^2\beta^2/H_o^2)(\varrho_H/\underline{r}^3)\sin^4(\omega_H\tau/2)$$

In the above equations, ω_H is the nuclear Larmor frequency, ϑ is the angle between the electron-nuclear distance $\underline{r}$ and the S_z quantization axis, ϱ_H is the proton mean number per volume unit.

The following relations were used:

$$D = gg_N\beta\beta_N/h\underline{r}^3$$

$$B = D\cos\vartheta\sin\vartheta$$

$$A = D(3\cos^2\vartheta-1) + \underline{a}$$

$$\omega_{\alpha,\beta}^2 = (\omega_H+A/2)^2+(B/2)^2$$

$$K = (\omega_H B/\omega_{\alpha}\omega_{\beta})^2$$

where $\underline{a}$ is the contact hyperfine constant.

Decay effects were determined from (19):

$$V(2\tau)_{\text{dec}} = N \exp\left\{-2\tau/T_1-(2\tau/T_2)^2\right\}$$

where the first term in the exponential represents effects due to the spectral diffusion resulting from S-S interactions, while the second term takes into account the decay due to S-I interactions with the matrix protons.

The procedure to obtain the best fit between experimental and simulated spectra involved a preliminary procedure to obtain approximate values of $N_i, \underline{r}_i,$ and ϱ_H. Subsequently a computer simulation is carried out to improve the above values including the contact hyperfine constants.

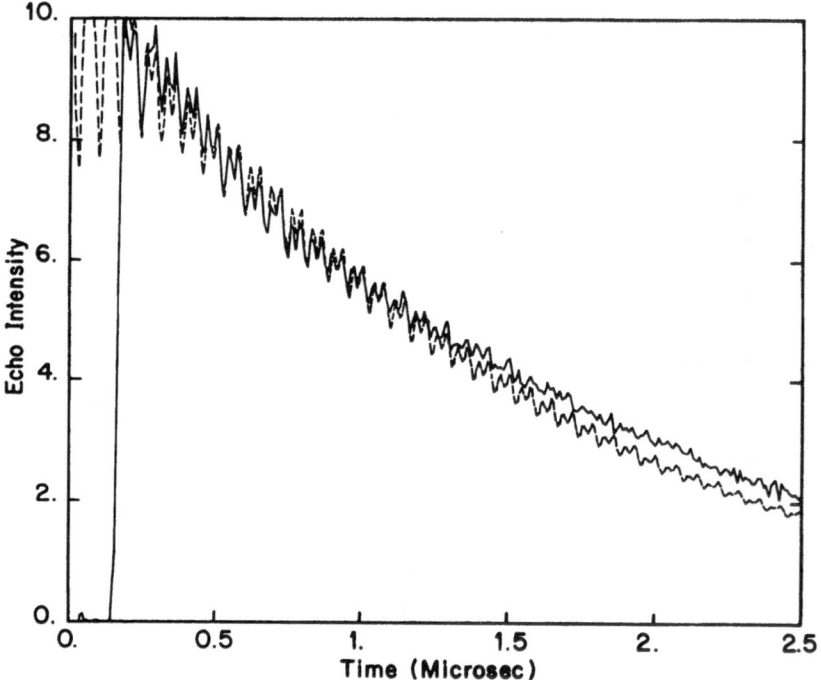

Figure 5 - Experimental (full line) and computed ESEM pattern of 2.5×10^{-4} M aqueous solution of TEMPYDO$^-$ adsorbed on S4; registration temperature: 4K

In the figures 5,6 and 7 a comparison is shown for two different systems and in table I are reported the values of some significant parameters as derived from the best simulation of the ESR and ESRM spectra.

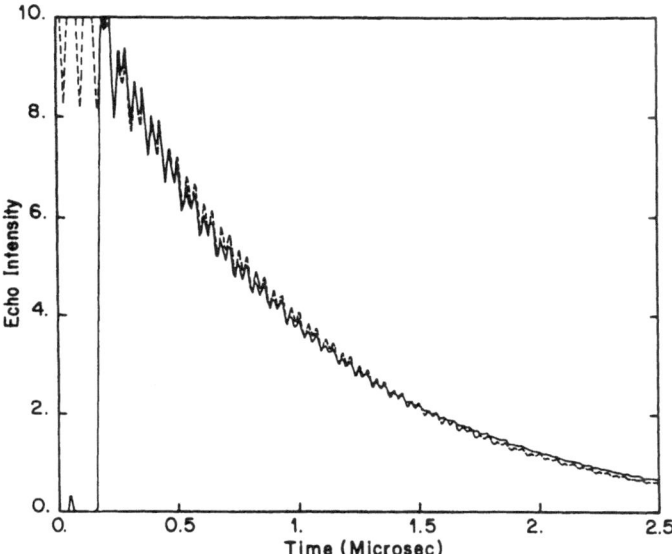

Figure 6 - Experimental (full line) and computed ESEM pattern of $1x10^{-2}$ M aqueous solution of TEMPYDO⁻ adsorbed on S4; registration temperature: 4K

Table I

sample	N	$\underline{r}(\text{Å})$	N	$\underline{r}(\text{Å})$	$\varrho_H(\text{Å}^{-3})$	$\underline{r}(\text{Å})$
S20-10^{-2}M	2	3.1	6	3.6	0.065	4.8
S4 -10^{-2}M	2	3.1	6	3.6	0.065	5.4
S4-2.5x10^{-4}M	2	2.9	6	3.6	0.065	5.4

sample	$T_1^*(\mu s)$	$T_2^*(\mu s)$	$\tau_z^{**}(ns)$	$\tau_x,\tau_y^{**}(ns)$
S20-10^{-2}M	0.49	13.	140	12
S4 -10^{-2}M	0.85	13.	210	12
S4-2.5x10^{-4}M	1.50	13.	210	12

*Values determined at 4 K;**Values determined at 160 K.

422

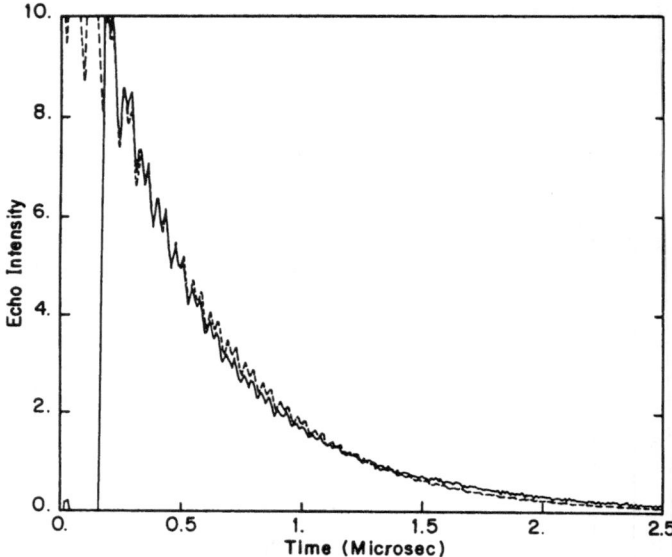

Figure 7 - Experimental (full line) and computed ESEM pattern of
1x10^{-2}M aqueous solution of TEMPYDO$^-$ adsorbed on S20;
registration temperature: 4K

The following conclusion can be drawn from the ESR and ESEM data:

a) different regions with different mobilities are confirmed by the
results reported in this paper on the water adsorbed on silica
gels with pore diameters 4, 10 and 20 nm;

b) the radical reorientation is clearly anisotropic probably due to
surface effects on the geometry of the solvation sphere;

c) the use of different radical concentration does not significantly alter the values of the correlation times;

d) the number of nearest nuclei surrounding the unpaired electron
agrees for two protons of water molecules and six equivalent
protons which, as suggested from distance and isotropic coupling
constant values (20), can be assigned to two methyl groups of
the radical;

e) the time for phase memory loss, T_1, depends, as expected, on both pore size and concentration, while T_2 results as independent from these variables.

ACKNOWLEDGMENTS

Thanks are due to Italian National Council of Researches (CNR), to Ministero della Pubblica Istruzione and to the U.S. National Science Foundation (LK) for financial support.

REFERENCES

1) A.C.Zettlemoyer,F.J.Micale and K.Klier, in "Water. A Comprehensive Treatise", (F.Franks ed.), vol. V, Plenum, New York, 1975, p. 249.

2) K.Klier and A.C.Zettlemoyer, J.Colloid Interface Sci. **58**, 216 (1977).

3) J.Clifford, in "Water. A Comprehensive Treatise", (F.Franks ed.), Vol. V, Plenum, New York, 1975, p. 75.

4) J.A.Texter, K.Klier and A.C.Zettlemoyer, Progr.Membr.Sci. **12**, 327 (1978).

5) V.Bassetti,L.Burlamacchi and G.Martini, J.Amer.Chem.Soc. **101**, 5471 (1979).

6) G.Martini, J.Colloid Interface Sci. **80**, 39 (1981).

7) C.J.Clark and M.B.McBride, Clays Clay Minerals **32**, 300 (1984).

8) M.B.McBride, Clays Clay Minerals **30**, 21 (1982).

9) J.C.Vedrine, E.G.Derouane and Y.Ben Taarit, J.Phys.Chem. **78**, 531 (1974).

10) A.von Zelewsky and J.Bemtgen, Inorg.Chem. **21**, 1771 (1982).

11) G.Martini,M.F.Ottaviani and M.Romanelli, J.Colloid Interface Sci. **94**, 105 (1983).

12) M.Romanelli,M.F.Ottaviani and G.Martini, J.Colloid Interface Sci. **96**, 373 (1983).

13) G.Martini, Colloids and Surfaces **11**, 409 (1984).

14) G.Martini and M.Bindi, J.Colloid Interface Sci., in the press.

15) G.Martini,M.Bindi,M.F.Ottaviani and M.Romanelli, J.Colloid Interface Sci., in the press.

16) J.H.Freed, in "Spin Labeling. Theory and Application", (L.J. Berliner ed.), Vol.1, Academic Press, New York, 1976, p. 33.

17) W.B.Mims,J.Peisach and J.L.Davis, J.Chem.Phys. **66**, 5536 (1977).

18) W.B.Mims and J.L.Davis, J.Chem.Phys. **64**, 4836 (1976).

19) I.M.Brown, J.Chem.Phys. **58**, 4242 (1973).

20) N.A.Sysoeva,V.I.Sheichenko and A.L.Buchachenko, Zh.Strukt.Khim. **9**, 1083 (1968).

PBB, Vol. 2
Advanced Magnetic Resonance Techniques
in Systems of High Molecular Complexity
© 1986 Birkhäuser Boston, Inc.

METAL—NUCLEOBASES COMPLEXATION EQUILIBRIA IN SOLUTION: AN ESR ANALYSIS

F.Laschi and C.Rossi

Department of Chemistry,University of Siena

Pian dei Mantellini 44, 53100 Siena, Italy

INTRODUCTION

The interactions between nucleotides and metal ions are of primary biological significance, the divalent metal ions playing a fundamental role
in several biological reactions (1-5). Metal ions are in fact strongly
involved in the action of nucleic acids and derivatives (1,2,4). Investigations of metal-nucleobase and -nucleoside interactions can offer
significant contributions for understanding chemical equilibria and molecular dynamics of larger biocomplexes in solution.

Magnetic resonance techniques are very useful for detecting the effects
of metal ions on biological systems (4-6)and the manganese(II) ion constitutes an useful paramagnetic relaxation probe that, in addition, is
itself involved as a cofactor in many relevant biological reactions
(7,8) or may replace other divalent metal ions in several biological
systems (1,2,9).

Electron spin resonance (ESR) parameters of Mn(II) are very sensitive to
the metal environment as well as to changes in chemical equilibria and
molecular dynamics.

In this paper, we report the X-band ESR analysis of Mn(II)-imidazole,

Mn(II)-pyridine and Mn(II)-purine bynary systems with the aim of defining the experimental conditions Mn(II) , L/M molar ratios, T, pH) that yield ESR evidence of metal-nitrogen bonding.

GENERAL APPROACH

High-spin Mn(II) complexes in solution are interpreted on the basis of the total spin Hamiltonian:

$$\mathcal{H} = \mathcal{H}_o + \mathcal{H}(t) \qquad (1)$$

where

$$\mathcal{H}_o = g\beta HS + aIS \qquad (2)$$

$$\mathcal{H}(t) = D\left[S_z^2 - S(S+1)\right] + E(S_x^2 - S_y^2) \qquad (3)$$

$\mathcal{H}_o$ is the time independent term in the Hamiltonian, while $\mathcal{H}(t)$ constitutes the dynamic part (10-12). $g\beta HS$ is the Zeeman interaction and aIS is the hyperfine interaction between electron and nuclear spins of Mn(II); both are isotropic and determine the position of ESR line; a, the isotropic coupling constant, is related to ligand electronegativity (14,15). $\mathcal{H}(t)$ is the zero field splitting (ZFS) energy that accounts for separation of electron spin levels in the presence of asymmetryc ligand fields around the Mn(II) ion. The ZFS term, the major structure-sensitive term in eq.(1), is an anisotropic second-rank traceless tensor and its time-dependent modulation acts as the dominant electron spin relaxation mechanism (10-12) for manganous complexes in solution. In particular, D and E are the ZFS structural parameters referring to axial and orthorombic distortions respectively within the first coordination shell.

Changes in molecular and structural dynamics of the Mn(II)-complexes induce large variations of ESR parameters such as linewidth (ΔH), spectral intensity (I), hyperfine coupling (a) and lineshape. Analysis of such variations yields a unique information on the nature and the extent of metal-ligand interactions.

Mn(II)-IMIDAZOLE SYSTEM

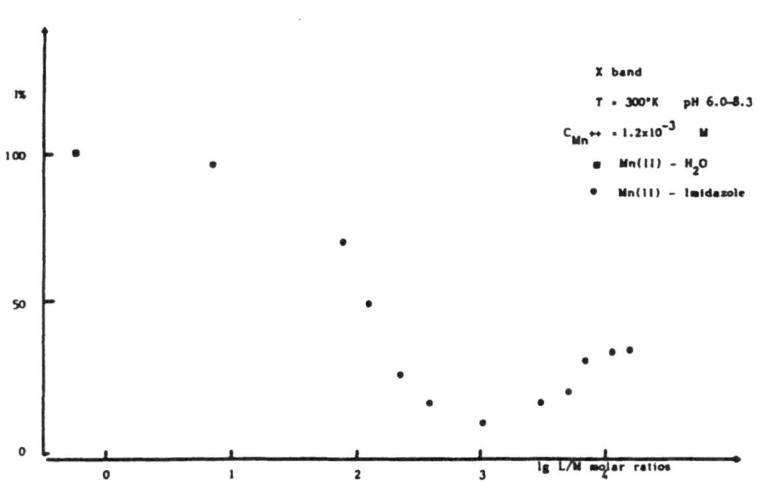

Figure 1 - X-band ESR intensity of the Mn(II)-Imidazole system vs L/M
molar ratios

Figure 1 shows the ESR intensity of the Mn(II)-imidazole system vs. L/M
molar ratios. the plot displays a minimum of 10% residual intensity for
L/M = 1.08×10^{3}; further ligand additions cause intensity recovery. Table
I shows the ΔH values vs. L/M; ΔH increases with ligand additions up to
L/M = 1.3×10^{3} (ΔH = 3.68 mT) whereas larger ligand concentrations cause
a large ΔH decrease at values lower than that of the hexaaquo ion.

The intensity decrease and the simultaneous ΔH increase can be explained
in terms of quantitative formation of the Mn(II)-H_2O-Im mixed species in
the following equilibrium (first step):

$$\text{Mn(II)-H}_2\text{O + Im} \overset{K_{mixed}}{\rightleftharpoons} \text{Mn(II)-H}_2\text{O-Im} \overset{K_{Im}}{\rightleftharpoons} \text{Mn(II)-Im + H}_2\text{O} \qquad (4)$$

$$\text{(a)} \qquad\qquad\qquad \text{(b)}$$

Table I

ESR linewidth (ΔH) vs.L/M ratios in the Mn(II)-imidazole system

$C_{Mn^{++}}$ = 1.2 mM; pH = 6.0-8.3; T = 305 K.

L/M molar ratios	H (mT)
0	1.93
650:1	2.03
950:1	2.85
1300:1	3.68
2900:1	3.02
3600:1	2.55
6000:1	1.83
8000:1	1.70

In fact mixed complexation alters the symmetry of the hexaaquo-Mn(II) complex, inducing a different molecular dynamics. In the presence of mixed species ZFS terms are large and hence ESR spectra become too broad to be detectable: modulation of such ZFS terms shortens the electron spin relaxation time, τ_s, yielding intensity loss (12,13). In these conditions, $\omega_o^2 \tau_c^2 \gg 1$, where ω_o is the ESR frequency and τ_c is the correlation time characteristic of ZFS modulation.

Step (b) of equilibrium (4) is favoured by high L/M and reflects into low ΔH values (Table I). The very low value of ΔH for L/M = 8.0×10^3 can be interpreted in terms of a highly symmetric arrangement of Im ligands around the metal ion.

Figure 2b shows the Mn(II)-Im spectrum (L/M = 8.0x10^3); the ESR line-shape is well resolved as expected for Mn(II) complexes with very small ZFS.

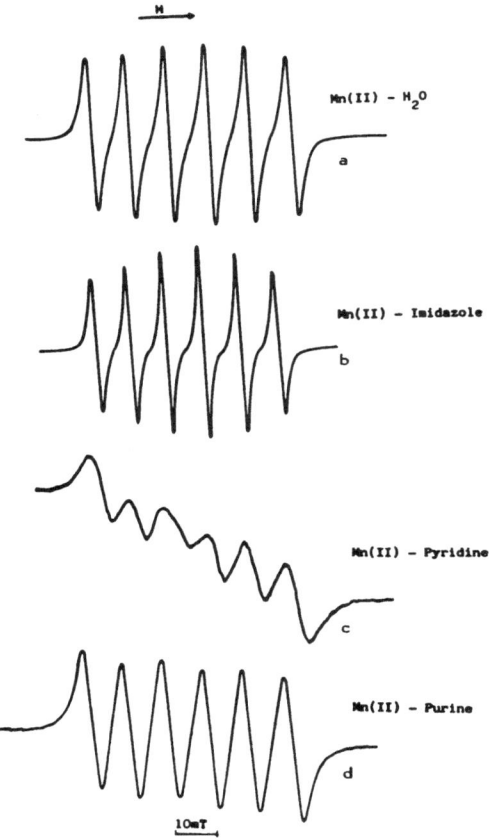

Figure 2 - (a) X-band ESR spectrum of the Mn(II)-H$_2$O system at pH=6.3 and T=290 K, $[\text{Mn}^{++}]$=1.2 mM; (b) X-band ESR spectrum of the Mn(II)-Imidazole system at pH=8.3 and T=300 K, $[\text{Mn}^{++}]$=1.2 mM; (c) X-band ESR spectrum of the Mn(II)-Pyridine system at pH=8.0 and T=290 K, $[\text{Mn}^{++}]$=1.5 mM; (d) X-band ESR spectrum of the Mn(II)-Purine system at pH=6.8 and T=290 K, $[\text{Mn}^{++}]$=1.2 mM

Figure 3 shows the plot of a vs. L/M. For L/M = 4.9x10^3 a=8.82 mT. For each L/M value, increase in pH does not alter the value of hyperfine constant. It is known that a depends on ligand electronegativity (14, 15); thus changes in a due to the presence of ligands different from solvent molecules agree with direct coordination of the ligand to Mn(II).

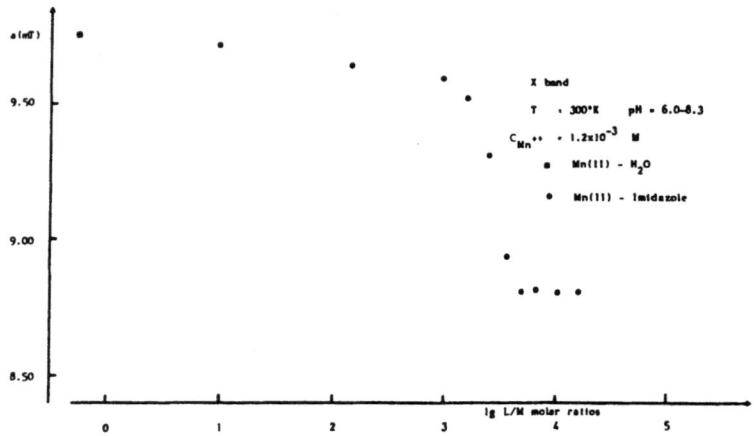

Figure 3 - Hyperfine coupling constant of the Mn(II)-Imidazole system vs L/M molar ratios

All these findings (I, ΔH, a and lineshape) can be taken to provide ESR evidence for a direct bond between the metal ion and the N-base within the first coordination shell. Moreover the large decrease of a together with that of ΔH suggest a tetrahedral arrangement of Im molecules around the Mn(II) ion.

Mn(II)-PYRIDINE SYSTEM

Figure 4 shows the ΔH behavior of the Mn(II)-Pyr system vs.L/M. A great increase in ΔH is evident at high pyr concentrations, underlining the τ_s variation due to metal-ligand interaction. These results can be related to mixed complexation, being the N-base present in the metal coordination sphere:

$$Mn(II)-H_2O + Pyr \xrightleftharpoons{K_{mixed}} Mn(II)-H_2O-Pyr \qquad (5)$$

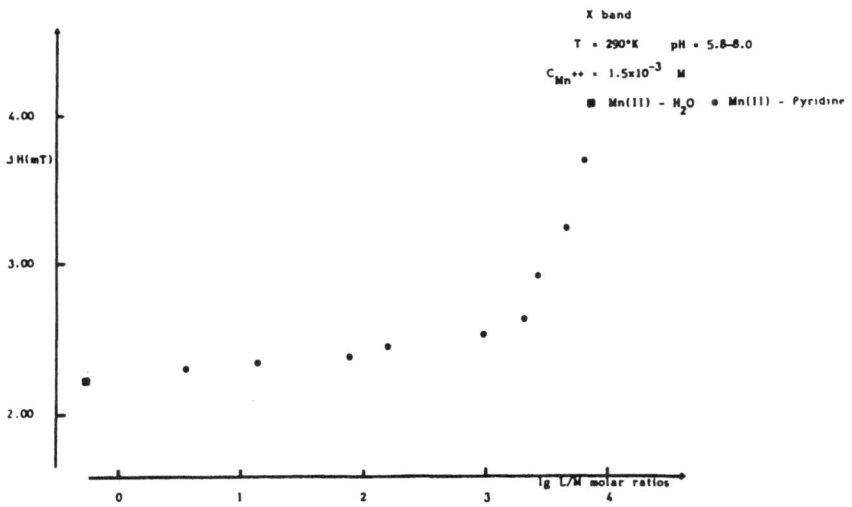

Figure 4 - Experimental ESR linewidth of Mn(II)-Pyridine system vs L/M molar ratios

Mixed coordination induces slowing down of molecular motions as compared with the hexaaquo complex, reflecting large ZFS contributions in the spin Hamiltonian (Eq.(1)). In such cases the ESR theory of Mn(II) solutions provides a second-order correction to the ESR linewidth (13):

$$T_{2el}^{-1} = K \left\{ 12.8 \ (a/\omega_o)^2 \left[I(I+1) - m^2 \right] J_o \right\} \qquad (6)$$

$$\text{with } K = K' \left[(D:D); J_o, J_1, J_2 \right]$$

In eq.(6), (D:D) is the inner product of the ZFS tensor and J_i is the i-th spectral density (13).

Figure 5 shows the ESR intensity of Mn(II)-Pyr vs. L/M. The decrease of I is monotone because of the quantitative Mn(II)-Pyr interaction and agrees with displacement of the equilibrium to the right (3). In the presence of Mn(II)-H_2O-Pyr mixed species, the cubic symmetry of the hexaaquo ion is lost and the unsymmetrical time-dependent ZFS terms strongly affect the overall lineshape and the ESR intensity (I% = 11.0 for L/M = 4.0×10^3 at pH = 8.0).

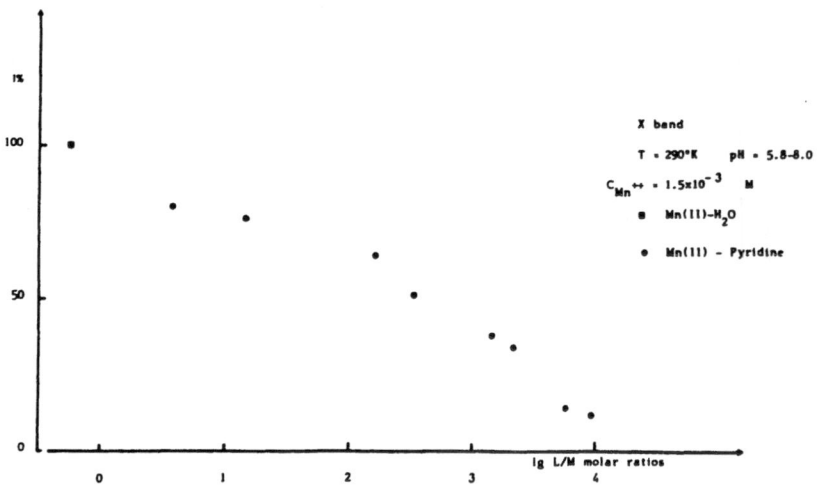

Figure 5 - X-band ESR intensity of the Mn(II)-Pyridine system vs L/M molar ratios

The Mn(II)-Pyr system is characterized by a slight but significant a decrease with increasing L/M (Table II). Such variation, more evident at higher L/M's, is consistent with direct interaction between the N-base and the Mn(II) ion (eq.(5)). The lineshape analysis of figure 2c shows the effects of the N-base on the metal ion: the six lines are quite broadened, especially the inner ones. This fact can be related to slow motion conditions within the mixed species such that $\omega_o^2 \tau_c^2 \gg 1$. We can state an upper limit for the ZFS correlation time at $\tau_c \geqslant 10^{-11}$ sec.

Table II

ESR hyperfine coupling constant (a) vs.L/M ratios

in the Mn(II)-pyridine system.

C_{Mn}^{++} = 1.5 mM; pH = 5.8-8.0; T = 290 K.

L/M molar ratios	a(mT)
0	9.78
100:1	9.76
400:1	9.70
1250:1	9.65
2000:1	9.62
3000:1	9.57
4000:1	9.50

It is worth noting that, at acidic pH's (pH = 4.6), ΔH does not exceed in any case the value of 2.6 mT while the intensity slightly drops. These findings, together the absence of changes in a vs.L/M are interpreted in terms of formation of outer-sphere complexes:

$$Mn(II)-(aq) + Pyr \underset{}{\overset{K_{outer}}{\rightleftharpoons}} Mn(II)(aq)-Pyr \qquad (7)$$

Mn(II)-PURINE SYSTEM

Figure 6 shows the ESR intensity of the Mn(II)-Pur system vs.L/M; the intensity slightly decreases and does not drop below the value of 50% residual intensity.

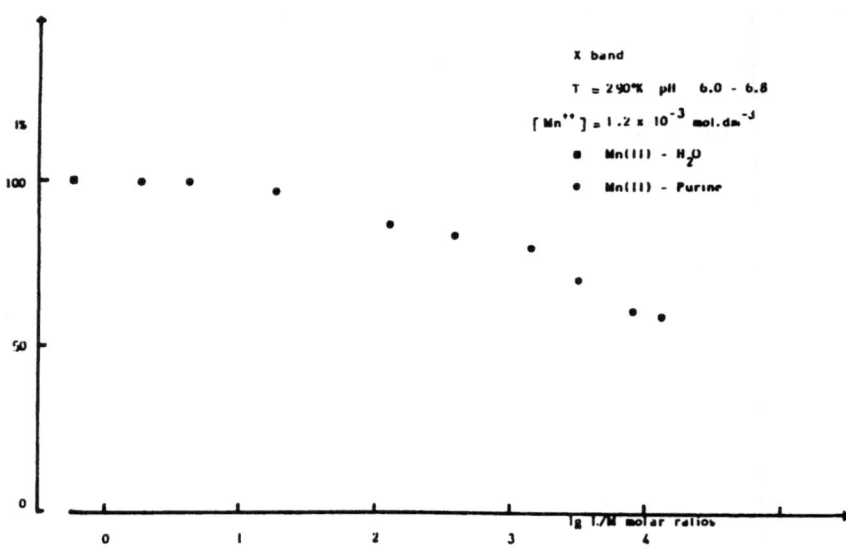

Figure 6 - X-band ESR intensity of the Mn(II)-Purine system vs L/M molar ratios

In Table III the ΔH values are reported for this system vs.L/M; the ΔH max is 2.82 mT for L/M = 4.0×10^3 at pH = 6.8. For the Mn(II)-Pur system, pH does not affect ΔH or I in the pH range 5.8-6.8, while Mn-hydroxycompounds precipitate at pH $\geqslant$ 7.2.

The a values do not vary with L/M and pH and are typical of the hexaaquo ion (a = 9.78 mT at T = 300 K). The lineshape analysis in figure 2d points out very small perturbations of the six lines.

Table III

ESR linewidth (ΔH) vs.L/M ratios in the Mn(II)-purine system

C_{Mn}^{++} = 1.2mM; pH = 6.0-6.8; T = 290 K.

L/M Molar ratios	H (mT)
0	2.05
30:1	2.10
130:1	2.20
300:1	2.39
700:1	2.45
1500:1	2.56

All these features can be interpreted on the basis of the quantitative presence of Mn(II)-Purine "outer sphere complexes":

$$Mn(II)(aq) + Pur \; \underset{}{\overset{K_{outer}}{\rightleftharpoons}} \; Mn(ii)(aq)-Pur \qquad (8)$$

$$(a) \hspace{6cm} (b)$$

It is known (17-19) that outer sphere complexes, i.e.hexaaquo-Mn(II) ions interacting with ligand molecules beyond the first coordination shell, display small random distortions from the cubic symmetry typical of the hexaaquo ion isolated from any environment. The random and fast ligand interactions on the Mn(II)-hexaaquo ions are able to yield small line broadening and intensity loss, while not affecting a. In fact dynamics of outer sphere coordination is usually assumed to be a diffusion-limited, which is a fast process in the ESR time scale.

For the Mn(II)-Pur system, the ESR spectrum (fig.2d) is resulting from the simultaneous contributions of species (a) and (b) of equilibrium (8) to the total lineshape, while ΔH is essentially affected by the outer sphere contribution. The a invariance strongly supports the presence of Mn(II)(aq)-Pur outer sphere species and impedes ESR evidence for the direct Mn(II)-N-base interaction.

REFERENCES

1) A.T.Tu and M.J.Heller, in "Metal Ions in Biological Systems" (H.Sigel Ed.),M.Dekker, New York, Vol.1, 1973 and references therein.

2) C.M.Frey and J.Stucher, in "Metal Ions in Biological Systems" (H.Sigel Ed.), M.Dekker, New York, Vol.1, 1973 and references therein.

3) R.M.Izatt,J.Christensen and J.H.Rytting, Chem.Rev. **71**, 439 (1971).

4) H.Pezzano and F.Podo, Chem.Rev. **80**, 365 (1980).

5) N.Niccolai,E.Tiezzi and G.Valensin, Chem.Rev. **82**, 359 (1982).

6) A.S.Mildvan and M.Cohn, Adv.Enzymol. **33**, 1 (1970).

7) Y.F.Lam,G.P.P.Kuntz and G.Kotowycz, J.Am.Chem.Soc. **96**, 1834 (1974).

8) Y.H.Chao and D.R.Kearns, J.Am.Chem.Soc. **99**, 6425 (1977).

9) A.Hudson and G.R.Luckhurst, Mol.Phys. **16**, 395 (1969).

10) B.R.McGarvey, J.Phys.Chem. **61**, 1232 (1957).

11) L.Burlamacchi,G.Martini,M.F.Ottaviani and M.Romanelli, Adv.Mol. Relax.Interact.Proc. **12**, 145 (1978).

12) G.R.Luckhurst and G.F.Pedulli, Chem.Phys.Letters **7**, 49 (1970).

13) S.I.Chan,B.M.Fung and H.Lutje, J.Chem.Phys. **47**, 2121 (1967) and references therein.

14) H.Levanon and Z.Luz, J.Chem.Phys. **49**, 2031 (1968).

15) L.Burlamacchi and E.Tiezzi, J.Mol.Struct. 2, 261 (1968).

16) L.Burlamacchi and E.Tiezzi, J.Phys.Chem. **73**, 1588 (1969).

17) L.Burlamacchi G.Martini and E.Tiezzi, J.Phys.Chem. **74**, 3980 (1970).

PBB, Vol. 2
Advanced Magnetic Resonance Techniques
in Systems of High Molecular Complexity
© 1986 Birkhäuser Boston, Inc.

COMPUTER ASSISTED EPR SPECTROSCOPY, WITH APPLICATIONS TO INTER-

AND INTRA-MOLECULAR DYNAMICS OF RADICALS IN SOLUTION

M.Barzaghi[a],M.Branca[b],A.Gamba[a],C.Oliva[a] and M.Simonetta[a]*

[a]CNR Center for the Study of Structure/Reactivity Relations and

Department of Physical Chemistry and Electrochemistry

University of Milan, 20133 Milan, Italy.

[b]Istituto Chimica Fisica, University of Sassari

INTRODUCTION

During the last years technical development induced considerable impact
of computer-techniques on electron spin resonance spectroscopy. There is
a surprising increase of application of EPR spectrometers, which are directly coupled to mini- and micro-computers (1). Most of these systems
have the ability to perform various types of digital data handling on
the raw spectral data in order to extract the maximum amount of information. Typical operations that are performed on spectra include signal
averaging, smoothing (2,3), differentiation (4), resolution enhancement
(3,5), data reduction (3,6), baseline flattening (7), double integration
(8), subtraction of spectra, creation of library of spectra. These
procedures will not be discussed in this paper. Most computer applications to EPR spectroscopy deal with the interpretation of experimental
spectra, based on spectral synthesis or simulation from a set of
judiciously guessed hyperfine coupling constants. Since the spectra can
be quite complex and large number of parameters may be required by
spectral synthesis, any procedure based on a simple trial-and-error
method is cumbersome, and its success is strongly dependent on the inve-
stigator's intuition, experience, diligence, and even luck. In this

paper we describe our experience in computer aided EPR spectral analysis. Illustrative applications will be taken from our investigation on hyndered internal rotation in radical anions as well as on the structure of ion pairs and triple ions.

AUTOMATED DECODING OF EPR FIRST–ORDER HYPERFINE STRUCTURE

The analysis of hyperfine patterns with EPR spectra is an obvious task to assign to a computer. Methods such as cepstral (9) and autocorrelation (10) analysis could be viable approaches to a completely automated decoding of EPR first-order hyperfine structure. Both techniques are transformations of an EPR spectrum which uncouple the nuclear hfs and provide a spectrum called a cepstrum and an autocorrelogram respectively, from which the hyperfine coupling constants can be readily identified and evaluated. The cepstrum of an EPR spectrum, taken as a continuous function $S(x)$ of magnetic field x, is given by

$$C(x) = F^{-1}\left[\ln\left|F\left[S(x)\right]\right|^2\right] = C(x,W) + 2\sum_k n_k C(x,a_k) \qquad (1)$$

Here F is the Fourier transformation operator and F^{-1} its inverse. Eq.(1) shows that the cepstrum of the EPR spectrum is a sum of the cepstrum of the lineshape $C(x,W)$ and of the cepstra of the hfs of each nuleus $C(x,a_k)$. $C(x,W)$ is a spike of width W centered on the origin and $C(x,a_k)$ is a series of impulses of decreasing strengths separated by the hyperfine coupling constant a_k. The negative pulses, which occur at $(2I_k+1)$ intervals of the hyperfine coupling constant, indicate the spin I_k of the k-th nucleus. The intensity of the pulses can be used to determine the number n_k of equivalent nuclei. Similar features are displayed by the autocorrelogram (10,11). The use and effectiveness of autocorrelogram and cepstral analysis in obtaining the hyperfine coupling constants is illustrated in figure 1 and is extensively discussed in ref.(11).

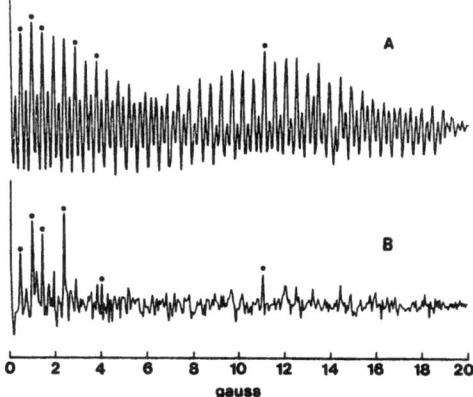

Figure 1 - Autocorrelogram (A) and cepstrum (B) of the EPR spectrum of
2-amino-4-methyl-5-nitropyridine (11). The lines corresponding to an hyperfine coupling constant are marked by a disk

A computer program (10,11) locates the maxima in the cepstrum and in the
autocorrelogram, assuming that the most intense peak corresponds to a
splitting constant. The remaining maxima are then scanned in order of
decreasing intensity and systematically examined for the integral relationships which indicate whether a maximum corresponds to a hyperfine
coupling constant or to a combination of previously determined hyperfine
coupling constants. In practice both overlap of lines and the presence
of nonequivalent nuclei with nearly equal hyperfine coupling constants
may prevent the extraction of all the hyperfine splitting constants from
the cepstrum or the autocorrelogram, and conversely, intense high-field
combination lines may be mistaken for true hyperfine coupling constant
peaks. The spectral noise usually prevents identification of the over-
tone patterns in the cepstra, so the spin information is lost. Also, the
lines for the largest coupling constants (> 5-6 gauss) are often not
much above noise, and they may not be located at the correct positions

(12). Second order hyperfine effects and linewidth broadenings can produce additional spectral noise and affect the determination of the hyperfine coupling constants. Despite of these drawbacks, the cepstrum and
the autocorrelogram are useful aids in the analysis of EPR spectra because they suggest reliable values of the hyperfine coupling constants.
The increased availability of fast and cheap micro-computers in line
with the spectrometers makes the cepstral and autocorrelation analyses a
quick and unexpensive technique to be routinely applied on any first-
order solution EPR spectrum. However a spectral synthesis is recommended
to verify the quality of cepstral and autocorrelation analyses.

LEAST-SQUARES LINESHAPE FITTING ANALYSIS

THE PROBLEM

Hyperfine coupling constants are not the only quantities which are relevant to the analysis of EPR spectra. Applications of EPR to kinetic and
thermodynamic problems as well as to the investigation of solute-solvent
interactions are well known (13). In these cases, the spectral linewidths are the most relevant features. From a general point of view the
maximum amount of information can be achieved by a synthetic approach of
the whole spectral trace. This procedure, in which theoretical trial
spectra, computed from guessed parameter values, are visually compared
with the experimental bandshapes, is in general adequate as long as the
hyperfine structure and the exchange scheme are sufficiently simple and
the assignments can be easily deduced from the spectra, but ceases to be
so in more complicated cases. In 1971 Heinzer published an efficient
iterative least-squares technique (14) which operates directly on the
raw digitized bandshapes and which optimizes approximated or guessed EPR
parameters such as coupling constants, scaling factors and linewidths.
His computer program ESRCON can deal with first-order isotropic solu-

tion spectra, by assuming all the lines equally broadened. Subsequently, the same author described the computer program ESRCEX (15) for the least-squares lineshape analysis of isotropic exchange-broadened EPR spectra based on the Liouville density matrix theory (16). This method, when applied to the fast motional region (i.e. at and above the coalescence temperature), presents severe limitations in obtaining reliable results and cannot be used without introducing arbitrary constraints on the parameters (15). In the fast motional region, the isotropic line broadening is better described by the results of the Redfield relaxation matrix theory, according to the relationship (13,17)

$$
W_{iso,\lambda} = \Sigma_\alpha F_\alpha {'M_{\lambda\alpha}}^2 + 2\Sigma_{\alpha<\beta} F_{\alpha\beta} {'M_{\lambda\alpha}} M_{\lambda\beta} + \Sigma_\alpha F_\alpha {''\zeta_\alpha}(m_{1\alpha}, m_{2\alpha}, \dots) +
$$

$$
+ \Sigma_{\alpha<\beta} F_{\alpha\beta} {''\zeta_{\alpha\beta}}(m_{1\alpha}, m_{2\alpha}, \dots; m_{1\beta}, m_{2\beta}, \dots) \tag{2}
$$

(For an explanation of the symbols in Eq.(2) see ref.17). The use of Eq.(2) for the fast motional region allows one a marked saving in computing time, if compared with the calculation afforded by the Liouville density matrix method (16,18). Most radicals which contain magnetic nuclei with large hyperfine coupling constants, such as nitrogroup nitrogen and ^{13}C, display anisotropic solution spectra whose lines are broadened according to the well-known relationship (13,17)

$$
W_{aniso,\lambda} = A + \Sigma_\alpha B_\alpha M_{\lambda\alpha} + \Sigma_\alpha C_\alpha n_\alpha (M_{\lambda\alpha}) + \Sigma_{\alpha<\beta} E_{\alpha\beta} M_{\lambda\alpha} M_{\lambda\beta} \tag{3}
$$

It was because we were often faced with EPR spectra displaying both isotropic and anisotropic line broadening in the fast motional region that in the mid-1970s we began to become interested in the development of an automated procedure which provides a quantitative, reliable and economical analysis of complex dynamic EPR spectra in the fast motional region, by taking full advantage of the results of the Redfield relaxation

matrix theory. Our efforts resulted in the implementation of the computer program ESR78 (19,20) and its descendant EPR80 (17,18) on the UNIVAC
1100/80 computer. A highly improved version of the latter program is
currently being implemented on our Gould CONCEPT 32/97 computer and it
will be referred to as program EPR83. The benefits to be derived from
spectral complexity when an iterative bandshape analysis is performed
are now well established (17,21). The true advantage of the least-squares method is that it is likely to produce in a very short time more accurate parameters than the standard simulation approach. Furthermore,
and even more importantly, it provides full error information (standard
errors, correlation matrix, etc.) which ensures a rigorous and quantitative evaluation of the agreement between experimental and calculated
spectra and which is often an indispensable prerequisite for a scrupo-
lous analysis of the temperature dependence of spectral parameters.

THE METHOD

An experimental EPR spectrum is regarded as a collection of digitized
signal intensities y_i (i = 1,2,...,N_o) represented in a computer by a
vector $\underline{y}$ in data space. In the same way, a theoretical EPR spectrum, no
matter how complicated, can always be expressed as a function $y(x,\underline{p})$ of
a discrete parameter vector $\underline{p}$ and a continuous variable x (field
strength), which can be discretized over a set of N_o points in data space. Then the task consists in finding a $\underline{p}$ such that it minimizes the error functional

$$f(\underline{p}) = \underline{s}^\dagger \underline{W} \underline{s} \qquad (4)$$

where the residual vector $\underline{s}$ is defined by elements

$$s_i = y_i - y(x_i,\underline{p}) \qquad (5)$$

and $\underline{W}$ is the weight matrix, i.e. the inverse of the variance-covariance matrix of the intensity data. Since in the case of interest $y(x,\underline{p})$ is non-linear in its parameters $\underline{p}$, an iterative minimization procedure must be used. The algorithm starts from an estimate $\underline{p}_k$ and chooses the next estimate recursively by means of

$$\underline{p}_{k+1} = \underline{p}_k + \triangle \underline{p}_{k+1} \tag{6}$$

where the displacement vector $\triangle \underline{p}_{k+1}$ is calculated by the Newton-Raphson formula

$$\triangle \underline{p} = -\underline{H}^{-1} g \tag{7}$$

The gradient vector g and the Hessian matrix $\underline{H}$ are given by

$$g = -2 \; \underline{J}^\dagger \underline{W} \underline{s} \tag{8}$$

$$\underline{H} = 2(\underline{J}^\dagger \underline{W} \underline{J} - \Sigma_{ij} \underline{K}_i \underline{W}_{ij} \underline{s}_j) \tag{9}$$

Here $\underline{J}$ is the Jacobian matrix of elements $J_{ir} = \partial y(x_i,\underline{p})/\partial p_r$ and $K_{i,rs} = \partial^2 y(x_i,\underline{p})/\partial p_r \partial p_s$. Several criteria can be applied to insure the stability and efficiency of the iterative procedure. Usually the Newton-Raphson step is damped by a scalar which can be estimated by some empirical formula such as $1+(\triangle \underline{p}^\dagger \triangle \underline{p})^{\frac{1}{2}} - 1$ or through a linear search along the Newton-Raphson direction $\triangle \underline{p}$. Alternatively, some interpolation methods between the Newton-Raphson direction and the gradient direction can be profitably applied (17,21b,22). The iteration loop is normally left when both the norm of the displacement vector and the maximum displacement are smaller than a given threshold. At this point the eigenvalues of the Hessian (Eq.(9)) are checked for their signs, because if a local minimum has been reached on the error hypersurface (Eq.4)), all the Hessian eigenvalues at that point must be positive.

AUTOMATED SPECTRAL ANALYSIS THROUGH A CONTINUOUSLY VARIABLE CORRELATION
BETWEEN THE RESIDUALS

Since the model function $y(x,\underline{p})$ is strongly non-linear in the parameters $\underline{p}$, the least-squares method to be efficient requires that the starting parameters $\underline{p}°$ approach the values $\underline{p}^*$ of minimum $f(\underline{p})$. Wrong guesses normally do not lead to convergence or they may give rise to local minima. A general iterative method for locating the global minimum of the error functional (Eq.4)) in the presence of a multitude of local minima has been published recently by Stephenson and Binsch and applied to NMR spectra (21d-g). The main advantage of this method is that it allows the starting parameters to be chosen randomly within liberally specified boundaries. In the standard least-squares lineshape fitting method of the previous section the weight matrix $\underline{W}$ is usually assumed to be a unit matrix, as required for an unbiased judgment of the quality of the fit. The novel idea of Stephenson and Binsch was to profit by the matrix $\underline{W}$ to induce some correlation between the residuals, in such a way it results an increasing function of the iteration distance to the global minimum. A sufficiently high correlation is expected to produce a region of local convexity which covers the entire domain of the parameter vector $\underline{p}$ (21d,f). The general procedure starts with a very high correlation and iteratively locates the minimum on the corresponding hypersurface. Then the correlation is stepped down by a constant factor, and another minimization cycle is performed. Such a procedure assures the existence of a unique pathway to the global minimum, which evades the multitude of traps (local minima) in the course of the minimization process (21d,f). We verified (17) that this computational strategy can be successfully applied to EPR spectra, so that the problem of the automated EPR spectral analysis may be regarded as solved, not only in principle but also in practice.

SYNTHETIC MODELS

Our computer programs can deal with first-order solution spectra both in the slow motional region (18,19) and in the fast motional region (17). Spectra due to two or more paramagnetic species can be as well analysed (17). This feature turns out to be of much benefit when investigating reactive systems (19,23). Experimental spectra are often recorded with relatively large modulation amplitude and/or microwave power, because of the beneficial effects on the signal-to-noise ratio. Since large modulation amplitude and microwave power are known to distort the lineshape, it is particularly useful to include modulation and saturation broadening in the synthetic models. Therefore efficient algorithms have been devised, which make programs EPR80 and EPR83 able to analyse EPR spectra which are saturated by microwave power up to 200 mW (provided that the modulation amplitude is relatively small, $B_m \leq 0.1$ G) or which are overmodulated by field up to 1 G (provided that the microwave power does not exceed 2-5 mW)(17). We will not discuss here the details of the synthetic models included in our programs. The interested reader should look them up in the quoted papers. Here we simply show some illustrative examples of their application in figures 2,5,6.

ERROR INFORMATION

The standard errors of the parameters are calculated as the square roots of the diagonal elements of $\underline{H}^{-1}$ multiplied by the standard error of the unit of weight, σ_0:

$$\sigma(p_r) = \sigma_0 \sqrt{(\underline{H}^{-1})_{rr}} \qquad (10)$$

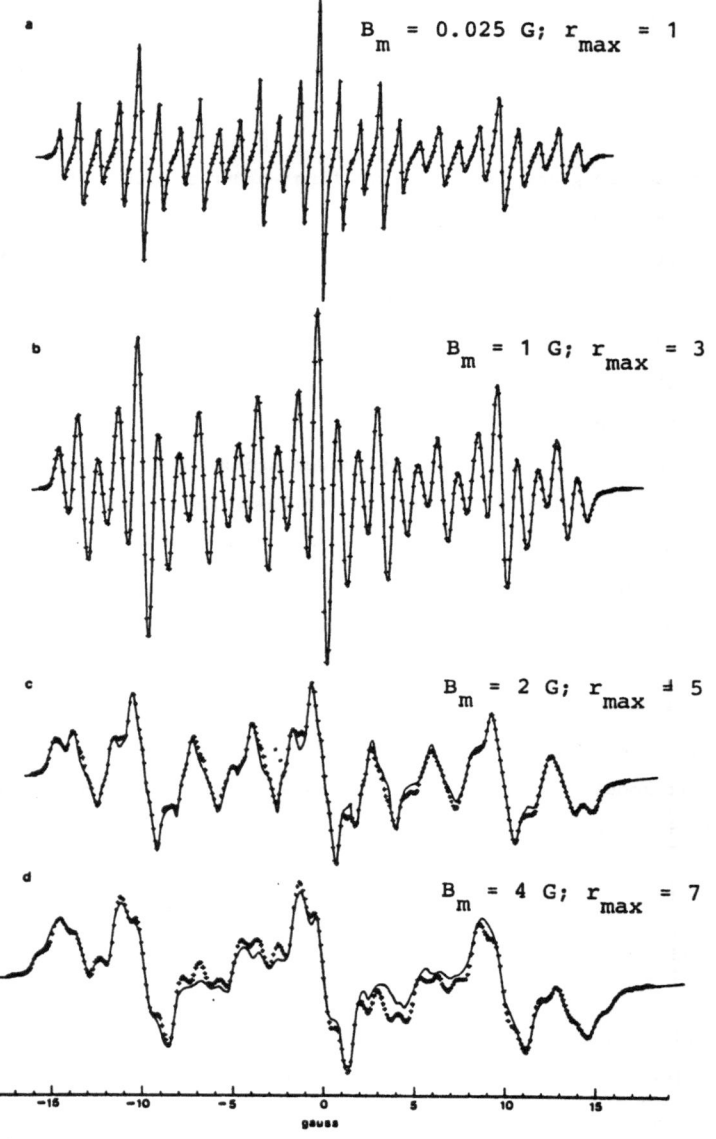

Figure 2 – EPR spectra of p-nitrobenzoate dianion radical in Me_2SO (counterion potassium) as a function of modulation amplitude (microwave power 2 mW). Solid lines: synthesized spectra; points: experimental digitized spectra. See ref. 17 for further details.

$$\sigma_o = \sqrt{(f(\underline{p})/\underline{I}^\dagger \underline{\underline{W}}\underline{I})(N_o/N_o-N_I)} \qquad (11)$$

N_I is the number of independent parameters which are varied. The elements C_{rs} of the correlation matrix are a measure for the interdepen-
dence of the parameters p_r and p_s, and are calculated from the Hessian
by means of the relation

$$C_{rs} = (\underline{\underline{H}}^{-1})_{rs} \left[(\underline{\underline{H}}^{-1})_{rr} (\underline{\underline{H}}^{-1})_{ss} \right]^{-\frac{1}{2}} \qquad (12)$$

In cases were fits of different spectra and/or different models are to
be compared, the dimensionless factor (24)

$$R = \sqrt{f(\underline{p})/\underline{y}^\dagger \underline{\underline{W}}\underline{y}} \qquad (13)$$

has proved to be a very useful error criterion (25) as shown in fig. 3

In the following sections we will discuss in some detail the results
obtained by applying the computational strategy described above to the
two main topics investigated in our EPR laboratory.

STRUCTURE AND DYNAMICS OF ION PAIRS AND TRIPLE IONS

Ion pair formation has been the subject of many investigations through
the detection of high resolution EPR spectra of anion radicals obtained
by alkali metal reduction in ethereal solvents (26-29). The main effects
of the alkali metal cations on the EPR spectra of anion radicals are the
following: (a) each line of the EPR spectrum is split in a multiplet of
lines due to the cation nuclear spin, (b) the redistribution of the odd
electron on the organic fragment induces a modification of the hyperfine
coupling constants of the anion radical magnetic nuclei. When an

electron withdrawing group is present in the organic fragment the cation

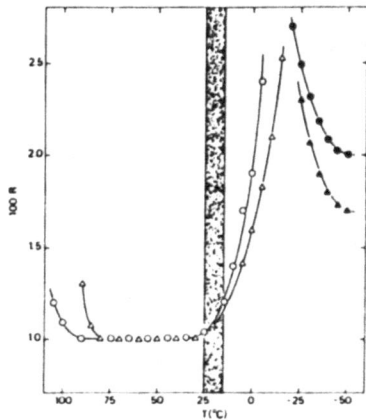

Figure 3 - Plot of R factor as a function of temperature for the EPR
spectra of 3-nitrobenzaldehyde anion radical in DMF, obtained by electrolytic reduction (circles) and by chemical
reduction with MeONa (triangles). Open symbols, fast motional model; blackened symbols, static model. The darkened
region corresponds to the coalescence temperature of the
hyperfine coupling constants.

places itself in its neighbouring. This is the case, for example, of ion

pairs formed by Li, Na, and K with 4-nitropyridine in dimethoxyethane

(DME) and tetrahydrofuran (THF) (30). Two competitive positions are fo-

und for the cation, near the nitrogroup and near the ring nitrogen

respectively, by INDO and ab initio calculations. The combined use of

experimental and theoretical information, including the evaluation of

the molecular electrostatic potential, suggests the localisation of the

cation in the region of the nitrogroup, in the σ_v plane at least 2.25 Å

over the molecular plane. Furthermore the splittings of both nitrogroup

nitrogen and metal cation increase with temperature, whilst the

splitting of the ring nitrogen is insensitive or decreases when metal splitting increases. According to Atherton and Weissman (31) this trend can be accounted for by the temperature dependence of the occupation probabilities of the vibrational states in which ion pairs can exist. A similar behaviour is observed in ion pairs formed by Li, Na, K, Rb and Cs with 4-nitropyridine-N-oxide in dimethoxyethane (32), and with other nitroderivatives (33,34). A theoretical investigation of the structure of both 4-nitropyridine and 4-nitropyridine-N-oxide ion pairs was carried out by the electrostatic potential model, using an ab initio wavefunction. The splitting constants were computed with both the UHF and RHF+ CI INDO methods and with the UHF ab initio methods. Perspective views of the potential surfaces for these anion radicals are compared in figure 4

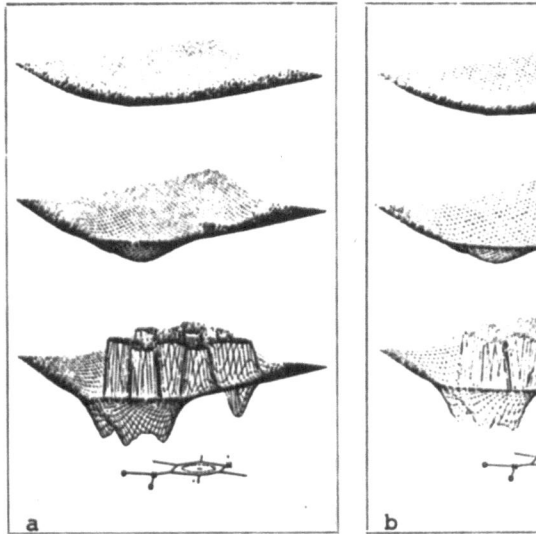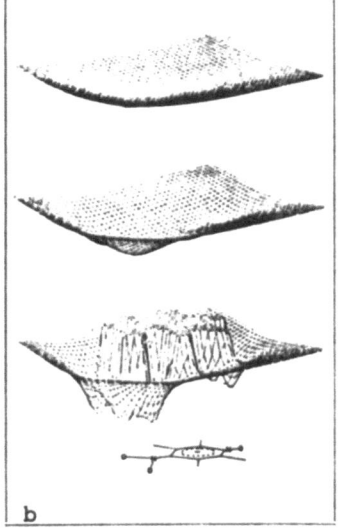

Figure 4 - Perspective views of potential surfaces for (a) 4-nitropyridine and (b) 4-nitropyridine-N-oxide anion radicals. Bottom: molecular plane; middle: plane at 1.5 Å; top: plane at 3 Å.

A completely different behaviour was observed for the ion pairs formed by Rb and Cs with 4-nitropyridine in DME, whose spectra show hyperfine patterns in which the metal splitting is absent. Their simulation requires twice the number of splitting constants as used to interpret the ESR spectrum of the free anion radical (32,35). This spectrum disappears by adding dimethylsulfoxide, i.e. on increasing the dielectric constant of the solvent, and a hyperfine pattern similar to that of the free anion appears, suggesting that an electrostatic binding was formerly linking two anionic fragments.

When two or more nitrogroups are bonded to an aromatic substrate, the alkali cation is localized near one nitrogroup, so that the symmetry of the ion pair is lowered with respect to that of the free anion radical. Correspondingly the electronic structure of the free anion is strongly changed upon complexation with alkali cations. This is true both when the nitrogroups are linked to the same aromatic ring, as in meta- and para-dinitrobenzene (36,37), 3,5-dinitropyridine (36), 3,5-dinitrobenzonitrile (38), 3,5-dinitrobenzamide (20), 1,3,5-trinitrobenzene (18), and when the nitrogroups are linked to two different aromatic rings, as in 4,4'-dinitrobenzophenone (33), in 1,1'-di-(p-nitrophenyl)ethylene (39) and in di-(p-nitrophenyl)methane (40). When a salt of a common cation is added to an ion pair solution the nitrogen coupling constant decreases and the lines broaden as the concentration of the added salt increases. These facts are ascribed to the formation of triple ions:

$$\bar{A}M^+ + M^+ \rightleftharpoons M^+\bar{A}M^+ \rightleftharpoons M^+A^- + M^+ \tag{14}$$

In the case of nitrobenzene and 4-nitrobenzonitrile the metal quartets are resolved and from their broadenings activation energies of 3.1 ± 0.7 and 2.2 ± 0.7 kcal mol^{-1} respectively were estimated (41). In the case of polynitro-derivatives the metal quartet is never resolved (18,36-38) and in general linewidth alternation affects the hyperfine pattern, because of the symmetry of the anion radical (see for example figure 5).

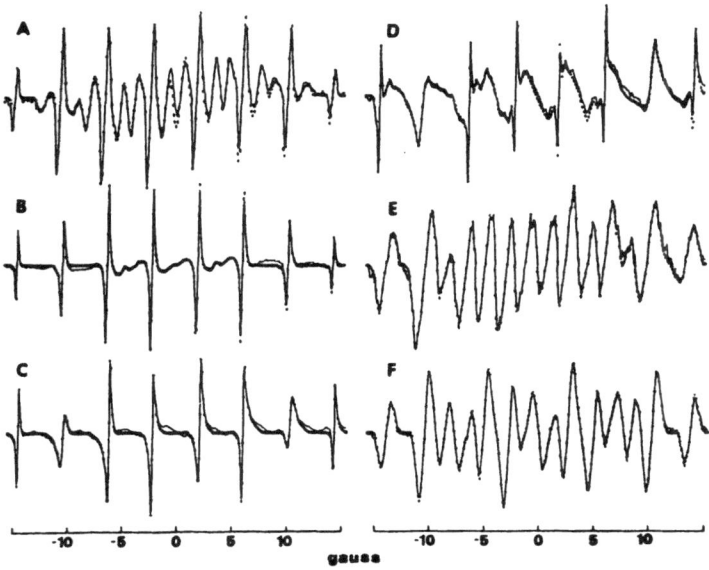

Figure 5 - Simulated (solid lines) and experimental (points) ESR spectra of 1,3,5-trinitrobenzene⁻,Na⁺ in THF at different temperatures and NaBPh₄ concentrations. (A) t=85°C; C=0.300 M; (B) t=35°C; C=0.206 M. (C) t=20°C; C=0.032 M; (D) t=-30°C; C=0.032 M (slow motion); (E) t=-75°C; C=0.206 M (static); (F) t=-65°C; C=0 (pure ion pair)

These features allow an analysis of the spectra according to a pseudo-intramolecular exchange model (18,36-38). In the case of m-dinitrobenzene⁻,Na⁺, after addition of sodium tetraphenylborate (NaBPh₄) intermolecular sodium exchange was observed, characterized by an activation energy of E_a=3.69 kcal mol^{-1} in THF and E_a=4.02 kcal mol^{-1} in DME (36). A similar behaviour was observed for 3,5-dinitropyridine⁻,Na⁺ (E_a=4.9$\pm$ 0.2 and 5.3$\pm$0.1 kcal mol^{-1} in THF and DME respectively) (36), 1,3,5-trinitrobenzene (E_a=4.16$\pm$0.02 kcal mol^{-1} in THF) (18), and 3,5-dinitrobenzonitrile (E_a=3.23$\pm$0.04 kcal mol^{-1} in THF) (38). This intermolecular exchange process has also been observed with p-dinitrobenzene⁻,Na⁺ (37).

In this case a value of $E_a = 4.9 \pm 0.2$ kcal mol^{-1} in DME is evaluated, very close to the value found for m-dinitrobenzene. However a more significant difference is found for the $\Delta G^{\ddagger}$ (2.9$\pm$0.2 and 4.2$\pm$0.2 kcal mol^{-1} for p-dinitrobenzene and m-dinitrobenzene in DME). By considering that $\Delta G^{\ddagger} = 0$ for o-dinitrobenzene (37), the activation free energy for the intermolecular sodium exchange decreases in the following order: $\Delta G^{\ddagger}$ (m-dinitrobenzene$^{\bar{}}$) > $\Delta G^{\ddagger}$(p-dinitrobenzene$^{\bar{}}$) > $\Delta G^{\ddagger}$(o-dinitrobenzene$^{\bar{}}$). The different extent of solvation in the ion pair and in the transition state for the three isomers seems to be the rate determining factor of the dynamic process. Since the ion pair and the triple ion have different hyperfine constants, these parameters can be exploited to determine the enthalpy and entropy change for the formation of triple ions from the corresponding ion pairs; in the case of 3,5-dinitrobenzo-nitrile (38) and in the case of 1,3,5-trinitrobenzene (18) we have obtained $\Delta H^{\circ} = 3.30 \pm 0.06$ and 5.2 ± 0.1 kcal mol^{-1} and $\Delta S^{\circ} = 18.4 \pm 0.2$ and 29.5 ± 0.5 cal K^{-1} mol^{-1} respectively. According to the Fuoss-Jagodzinski model of ionic association (42) a smaller distance between the two interacting ions may be expected when the charge of the unpaired electron is distributed between more nitrogroups on the same organic fragment, probably due to different solvation extent.

A completely different behaviour has been observed when NaBPh$_4$ is added to solutions of ion pairs in which nitrogroups are linked to two different aromatic rings, as in the case of dinitrobenzophenone (43) and dinitrodiphenylethylene (39). No line-width alternation by intermolecular cation exchange was observed in the ESR spectra of these compounds. However, the hyperfine lines of the cation multiplet have disappeared, indicating that the cation is involved in the intermolecular process. Furthermore the splitting constant of the nitrogroup nitrogen decreases, as usually observed when triple ion formation occurs. An approximate evaluation of $\Delta H^{\circ} = 7.7$ kcal mol^{-1} and $\Delta S^{\circ} = 38.7$ cal K^{-1} mol^{-1}

is given (43) for the formation of the triple ion of the dinitrobenzo-phenone anion radical with two sodium cations. It is worthnoting that the alternation linewidth observed for the dinitrobenzophenone$^-$,Na$^+$ ion pair due to ring rotation (see next Section) disappears when a triple ion is formed. These findings suggest that the bond order C_1-C_α (between the carbonyl and the rotating rings) decreases upon triple ion for-mation, and the two rings can rotate freely. This fact also excludes that the second cation could be localized near the carbonyl group.

HINDERED INTERNAL ROTATIONS

The least-squares fitting to EPR spectra affected by linewidth broa-dening can provide information on the barriers to internal rotation in paramagnetic species. In this section we focus our attention on a few examples related to the rotation about the benzene-to-carbonyl bond of anion radicals generated by electrochemical and chemical reduction in solution.

A first group of investigated systems includes 3-nitrobenzamide, 4-nitrobenzamide and 3,5-dinitrobenzamide anion radicals generated by electrochemical reduction in N,N-dimethylformamide (DMF)(20). Free rotation of the amido group is deduced by the analysis of the EPR spectra of 3- and 4-nitrobenzamide anion radicals in the whole range of the examined temperatures. The EPR spectra of 3,5-dinitrobenzamide anion radical show pronounced line-width variations (figure 6) due to the out-of-phase modulation of two pairs of coupling constants (a_{2H}, a_{6H} and $a_{3N(NO2)}$, $a_{5N(NO2)}$).

The Arrhenius plot gives an activation energy of 3.31 kcal mol^{-1} for the rotation of the amido group. The different behaviour of the mononitro- and dinitro-benzamide anion radicals can be explained by the analysis of

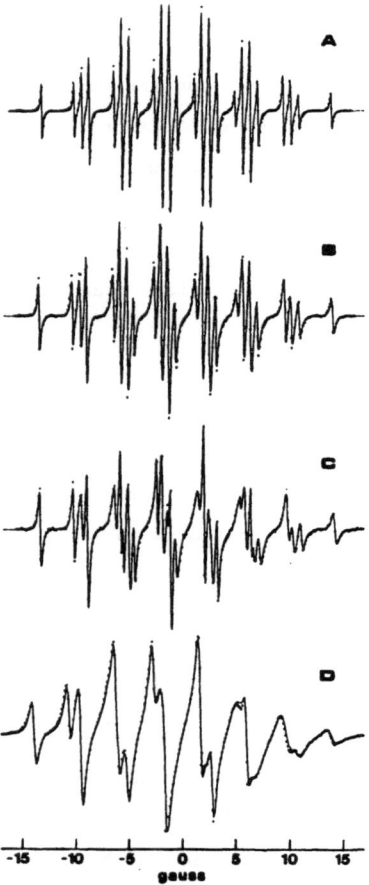

Figure 6 – Simulated (solid lines) and experimental (points) ESR spectra of 3,5-dinitrobenzamide radical anion in DMF at 20°C (A), -35°C (B), -55°C (C) and -85°C (D)

the viscosity-dependent line broadening, which is due to the modulation of the anisotropic nitrogroup nitrogen splitting in the tumbling anion radicals in solution. The barrier to internal rotation of the amido group in 3,5-dinitrobenzamide$^{-\cdot}$ is comparable to the activation energy for the rotational diffusion of the whole radical in the same solvent (3.20 kcal mol^{-1}). This fact allows us to conclude that the internal rotation of the amido group in 3,5-dinitrobenzamide$^{-\cdot}$ is controlled by solvent interactions, whereas in the remaining benzamide anion radicals is controlled by the intramolecular torsional potential (their rotational diffusion activation energies are quite close to that of the pure solvent).

Another group of radicals was generated by chemical and electrochemical reduction of benzophenone, 4,4'-dinitrobenzophenone and 4-nitrobenzophenone (33). The phenyl rings of both the free radicals and the ion pairs undergo hindered rotation detectable on the EPR time scale in the range of explored temperatures. The fit to the experimental spectra was obtained by assuming a four-site exchange model for the systems with a symmetrically delocalized spin distribution (benzophenone$^{-\cdot}$ and 4,4'-dinitrobenzophenone$^{-\cdot}$), while a two-site exchange model was adopted for the benzophenone nitro-derivative ion pairs, whose spin distribution is localized on one nitrophenyl ring only. The thermodynamic parameters which control the ring rotation of both the free and associated anion radicals are collected in Table I. The formation of ion pairs lowers the energy barrier to the rotation of the nitrophenyl group about the benzene-to-carbonyl bond. Furthermore the barrier height depends significantly on the strength of ionic association, as it increases with increasing the ionic radius of the cation. In the case of 4,4'-dinitrobenzophenone$^{-\cdot}$ the unexpected lower value of the enthalpy for the free anion with respect to that for the ion pair is understood by considering their very different electronic structure: in the free anion the odd electron is symmetrically delocalized over the whole radical whereas in the ion pairs the odd electron lies on one nitrophenyl ring only.

Nitrosubstituted benzaldehyde (25) and acetophenone (34) anion radicals are the last group of examined paramagnetic species. Para- and ortho-nitrosubstituted anion radicals show internal rotation controlled

TABLE I

Activation data of internal rotation and rotational diffusion

Compound	Solvent	Counterion	$\Delta G^{\ddagger}$ (kcal/m)	$\Delta H^{\ddagger}$ (kcal/m)	$\Delta S^{\ddagger}$ (cal/K.m)	$-E_a$ [a] (kcal/m)
3-nitro-benzamide	DMF	TBA				1.8 ± 0.1 [b]
4-nitro-benzamide	DMF	TBA				2.3 ± 0.8 [b]
3,5-dinitro-benzamide	DMF	TBA		3.31 ± 0.04		3.2 ± 0.2
	THF	Na				2.3 ± 0.3
benzophenone	DMF	TBA	[c] 5.00 ± 0.09	1.2 ± 0.2	-19 ± 1	2.2 ± 0.4
4-nitro-benzophenone	DMF	TBA	[c] 6.35 ± 0.03	5.2 ± 0.2	-5 ± 0.4	2.0 ± 0.2
4,4'-dinitro-benzophenone	DMF	TBA	[c] 5.8 ± 0.1	2.5 ± 0.4	-17 ± 2	
4-nitro-benzophenone	THF	Li	[c] 5.09 ± 0.06	5.6 ± 0.3	0	1.1 ± 0.1
	THF	Na	[c] 5.59 ± 0.04	4.9 ± 0.2	-4 ± 1	2.5 ± 0.1
	THF	K	[c] 5.29 ± 0.01	3 ± 1	-11 ± 7	1.7 ± 0.1
	THF	Cs	[c] 5.60 ± 0.06	2.4 ± 0.2	-16 ± 8	1.1 ± 0.1
4,4'-dinitro-benzophenone	THF	Li	[d] 6.34 ± 0.04	6.0 ± 0.1	-1.9 ± 0.5	1.2 ± 0.1
	THF	Na	[d] 6.65 ± 0.06	4.9 ± 0.6	-7 ± 2	1.9 ± 0.1
	THF	K	[d] 6.63 ± 0.06	4.9 ± 0.6	-8 ± 2	1.7 ± 0.1
	THF	Cs	[d] 6.57 ± 0.04	5.6 ± 0.6	-4 ± 2	1.5 ± 0.2
3-nitro-acetophenone [e]	DMF	TBA		4.8 ± 0.4		3.0 ± 0.6
	DME	Na		3.7 ± 0.6		3.0 ± 0.1
4-nitro-acetophenone	DMF	TBA		high		3.0 ± 0.2
	DME	Na		high		2.8 ± 0.2
3-nitro-benzaldehyde [e]	DMF	TBA	[f] 7.08 ± 0.01	6.01 ± 0.01	-3.63 ± 0.02	3.2 ± 0.3
	Me_2SO	TBA	[f] 6.47 ± 0.02			
	Me_2SO/t-BuOH	TBA	[f] 6.29 ± 0.03			
	MeOH	TBA	[f] 5.33 ± 0.02			
	H_2O	TBA	$\cong0$			
	DMF	Na	[f] 6.98	3.89	-10.40	3.3 ± 0.3

TABLE I (continued)

4-nitro-				
benzaldehyde	DMF	TBA	high	3.3±0.2
2-nitro-				
benzaldehyde	DMF	TBA	high	3.4±0.1
	THF	Na	high	2.5±0.1

(a) Determined by $\ln(B_N T)$ vs $1/T$ if not otherwise stated.
(b) Determined by $\ln(C_N T)$ vs $1/T$.
(c) At $-70°C$.
(d) At $-40°C$.
(e) The data refer to the overall kinetic process described by (25)

$$k_{obs} = k_{tc}^{-1} k_{ct}^{-1} (k_{ct} + k_{tc})^3$$

(f) At $25°C$.

---

by barrier heights too high to be detected on the EPR time scale. Spectral fittings by a static model confirm unequivocally this suggestion. On the contrary hindered rotation is deduced from the EPR spectra of the meta-nitrosubstituted anion radicals as they are affected by linewidth alternation. Activation enthalpies of 6.1 and 4.8 kcal mol^{-1} were obtained for 3-nitrobenzaldehyde and 3-nitroacetophenone anion radicals in DMF/TBA respectively. Ion pairing lowers these energies to 3.9 (DMF/Na) and 3.7 (DME/Na) kcal mol^{-1}, respectively. This effect was already observed for nitrobenzophenone anion radical (see Table I) (33). In the EPR spectra of 3-nitrobenzaldehyde$^{-\cdot}$ in DMF, the hfs patterns of two different paramagnetic species, shifted by $\Delta g = 8.6 \times 10^{-5}$, are easily recognized below 20°C and assigned to trans and cis rotational isomers. At higher temperatures averaged spectra are present, whose hfs patterns can be analyzed in terms of averaged hfs constants $\bar{a} = a_{cis} w_{cis} + a_{trans} w_{trans}$ (w_i = equilibrium population of isomer i). A complete picture of the energies involved in the internal rotation of the formyl group of 3-nitrobenzaldehyde radical anion has been obtained by the analysis of temperature dependence of both linewidth alternation and hyperfine coupling constants. It appears from the free energy profile that the cis isomer is 0.98 kcal mol^{-1} more stable than the trans isomer and that the barrier for the cis-to-trans conversion is 9.35 kcal mol^{-1}. On the other hand, the enthalphy profile indicates that the trans isomer is 1.13 kcal mol^{-1} less energetic than the cis isomer. Solvent interactions stabilize the cis form through a relevant entropy contribution to the free energy of isomerization ($\Delta S° = -7.1$ cal mol^{-1}K^{-1}). The activated complex is more favorable to efficient solvation and ionic interaction than both the cis isomer ($\Delta S^{\ddagger} = -18.6$ cal mol^{-1}K^{-1}) and the trans isomer ($\Delta S^{\ddagger} = -10.3$ cal mol^{-1}K^{-1}).

REFERENCES

1) (a) B.Johnson,T.Kuga and H.M.Gladney, IBM J.Res.Develop. **13**, 36 (1969).

(b) S.E.O'Conner,T.A.Spraggins and C.M.Grisham, Comp.Chem. **1**, 181 (1981).

(c) E.Klopfenstein,P.Jost and D.H.Griffith, in "Computers in Chemical and Biochemical Research 1", (C.E. Klopfenstein and L.L. Wilkins, Eds), Academic Press, New York, 1972, p. 176.

(d) T.Sasaki,Y.Kanoaka,T.Watanabe and S.Fujiwara, J.Magn.Reson. **38**, 385 (1980).

(e) N.G.Rudie,M.G.Mulkerrin and J.E.Wampler, Biochemistry **20**, 344 (1981).

(f) G.Grampp and C.A.Schiller, Anal.Chem. **53**, 560 (1981).

(g) E.S.Rich and J.E.Wampler, Int.Lab. 32 (1982).

(h) F.Momo,G.A.Ranieri and A.Sotgiu, Comp.Enh.Spectr. **1**, 79 (1983).

2) (a) A.Savitzsky and M.J.E.Golay, Anal.Chem. **36**, 1627 (1964).

(b) J.W.Hayes,D.E.Glover,D.E.Smith, and MacW.Overton, Anal.Chem.

45, 277 (1973).

3) T.A.Maldacker,J.E.Davis and L.B.Rogers, Anal.Chem. **46,** 637 (1974).

4) (a) A.E.Martin, Spectrochim.Acta **14,** 97 (1959).

    (b) J.R.Morrey, Anal.Chem. **40,** 905 (1968).

5) (a) R.N.Jones,R.Venkataragharan and J.W.Hopkins, Spectrochim.Acta **23A,** 925 (1967).

    (b) L.C.Allen,H.M.Gladney and S.H.Glarum, J.Chem.Phys. **48,** 4822 (1968).

    (c) Yu.G.Tkach,S.N.Dobryakov and Ya.S.Lebedev, Zh.Fiz.Khim. **51,** 1920 (1975).

6) T.Nishikawa and K.Someno, Anal.Chem. **47,** 1290 (1975).

7) G.A.Pearson, J.Magn.Reson. **27,** 265 (1976).

8) R.T.Vollmer and W.J.Caspary, J.Magn.Reson. **27,** 181 (1977).

9) D.W.Kirsme, J.Magn.Reson. **11,** 1 (1973).

10) K.D.Bieber and T.E.Gough, J.Magn.Reson. **21,** 285 (1976).

11) M.Barzaghi,A.Gamba,G.Morosi and M.Simonetta, J.Phys.Chem. **82,** 2105 (1978).

12) G.A.Pearson,M.Rocek and R.I.Walter, J.Phys.Chem. **82,** 1185 (1978).

13) (a) J.H.Freed and G.K.Fraenkel, J.Chem.Phys. **39,** 326 (1963).

    (b) G.K.Fraenkel, J.Phys.Chem. **71,** 139 (1967).

    (c) A.Hudson and G.R.Luckhurst, Chem.Rev. **69,** 191 (1969).

    (d) L.T.Muus and P.W.Atkins,Eds, "Electron Spin Resonance in Liquids", Plenum, New York, 1972.

    (e) P.D.Sullivan and J.R.Bolton, Advan.Magn.Reson. **4,** 39 (1975).

14) (a) J.Heinzer, QCPE **11,** 197 (1971).

    (b) Illustrative applications of the QCPE program ESRCON can be found in: F.Gerson,R.Gleiter,J.Heinzer and H.Behringer, Angew. Chem. **82,** 294 (1970); F.Gerson,J.Heinzer and E.Vogel, Helv. Chim.Acta **53,** 103 (1970); F.Gerson,G.Moshuk and M.Schwyzer, Helv.Chim.Acta **54,** 361 (1971); M.Barzaghi,A.Gamba,G.Morosi and M.Simonetta, J.Phys.Chem. **78,** 49 (1974).

15) J.Heinzer, J.Magn.Reson. **13**, 124 (1974).

16) (a) J.Heinzer, Mol.Phys. **22**, 167 (1971).

    (b) J. Heinzer, QCPE **11**, 209 (1972).

17) M.Barzaghi and M.Simonetta, J.Magn.Reson. **51**, 175 (1983).

18) M.Barzaghi,C.Oliva and M.Simonetta, J.Phys.Chem. **85**, 1799 (1981).

19) M.Barzaghi,P.L.Beltrame,A.Gamba and M. Simonetta, J.Am.Chem.Soc. **100**, 251 (1978).

20) M.Barzaghi,C.Oliva and M.Simonetta, J.Phys.Chem. **84**, 1959 (1980).

21) Least-squares bandshape analysis has been widely and successfully applied also to NMR spectra in recent years.

    (a) J.Heinzer, J.Magn.Reson. **26**, 301 (1977);

    (b) D.S.Stephenson and G.Binsch, J.Magn.Reson. **32**, 145 (1978);

    (c) E.E.Wille,D.S.Stephenson,P.Capriel and G.Binsch, J.Am.Chem. Soc. **104**, 405 (1982);

    (d) D.S.Stephenson and G.Binsch, J.Magn.Reson. **37**, 395 (1980);

    (e) D.S.Stephenson and G.Binsch, J.Magn.Reson. **37**, 409 (1980);

    (f) G.Binsch, in "Computational Methods in Chemistry", Plenum, New York, 1980, p.15;

    (g) D.S.Stephenson and G.Binsch, Mol.Phys. **43**, 697 (1981).

22) (a) K.Levenberg, Quart.Appl.Math. 2, 164 (1944).

    (b) D.W.Marquardt, J.Soc.Ind.Appl.Math. **11**, 431 (1963).

    (c) J.Simons,P.Jorgensen,H.Taylor and J.Ozment, J.Phys.Chem. **87**, 2745 (1983).

23) (a) M.Barzaghi,C.Oliva,A.Gamba and A.Saba, J.Chem.Soc. Perkin II, 1617 (1980).

    (b) M.Branca,A.Gamba,A.Saba,M.Barzaghi and M.Simonetta, J.Chem. Soc. Perkin II, 349 (1982).

24) W.C.Hamilton, Acta Crystallogr. **18**, 502 (1965).

25) M.Branca,A.Gamba,M.Barzaghi and M.Simonetta, J.Am.Chem.Soc. **104**, 6506 (1982).

26) The Chemical Society, Ed. "Electron Spin Resonance", London, Vol.

1-8.

27) S.Bonu,M.L.Lorenzetti and A.Gamba, Boll.Soc.Sarda Sci.Nat. **18**, 121 (1979).

28) M.Simonetta, Intern.Reviews in Phys.Chem. **1**, 31 (1981).

29) A.Gamba,C.Oliva and M.Barzaghi, Boll.Soc.Sarda Sci.Nat. **21**, 79 (1982).

30) P.Cremaschi,A.Gamba,G.Morosi,C.Oliva and M.Simonetta, J.Chem.Soc. Faraday II **71**, 1829 (1975).

31) N.M.Atherton and S.I.Weissman, J.Am.Chem.Soc. **83**, 1330 (1961).

32) A.Gamba,P.Cremaschi,G.Morosi,C.Oliva and M.Simonetta, Gazz.Chim. Ital. **106**, 337 (1976).

33) M.Barzaghi,S.Miertus,C.Oliva,E.Ortoleva and M.Simonetta, J.Phys. Chem. **87**, 881 (1983).

34) M.Branca,A.Gamba,C.Oliva and M.Simonetta, Chem.Phys. **75**, 253 (1983).

35) A.Gamba,C.Oliva and M.Simonetta, Chem.Phys.Letters **36**, 88 (1975).

36) M.Barzaghi,P.Cremaschi,A.Gamba,C.Oliva and M.Simonetta, J.Am. Chem. Soc. **100**, 3132 (1978).

37) M.Branca,A.Gamba and C.Oliva, J.Magn.Reson. **54**, 216 (1983).

38) M.Barzaghi,C.Oliva and M.Simonetta, J.Phys.Chem. **84**,1717 (1980).

39) C.Oliva, Istituto Lombardo (Rend.Sci.) **B 115**, 71 (1981).

40) M.Barzaghi,C.Oliva and M.Simonetta, unpublished results.

41) N.M.Atherton,P.M.Blustin,C.A.Humphreys and A.S.Shalaby, Org.Magn. Reson. **22**, 456 (1984).

42) (a) R.M.Fuoss and C.A.Krauss, J.Am.Chem.Soc. **55**, 2387 (1933).
    (b) P.Jagodzinski and S.Petrucci, J.Phys.Chem. **78**, 917 (1974).

43) C.Oliva, Spectrochim.Acta **39A**, 191 (1983).

PBB, Vol. 2
Advanced Magnetic Resonance Techniques
in Systems of High Molecular Complexity
(c) 1986 Birkhäuser Boston, Inc.

PHOTOREACTIVITY OF AROYL SILANES AS STUDIED BY ESR SPECTROSCPY

A.Alberti[a] and G.F.Pedulli[b]

[a]C.N.R., Istituto dei Composti del Carbonio Contenenti

Eteroatomi e loro Applicazioni, 40064 OZZANO EMILIA, Italy

[b]Università di Cagliari, Istituto di Chimica Organica

09100 CAGLIARI, Italy

The photoreactivity of acyl and aroylsilanes has been matter of debate for a long time, the point being whether their reactivity pattern is dictated by homolysis of the bond between silicon and the carbonyl carbon, or by initial 1,2 carbon to oxygen silicon shift with formation of a siloxycarbene.

Thus the formation of solvent derived products upon photolysis of acylsilane in carbon tetrachloride has been explained by admitting a Norrish I cleavage of the silicon-acyl bond with formation of $R_3Si^{\cdot}$ and $R\overset{\cdot}{C}O$ radicals which may subsequently react with solvent (1). Evidence for the formation of silicon radicals in the photodecomposition of acylsilanes was also obtained through product analysis (2,3) and spectroscopic studies (4).

On the other hand the formation of insertion products upon photolysis of these compounds in alcoholic solutions has been postulated to occur via a siloxycarbene intermediate (2,5), as outlined in the following reaction sequence:

464

$$R_3SiC(O)R' \xrightarrow{h\nu} \left[R_3SiC(O)R'\right]^* \longrightarrow R_3SiOCR' \xrightarrow{R''OH} R_3SiOCH(OR'')R'$$

We have carried out an ESR investigation of the photoreaction of benzoyltrimethylsilane and benzoyltriphenylsilane in hydrocarbon solvents with a number of phosphorus compounds: the results indicate that initial homolysis of the aroyl-silicon bond satisfactorily explains the formation of the observed radical intermediates.

Table I

ESR spectral parameters for radical adducts of general structure Ph(X)ĊOY (Coupling constants in Gauss)

X	Y	a_{Ho}	a_{Hm}	a_{Hp}	$a13_C$	a_X	a_Y	g
SiPh$_3$	SiPh$_3$	3.88	1.45	4.44	27.08	10.81*	6.79*	2.0030
SiPh$_3$	P(O)Ph$_2$	4.35	1.49	5.04	---	10.15*	32.13§	2.0030
SiMe$_3$	P(O)Ph$_2$	4.47	1.52	5.10	---	---	32.26§	2.0029
SiPh$_3$	P(O)(OEt)$_2$	4.49	1.58	5.22	---	---	35.01§	2.0029
SiMe$_3$	P(O)(OEt)$_2$	4.54	1.56	5.23	---	---	34.66§	2.0029
P(O)(OEt)$_2$	SiPh$_3$	4.16	1.47	4.83	---	27.53§	---	2.0029
P(O)(OEt)$_2$	SiMe$_3$	4.26	1.44	4.82	---	27.62§	7.14*	2.0029
P(O)(OEt)$_2$	P(O)(OEt)$_2$	4.31	1.42	4.91	---	26.90§	14.54§	2.0029

*coupling with ^{29}Si; §coupling with ^{31}P.

In agreement with previous results, photolysis of benzene solutions of benzoyltriphenylsilane led to the detection of the self-adduct, Ph(Ph$_3$Si)ĊOSiPh$_3$, whose ESR spectral parameters are reported in Table I.

When however photolysis was carried out in the presence of diphenyl-phosphine oxide or diethylphosphite and di-tert-butylperoxide, different radical species were observed, whose ESR spectra showed coupling of the unpaired electron with a phosphorus nucleus (see Table I). These radicals have been assigned the general structure $Ph(R_3Si)\dot{C}OP(O)(R')$, where R=Ph,Me and R'=Ph or OEt. Their formation is explained with the reaction sequence outlined below, where photolytically generated butoxyl radicals

$$^tBuOO^tBu \xrightarrow{h\nu} 2\ ^tBuO^\cdot$$

$$^tBuO^\cdot + HP(O)(R')_2 \longrightarrow {}^tBuOH + {}^\cdot P(O)(R')_2$$

$$PhC(O)SiR_3 + {}^\cdot P(O)(R')_2 \longrightarrow Ph(R_3Si)\dot{C}OP(O)(R')_2$$

react with the phosphorus compounds to give phosphonyl radicals which then add to the carbonyl group of the benzoylsilane. The 10.15 G value of the ^{29}silicon splitting determined for the adduct with R=R'=Ph (value typical of a silicon α to a radical centre) (6) as well as the marked temperature dependence of the ^{31}P splitting (typical of a nucleus ß to a radical centre) leave little doubt about the correctness of the assignment. As the adducts $Ph(R_3Si)\dot{C}OP(O)(R')_2$ were the only detectable species, it appears that the 1,2 silicon shift, that is the formation of a siloxycarbene, is not important under these conditions.

When benzoylsilanes were photolysed in the presence of tetraethyl-pyrophosphite, three different radical species were detected, whose relative amounts depended on temperature, concentration of pyrophosphite, and time of photolysis. Two of these species were formed as soon as irradiation was started: one could be unambiguosly identified as the self-adduct on the basis of its spectral parameters, while the other, which showed coupling of the unpaired electron with the protons of a phenyl ring and with a phosphorus atom, was assigned structure (A) because i) the spectral parameters were similar to those of the adduct observed by Neumann and coworkers (7) upon addition of silyl radicals to the phenyl diethoxyphosphonyl ketone, ii) the silicon splitting is typical of a silicon atom β to the radical centre, and iii) the phosphorus splitting is practically temperature independent, as expected for a nucleus α to a radical centre.

$$
\begin{array}{cc}
\text{Ph} \diagdown \underset{\overset{|}{\text{OSiR}_3}}{\overset{\bullet}{\text{C}}} \diagup \text{P(O)(OEt)}_2 & \text{Ph} \diagdown \underset{\overset{|}{\text{OP(OEt)}_2}}{\overset{\bullet}{\text{C}}} \diagup \text{P(O)(OEt)}_2 \\
\text{(A)} & \text{(B)}
\end{array}
$$

The species detected after prolonged photolysis showed the coupling of the unpaired electron with two different phosphorus atoms, and was assigned structure (B).

In view of the photolability of aroylsilanes, we believe that the following reaction sequence is adequate to account for the formation of all the observed species:

$$
\begin{array}{l}
\text{ArC(O)SiR}_3 \xrightarrow{\ h\nu\ } \text{Ar}\overset{\bullet}{\text{C}}\text{O} + {}^{\bullet}\text{SiR}_3 \\
\text{Ar}\overset{\bullet}{\text{C}}\text{O} + \big[\text{(EtO)}_2\text{P}\big]_2\text{O} \longrightarrow \text{ArC(O)P(O)(OEt)}_2 + {}^{\bullet}\text{P(OEt)}_2 \\
{}^{\bullet}\text{SiR}_3 + \text{ArC(O)SiR}_3 \longrightarrow \text{Ar(R}_3\text{Si})\overset{\bullet}{\text{C}}\text{OSiR}_3 \\
{}^{\bullet}\text{SiR}_3 + \text{ArC(O)P(O)(OEt)}_2 \longrightarrow \text{Ar}\big[\text{(EtO)}_2\text{(O)P}\big]\overset{\bullet}{\text{C}}\text{OSiR}_3 \\
{}^{\bullet}\text{P(OEt)}_2 + \text{ArC(O)P(O)(OEt)}_2 \longrightarrow \text{Ar}\big[\text{(EtO)}_2\text{(O)P}\big]\overset{\bullet}{\text{C}}\text{OP(OEt)}_2
\end{array}
$$

This scheme gains additional support by the fact that the spectrum observed by photolysis of solutions of phenyl tert-butyl ketone in the presence of pyrophosphite was identical with that of radical (B). The formation of benzoyl and silyl radicals was further evidentiated by trapping these two species with suitable scavengers; thus when 2-methyl-2-nitrosopropane was added to the system undergoing photolysis, the overimposed spectra due to benzoyl tert-butyl nitroxide and di-tert-butyl nitroxide were observed, whereas 3,5-dinitrophenyl triphenylsilyloxy nitroxide was the only observable species upon addition of 1,3,5-trinitrobenzene.

In the light of these results we may safely conclude that the homolytic cleavage of the benzoyl-silicon bond represents the key step in the photochemistry of benzoylsilanes under the experimental conditions employed.

REFERENCES

1) A.G.Brook,P.J.Dillon and R.Pearce, Can.J.Chem. **49**, 133 (1971).

2) A.G.Brook and J.M.Duff, Can.J.Chem. **51**, 352 (1973).

3) A.G.Brook and J.M.Duff, J.Amer.Chem.Soc. **89**, 454 (1967).

4) A.G.Brook, J.W.Harris, J.Lennon and M.El Sheiker, J.Amer.Chem.Soc. **101**, 83 (1979).

5) J.M.Duff and A.G.Brook, Can.J.Chem. **51**, 2869 (1973).

6) A.Alberti, G.Seconi, G.F.Pedulli and A.Degl'Innocenti, J.Organomet. Chem. **253**, 291 (1983).

7) W.Schulten and W.P.Numann, in "Landolt-Börnstein", Springer-Verlag, Berlin, 1977; series II/9b.

PBB, Vol. 2
Advanced Magnetic Resonance Techniques
in Systems of High Molecular Complexity
©️ 1986 Birkhäuser Boston, Inc.

FIELD-SWEPT AND FREQUENCY-SWEPT EPR SPECTRA FOR SPIN ½:

COMPUTER SIMULATION IN THE PRESENCE OF g AND A STRAIN

S.Cannistraro and G.Giugliarelli

Gruppo di Biofisica Molecolare

Dipartimento di Fisica dell'Università

06100 Perugia (Italy)

INTRODUCTION

Randomly oriented tetragonal complexes of copper ions $(S=½)$ show in the low field region ($g_{\parallel}$ region) of their Electron Paramagnetic Resonance (EPR) spectrum, four lines centered at $g_{\parallel}$ and separated by $A_{\parallel}$ which arise from the hyperfine coupling to the copper nucleus ($I=3/2$). The width of these lines has been reported to exhibit a dependence on the nuclear quantum number, m_I, of copper (1-4) and on the microwave frequency (1). Moreover we found the existence of an unequal spacing between the four parallel hyperfine lines, occurring to a variable extent from complex to complex (5). Such an effect was particularly evident in the EPR spectrum of the Cu^{++}-tRNA complex where an appreciable shift of the $m_I=½$ hyperfine lines was observed (6). To take into account the dependence of both the width and the resonant field position of the parallel hyperfine lines, we have recently developed a theoretical model for the simulation of the EPR spectra of copper complexes (5). The model, which assumes a Gaussian distribution of $g_{\parallel}$ and $A_{\parallel}$ values as due to strain occurring during freezing, has been successfully used to obtain the best fit of several X-band tetragonal copper spectra arising from complexes in the frozen state. The quality of the fits (as estimated by the chi-square test) was found to be greatly improved as compared with the case where the simulation model does not take account the two above mentioned effects. However, the theoretical EPR lineshape used in our method (as in most of the simulation methods present in literature) was expressed following a formalism derived from frequency-swept spectra while, in actual fact, field-swept EPR is a fixed frequency form of spectroscopy. This problem and all the consequences connected has very recently been raised by Pilbrow (7). In his very perceptive paper, Pilbrow demonstrated that the following expression should be used for a correct simulation of $S=\frac{1}{2}$ EPR absorption spectra:

$$S(\nu,H) = C\nu f\left[(\nu-\nu_o)^2,\sigma_\nu\right] \tag{1}$$

where S expresses the net absorption of microwave radiation, C is a constant encompassing all the instrumental parameters and the transition probability and $f\left[(\nu-\nu_o)^2,\sigma_\nu\right]$ represents the frequency-swept lineshape. Eq.(1) is to be preferred to the currently used expression:

$$S(\nu_c,H) = Ch\nu_c/\beta g \cdot f\left[(H-H_o)^2,\sigma_H\right] \tag{2}$$

where $f\left[(H-H_o)^2,\sigma_H\right]$ is the field-swept lineshape. Use of Eq.(1) may also take account of lineshape asymmetries deriving from the presence of strain (7). In this paper we present a reformulation of our theoretical model in the light of Pilbrow's paper. Moreover, we report some applications of the simulation model to the best fitting of some experimental Cu^{++}-complex EPR spectra, the results obtained being discussed in connection with previous results.

THEORETICAL MODEL

If we take into account eq.(1), the frequency-swept EPR absorption spectrum of randomly oriented axially symmetric copper complexes ($S=\frac{1}{2}$) can be generated by an expression of the type (8,9):

$$S(\nu,H) = N\nu \int_0^{\pi/2} g_1^2 f\left[(\nu-\nu_o)^2, \sigma_\nu^R\right] \sin\vartheta d\vartheta \qquad (3)$$

where:

$$f\left[(\nu-\nu_o)^2, \sigma_\nu^R\right] = (\sqrt{2\pi}\sigma_\nu^R)^{-1} \exp\left[-\tfrac{1}{2}((\nu-\nu_o)/\sigma_\nu^R)^2\right] \qquad (4)$$

is the Gaussian frequency-swept lineshape, σ_ν^R indicating the half peak-peak width of the first derivative of f and is mainly determined by the unresolved coupling of electronic spin to ligand nuclei (2); g_1^2 expresses the orientation dependence of the transition probability (8); ϑ is the angle between the magnetic field direction and the molecular axis and the normalization constant N takes account of the instrumental parameters.

The resonant frequencies of the copper EPR lines are obtained by solving to the second perturbative order the spin Hamiltonian:

$$\mathcal{H} = \beta\left[g_{\parallel}H_z S_z + g_{\perp}(H_x S_x + H_y S_y)\right] + A_{\parallel}I_z S_z + A_{\perp}(I_x S_x + I_y S_y) \qquad (5)$$

where nuclear Zeeman and quadrupole terms are neglected, then:

$$\nu_o = h^{-1}\left[g\beta H + Km_I + K'/g\beta H\right] \qquad (6)$$

where

$$K' = A_{\perp}^2(A_{\parallel}^2 + K^2)/4K^2\left[I(I+1) - m_I^2\right] + g_{\parallel}^2 g_{\perp}^2/2K^2 g^4 \cdot (A_{\parallel}^2 - A_{\perp}^2)2\sin^2\vartheta\cos^2\vartheta\, m_I^2 \qquad (7)$$

$$g^2 = g_{\parallel}^2\cos^2\vartheta + g_{\perp}^2\sin^2\vartheta \qquad (8)$$

$$K^2 g^2 = g_{\parallel}^2 A_{\parallel}^2 \cos^2\vartheta + g_{\perp}^2 A_{\perp}^2 \sin^2\vartheta \qquad (9)$$

It should be noted that K bears the same sign as A in eqs.(6) to (9). Now, in a general way, we make the hypothesis that strains occurring during freezing give rise to a Gaussian distribution of the spin Hamiltonian parameters λ_i which appear in eq.(6). This will result in a fluctuation $\Delta\nu_0$ of the resonant frequency around ν_0. Under the hypothesis of small changes, $\Delta\lambda_i$, of the λ_i parameters this fluctuation can be expressed as

$$\Delta\nu_0 = \sum_i (\partial\nu_0/\partial\lambda_i)\Delta\lambda_i \qquad (10)$$

According to some authors (1,2) we have taken into account only the fluctuations, $\Delta g_{\parallel}$ and $\Delta A_{\parallel}$, in the parameters $g_{\parallel}$ and $A_{\parallel}$. On the other hand, inclusion in the model of fluctuations in $g_{\perp}$ and $A_{\perp}$ resulted in only a very small change in the computer spectra (5). Then eq.(10) takes the form:

$$\Delta\nu_0 = (\partial\nu_0/\partial g_{\parallel})\Delta g_{\parallel} + (\partial\nu_0/\partial A_{\parallel})\Delta A_{\parallel} \qquad (11)$$

To take into account the strain distribution of $g_{\parallel}$ and $A_{\parallel}$ spin Hamiltonian parameters in the generation of the EPR absorption spectrum eq.(11) must be integrated over these distributed values:

$$S_{strain}(\nu,H) = \int S(\nu,H)F(g_{\parallel},A_{\parallel})dg_{\parallel}dA_{\parallel} \qquad (12)$$

where $F(g_{\parallel},A_{\parallel})$ is the probability density function for the random distribution of $g_{\parallel}$ and $A_{\parallel}$.

In a general way, we have chosen a bivariate normal density function for F of the form:

$$F(g_{\parallel}, A_{\parallel}) = \left(2\pi\sigma g_{\parallel}\sigma A_{\parallel}(1-\varrho^2)^{\frac{1}{2}}\right)^{-1}.$$

$$\cdot\exp{-\tfrac{1}{2}(1-\varrho^2)^{-1}\left[((g-g_{\parallel}^o)/\sigma g_{\parallel})^2 - 2\varrho(g-g_{\parallel}^o)(A-A_{\parallel}^o)/\sigma g\sigma A_{\parallel} + ((A-A_{\parallel}^o)/\sigma A)^2\right]} \tag{13}$$

where $(\sigma g_{\parallel})^2$ and $(\sigma A_{\parallel})^2$ are the variances of the Gaussian distributions of $g_{\parallel}$ and $A_{\parallel}$ around the mean values $g_{\parallel}^o$ and $A_{\parallel}^o$; ϱ being a correlation coefficient $-1 \leqslant \varrho \leqslant 1$.

Recalling eqs.(3) and (4) and integrating eq.(12) we obtain:

$$S_s(\nu, H) = N\nu \int_0^{\pi/2} g_1^2 f\left[(\nu - \nu_o)^2, \sigma_\nu^T\right]\sin\vartheta d\vartheta \tag{14}$$

where

$$(\sigma_\nu^T)^2 = (\sigma_\nu^R)^2 + (\sigma_\nu^S)^2 \tag{15}$$

and

$$(\sigma_\nu^S)^2 = (\partial\nu_o/\partial g_{\parallel})^2(\sigma g_{\parallel})^2 + (\partial\nu_o/\partial A_{\parallel})^2(\sigma A_{\parallel})^2 + 2\varrho(\partial\nu_o/\partial g_{\parallel})(\partial\nu_o/\partial A_{\parallel})\sigma g_{\parallel}\sigma A_{\parallel} \tag{16}$$

Strictly speaking eq.(9) was obtained in the present form under the hypothesis that the fluctuation of g_1^2, as due to its dependence on $g_{\parallel}$ (9), is negligible and by assuming that the distribution of $g_{\parallel}$ and $A_{\parallel}$ are not orientation-dependent. Eq.(16) expresses the additional linewidth introduced by the strain-induced random distribution of $g_{\parallel}$ and $A_{\parallel}$. Considering eq.(6), this last term can be expressed as a power expansion of m_I and H as follows:

$$(\sigma_\nu^S)^2 = (g\beta/h)^2(Am_I^2 + Bm_I H + CH^2) \tag{17}$$

where:

$$A=1/\beta^2 K^2 g^2 \left[(A_\parallel^2-K^2)^2 \sigma g_\parallel^2 + A_\parallel^2 g_\parallel^2 \sigma A_\parallel^2 + 2\varrho(A_\parallel^2-K^2)A_\parallel g_\parallel \sigma g_\parallel \sigma A_\parallel\right] g_\parallel^2 \cos^4\vartheta/g^4 \quad (18)$$

$$B = -2/\beta Kg \cdot \left[(A_\parallel^2-K^2)\sigma g_\parallel^2 + \varrho A_\parallel g_\parallel \sigma g_\parallel \sigma A_\parallel\right] g_\parallel^2 \cos^4\vartheta/g^4 \quad (19)$$

$$C = \sigma g_\parallel^2 g_\parallel^2 \cos^4\vartheta/g^4 \quad (20)$$

Eq.(17) shows two important facts the first of which is that the strain-induced additional line broadening is dependent on m_I. Such an analytical dependence underlies the explanation of the observed m_I dependence of the hyperfine linewidth. On the other hand the line broadening is dependent on the magnetic field value and this necessarily leads to an EPR line asymmetry when strain effects are present. This point has been analysed only qualitatively by other authors (3,4).

Finally to compute the field swept EPR spectra which are usually recorded at a fixed frequency ν_c and displayed as the first derivative of the absorption, it is necessary to differentiate eq.(14), taken at the frequency ν_c, with respect to H, after having expressed the resonant frequency ν_0 as a function of H through eq.(16). The following final expression is thus obtained:

$$I_S(H_i) = dS_S(\nu_c,H)/dH = N\nu_c h/\beta \int_0^{\pi/2} g_1^2/g(\sigma_H^T)^3 \left\{\left[1-K'/(g\beta H)^2\right](H_0'-H) + \right.$$
$$+\left[((H-H_0')^2/2(\sigma_H^T)^2)-\tfrac{1}{2}\right](Bm_I+2CH)\bigg\} \cdot$$
$$\cdot \exp\left[-(H-H_0')^2/2(\sigma_H^T)^2\right]\sin\vartheta d\vartheta \quad (21)$$

where:

$$H_0' = (1/g\beta)(h\nu_c - Km_I - K'/g\beta H) \quad (22)$$

and

$$\sigma_H^T = h\sigma_\nu^T/g\beta \tag{23}$$

In this expression the above mentioned field dependence of σ_ν^S results in a non-zero $\partial\sigma_H^T/\partial H$ term, hence in an overall asymmetry of the total lineshape. We may therefore observe the appearance of the g_1^2/g ratio for the orientation-dependent transition probability. Such a ratio was qualitatively introduced by Aasa and Vanngard (10) in the simulation of $S=\frac{1}{2}$ EPR powder-like spectra but only in Pilbrow's paper is its analytical derivation provided.

MATERIALS AND METHODS

All chemicals used were of analytical reagent grade and where used without additional purification. [65]Cu was purchased from the Oak Ridge National Laboratories, digested in HCl and diluted in phosphate buffer. EPR sample preparation was performed according to the procedure reported in ref.(9). EPR spectra were recorded by an X-band Varian E-109 spectrometer at 77 K. To calculate the experimentally observed g-factors, a magnetic field calibration was performed with a Magnion Precision NMR Gaussmeter Mod. 6-502, the microwave frequency being measured with a Marconi 2440 counter. EPR data acquisition was performed on an HP 86A personal computer, through a home-made interface connected to an IEEE-488 bus. To run simulation and best-fit programs, the same microcomputer was switched to an intelligent terminal of the main frame computer (a Prime 550/1) through an RS-232C serial interface and an HP terminal emulator. The chi-square test was used to evaluate the goodness of fits utilizing the expression:

$$\chi^2 = \sum_{i=1}^{N} \left| (I^{exp}(H_i) - I_S(H_i))/\sigma_i \right|^2 \tag{24}$$

where $I^{exp}(H_i)$ is the derivative of the experimental EPR absorption spectrum sampled at N=160 discrete points of the field, $I_S(H_i)$ is the model function of eq.(21), and σ_i is the standard deviation calculated for the i-th experimental point of the EPR spectrum by repeated runs (9).

RESULTS AND DISCUSSION

Contrary to our previous work, where we were mainly interested in the simulation of experimental spectra arising from copper complexes normally containing a mixture of the two copper isotopes (^{63}Cu and ^{65}Cu, as in native copper proteins), in the present study we have only used ^{65}Cu in the preparation of the paramagnetic complexes whose low temperature EPR spectra were to be fitted. This resulted in an improved resolution of the hyperfine lines. In the course of the present study we simulated the

---

Table I

Spin-Hamiltonian parameters used for the simulation of
EPR spectra b and c of fig.1 (hyperfine couplings and
linewidths are expressed in gauss)

	$g_\parallel$	$g_\perp$	$A_\parallel$	$A_\perp$	$\sigma_{H_\parallel}^R$	$\sigma_{H_\perp}^R$	$\sigma g_\parallel$	$\sigma A_\parallel$	ϱ
spectrum b	2.394	2.073	127.0	9.0	8.0	18.0	0.018	8.5	1.0
spectrum c	2.394	2.073	127.0	9.0	8.0	20.8	0.018	8.5	1.0

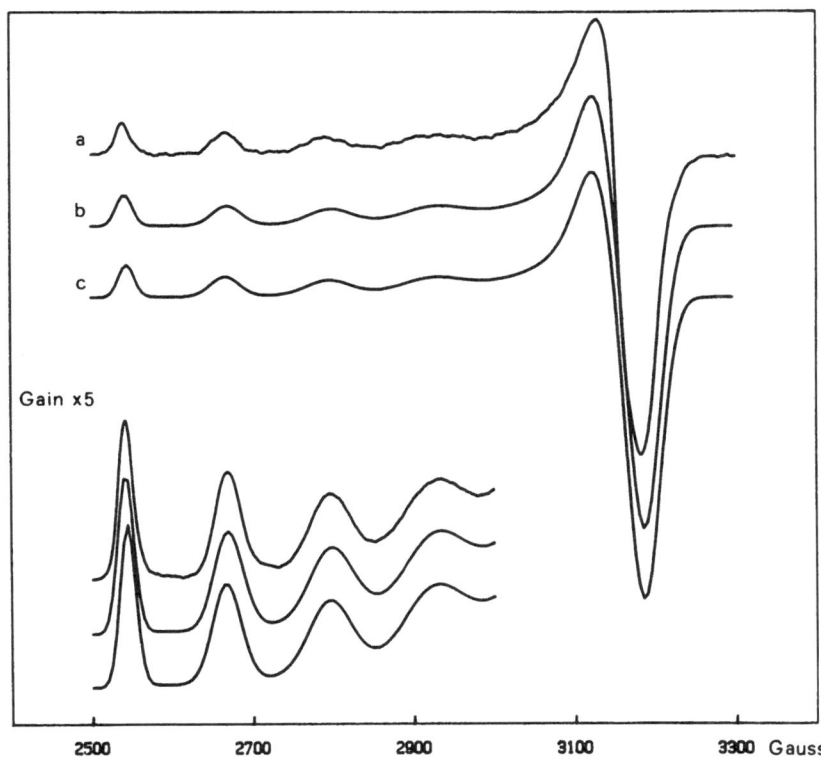

Figure 1 – Experimental (a) and computer-synthesized ((b) and (c)) EPR spectra of aqueous solution of $^{65}Cu^{++}$ (1mM) in the presence of an excess sodium triphosphate, pH=7. The experimental spectrum was recorded at 77 K, the other spectrometer settings being: microwave frequency=9.090 GHz, microwave power =20mW, magnetic field sweep rate = 2000 Gauss in 8 min, time constant = 0.5 s, modulation amplitude = 5 G. Spectra (b) and (c) were simulated with the parameters reported in Table I.

X-band EPR spectra of several copper complexes having different ion ligand sets. Figure 1 shows an example of these simulations performed with the theorical model presented in this paper. Pattern a shows the experimental EPR spectrum, recorded at 77K, of an aqueous solution of $^{65}Cu^{++}$ ions in the presence of sodium triphosphate at pH=7. We chose this spectrum because strain effects on hyperfine lines are more evident when oxygen atoms are coordinated to the paramagnetic ion (2). In fact, m_I-dependent parallel hyperfine line broadening is clearly observed; moreover a careful examination of the $g_{||}$ region (see insert of fig.1) shows a slight increase in the spacing between pairs of adjacent lines toward higher fields. These two effects could not be simulated by using the ordinary simulation methods (see ref (5) and references cited therein). Pattern b of fig.1 represents the computer generated spectrum which best fits the experimental one. This spectrum has been obtained through the use of eq.(21) provided that ϱ is taken equal to 1. As can be seen the agreement between the two patterns, a and b, is very good; moreover the χ^2 value calculated was 200, indicating a very good fit. As a comparison, curve c shows the computer synthesized spectrum which again best fits the experimental one but this time arising from our previous simulation model and essentially generated through eq.(14) of ref.(5). It should be noted that no appreciable differences are observed between pattern b and c and also that the goodness of the fits was essentially the same. In table I we report the spin Hamiltonian parameters used in the simulation of spectra b and c shown in fig.1. All the parameters are essentially the same except for σ_H^R which takes a lower value in the $g_\perp$ region ($\sigma_{H\perp}^R$) in the synthesis of spectrum b. This could however be inferred from a comparison of the relationship expressing the angular dependence of the residual linewidth (in field swept domain; i.e. eq.(9) of ref.(5) with the equivalent expression in the present paper, eq.(23).

As in the previous work most of our experimental spectra could be simulated by putting $\varrho=1$ in eq.(13). Then again it can be shown that (5):

$$g_{||} = aA_{||} + b \qquad (25)$$

and all the considerations made in that occasion hold in this case.

REFERENCES

1)   W.B.Lewis,M.Alei and L.D.Morgan, J.Chem.Phys. **44**, 2469 (1966).

2)   W.Froncisz and J.S.Hyde, J.Chem.Phys. **73**, 3123 (1980) and referen-
     ces therein.

3)   W.R.Hagen, J.Magn.Reson. **44**, 447 (1981).

4)   A.S.Brill, "Transition Metals in Biochemistry", Springer Verlag,
     Berlin, 1977.

5)   G.Giugliarelli and S.Cannistraro, Chem.Phys. **98**, 115 (1985).

6)   S.Cannistraro,G.Giugliarelli,G.Onori and E.Rongoni, Biophys.Chem.
     22, 107 (1985).

7)   J.R.Pilbrow, J.Magn.Reson. **58**, 186 (1984).

8)   A.Abragam and B.Bleaney, "EPR of Transition Ions", Oxford Univ.
     Press, London, 1970.

9)   G.Giugliarelli and S.Cannistraro, Il Nuovo Cimento **4D**, 194 (1984).

10) R.Aasa and T.Vanngard, J.Magn.Reson. **19**, 308 (1975).

PBB, Vol. 2
Advanced Magnetic Resonance Techniques
in Systems of High Molecular Complexity
© 1986 Birkhäuser Boston, Inc.

MAGNETIC RESONANCE LINESHAPES FOR PROBES UNDERGOING

NON-MARKOFFIAN MOTIONS

U. Segre

Dipartimento di Chimica Fisica

Università di Padova

Via Loredan 2, 35131 Padova

Molecular reorientation in liquids is usually tested by measuring or computing the correlation function of some angular dependent molecular property $f(\Omega)$. If the orientational equilibrium distribution function is $p(\Omega)$, the time correlation function of f is given by:

$$g(t) = <f^*(t)f(0)>$$
$$= <f\ p^{\frac{1}{2}}|\exp(-\hat{R}t)|p^{\frac{1}{2}}\ f> \qquad (1)$$

(we assume $<f> = 0$). $\hat{R}$ is the symmetrized diffusion operator (1):

$$\hat{R} = p^{-\frac{1}{2}}\ \underset{\sim}{\nabla}_\Omega\ \underset{\sim}{Dp}\ \underset{\sim}{\nabla}_\Omega\ p^{-\frac{1}{2}} \qquad (2)$$

for the sake of compactness we define:

$$|u> = p^{\frac{1}{2}}|f>$$
$$|v(t)> = \exp(-\hat{R}t)\ |u> \qquad (3)$$
$$|z(\omega)> = \int_0^\infty dt\ \exp(-i\omega t)\ |v(t)>$$

The Fourier-Laplace transform of the diffusion equation is:

$$(i\omega + \hat{R})\ |z(\omega)> = |u> \qquad (4)$$

Using these notations, the correlation function $g(t)$ and its spectral density $j(\omega)$ are given by:

$$g(t) = <u|v(t)> \qquad (5)$$

$$j(\omega) = <u|z(\omega)> = <u|(i\omega+\hat{R})-^{1}|u> \qquad (6)$$

Molecular reorientation is a Markov process as long as the diffusion tensor is a constant. However, if $\underset{\sim}{D}$ fluctuates in time, the reorientation process is not longer Markoffian, but it is just the projection of a composite process which concerns the orientation of the molecule and its internal degrees of freedom.

Let us suppose that N different conformations are allowed for the molecules, with weights π_μ and diffusion tensors $\underset{\sim}{D}_\mu$. Very often the transition rate between different conformations is given by the (symmetrized) random jump expression:

$$\Gamma_{\mu\nu} = k \left\{ (\pi_\mu \pi_\nu)^{\frac{1}{2}} - \delta_{\mu\nu} \right\} \qquad (7)$$

Then Eq.(4) becomes:

$$(i\omega+\hat{R}_\mu)|z_\mu> - \Sigma_\nu \Gamma_{\mu\nu}|z_\nu> = \pi_\mu^{\frac{1}{2}}|u> \qquad (8)$$

Its solution is:

$$j(\omega) = \Sigma_\mu <u|\pi_\mu^{\frac{1}{2}}z_\mu(\omega)>$$

$$= <u|\hat{Q}(\omega)\left\{1-k\hat{Q}(\omega)\right\}^{-1}|u> \qquad (9)$$

where:

$$\hat{Q}(\omega) = \Sigma_\mu \pi_\mu (i\omega + k + \hat{R}_\mu)^{-1} \qquad (10)$$

The EPR lineshape can be obtained from the above expressions, by replacing f with the spin operator S_x and R_μ with $-iL+\hat{R}_\mu$. The absorption is given by:

$$I(\omega) = \text{Re } j(\omega) \qquad (11)$$

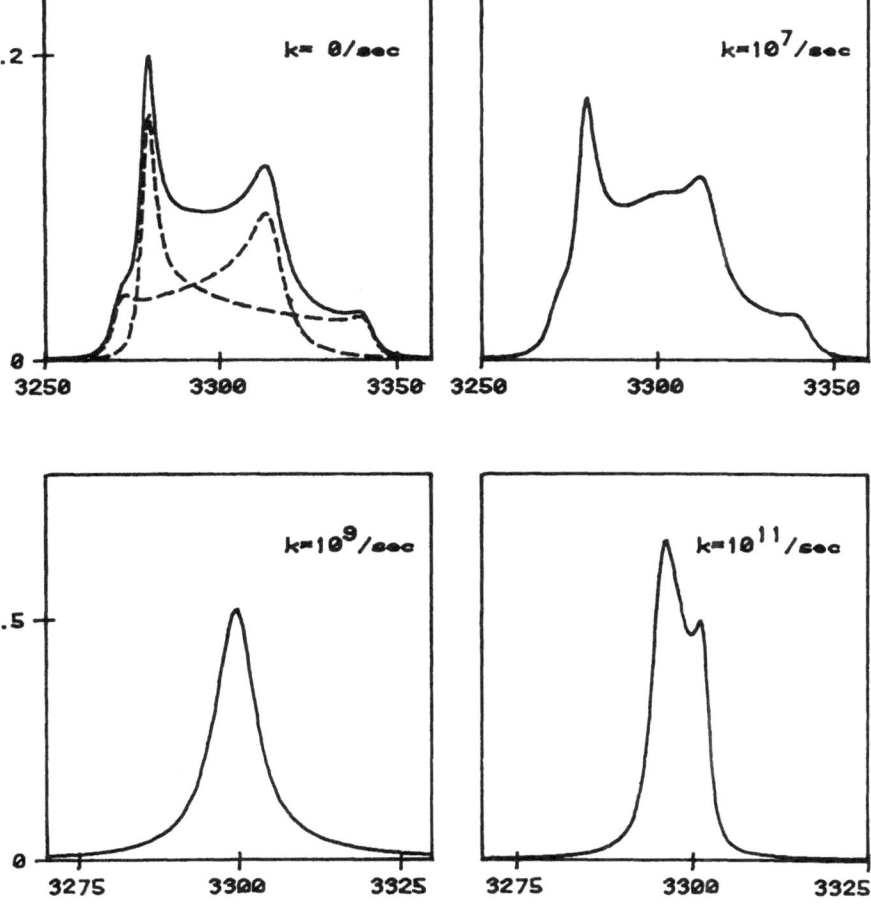

The EPR lineshapes for a probe with non axial g-tensor, exchanging between two molecular conformations, are shown in the figures. The Lanczos algorithm (3) has been used to solve the stochastic Liouville equation. The $\underset{\sim}{g}$ and $\underset{\sim}{D}_\mu$ tensor components used in the computations are given in the Table.

	$\underset{\sim}{g}$	$\underset{\sim}{D}_1 (\text{sec}^{-1})$	$\underset{\sim}{D}_2 (\text{sec}^{-1})$
x	1.972	2.5×10^6	5.0×10^8
y	2.008	2.5×10^6	2.5×10^6
z	2.020	5.0×10^8	2.5×10^6

REFERENCES

1) G.Moro,U.Segre and P.L.Nordio, in "Nuclear Magnetic Resonance of Liquid Crystals", (J.W. Emsley ed.), Reidel, (1985).

2) H.Sillescu, J.Chem.Phys. **54**, 2110 (1971).

3) G.Moro and J.H.Freed, J.Chem.Phys. **74**, 3757 (1981).

PBB, Vol. 2
Advanced Magnetic Resonance Techniques
in Systems of High Molecular Complexity
© 1986 Birkhäuser Boston, Inc.

ARRANGEMENT OF GUEST DONORS IN THE NAPHTHALENE-TCNB CRYSTAL

AN ENDOR STUDY

C.Corvaja,A.L.Maniero and L.Pasimeni

Physical Chemistry Department, University of Padova

Via Loredan, 2 - 35131 Padova (Italy)

INTRODUCTION

Crystals of weak CT complexes are frequently distinguished by the feature that there exist large degrees of freedom for in-plane librational
motions of donors. The Naphthalene (N)-1,2,4,5 Tetracyanobenzene (TCNB)
CT crystal consists of stacks of alternating donor and acceptor
molecules. While the TCNB molecules occupy a fixed position in the crystal lattice, N molecules are allowed to librate around the crystallographic c axis. Freezing-in of these librations is accompanied often by
phase transitions at 63 K. EPR and ENDOR data of X-traps in their first
excited triplet state at 4.2 K have shown that two inequivalent sites
are present in the crystal lattice which differ for the orientation of
the N molecules (1). The long in-plane axis of N forms an angle of +16°
with respect to the $\mathbf{a'}(\mathbf{a'}=\mathbf{b}\times\mathbf{c})$ crystallographic axis. Also the normal to
the N molecular plane is tilted by 23° with respect to the stacking axis
c. The N-TCNB crystal can accomodate in its crystal lattice several
guest donors.

The host crystal N-TCNB doped with the guest donors Phenanthrene (Ph)
and Pyrene (Py) has been studied by EPR (2,3). Measurements of the ZFS
tensor were performed below and above the temperature T_c of phase transition and the orientation of the symmetry axes of the guest donors with

respect to the crystallographic axes was determined. The angles between their long axes **x** and **a'**-direction and between plane normals **z** and stack axes **c**, as shown in Table I, are quite different from one guest donor to another. Pyrene is a particularly interesting donor since it shows the largest tilting angles in the low temperature phase.

Table I

Angles between long axes x and a'-direction and between plane normals z and stack axes c of traps in the N-TCNB crystal host at 30 K

Molecule	(x,a')	(z,c)	Ref.
Naphthalene	18°	10°	1
Pyrene	40°	10°	3
Phenanthrene	0°	0°	3

Unlike ZFS parameters ENDOR data can give valuable information on donor librational motion when passage through the phase transition is accompanied by spin redistribution in the Py-TCNB complex. This paper presents the results of an ENDOR study of the Py-TCNB complex in its first excited state.

HYPERFINE PROTON TENSORS OF ORGANIC TRIPLETS FROM ENDOR SPLITTINGS

The nuclear part of the spin hamiltonian for a triplet state molecule is

$$H_{SI} + H_I = \langle \overline{S} \rangle \; A_k \overline{I}_k - g_p |\beta_n| \overline{B}_o \sum \overline{I}_k \quad (I_k = \tfrac{1}{2}, \text{ all } k)$$

Where $\langle \overline{S} \rangle$ is the expectation value of the electronic spin moment.

ENDOR transition frequency ν_k of the k-th proton is:

$$h\nu_k = |\langle S \rangle A_k - g_p |\beta_n| B_o|$$

$$h\nu_k = \left[\left(\langle S \rangle_x A_{kxx} + \langle S \rangle_y A_{kxy} + \langle S \rangle_z A_{kxz} - lh\nu_p \right)^2 + \right.$$
$$\left(\langle S \rangle_x A_{kxy} + \langle S \rangle_y A_{kyy} + \langle S \rangle_z A_{kyz} - mh\nu_p \right)^2 +$$
$$\left. \left(\langle S \rangle_x A_{kxz} + \langle S \rangle_y A_{kyz} + \langle S \rangle_z A_{kzz} - nh\nu_p \right)^2 \right]^{\frac{1}{2}}$$

The analysis of the experimental curves reported in figure 1 gives the hyperfine tensors of Table II.

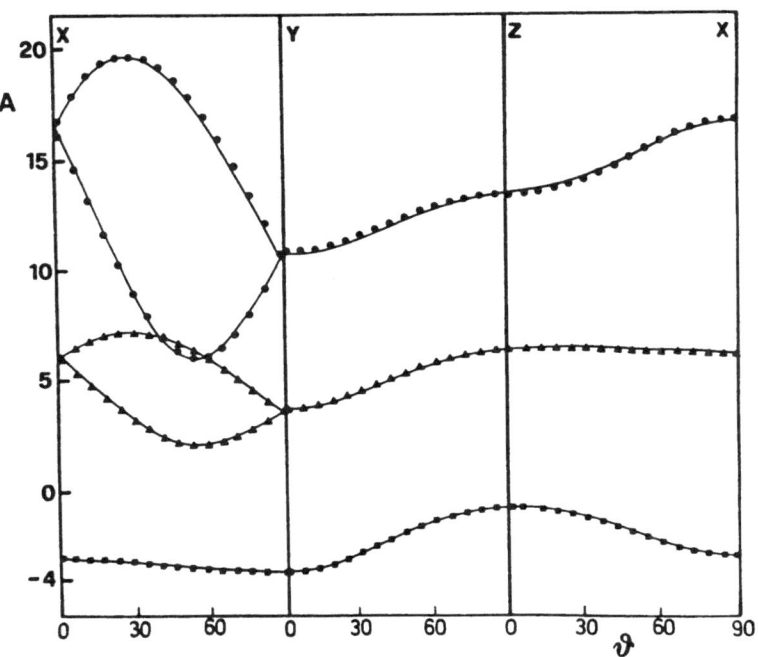

Figure 1

Table II

Hyperfine protons	Protons			
	$\alpha(1,3,6,8)$	$\beta(4,5,9,10)$	$\gamma(2,7)$	$\delta(TCNB)$
A_{xx}/h	-15.40	-5.70	3.07	0.11
A_{xy}/h	$\pm\,5.99$	± 2.23	0.01	0.01
A_{xz}/h	0.01	0.01	0.01	0.01
A_{yy}/h	$-\,9.52$	-3.30	3.74	-0.19
A_{yz}/h	0.01	0.01	0.01	± 0.43
A_{zz}/h	-12.91	-6.01	0.88	0.73

DISCUSSION

The CT character of the Py-TCNB complex in the N-TCNB complex can be inferred from the ZFS parameters and from the hyperfine data.

The triplet state of a CT complex is described as a superposition of a CT state and locally excited donor (and acceptor) triplet states. In the case of Py-TCNB the acceptor state can be neglected

$$\psi_c = (1-b^2)\,\psi_{loc}(D*A) + b^2\,\psi_{CT}(D^+A^-)$$

b^2 is the degree of CT character.

One obtains the following relations for the principal values X_i of the fine tensor

$$X_i = b^2 X^{ion}(D^+A^-) + (1-b^2)X^{loc}(D*A)$$

and for one element A_{ij} of the hyperfine tensor of the complex:

$$A_{ii}(exp) = b^2 A_{ij}(D^+A^-) + (1-b^2)A_{ij}(D*A)$$

with

$$A_{ij}(D*A) = \frac{1}{2}A_{ij}(D*A)$$

The ZFS parameters of the ionic state are easily calculated and those of the locally excited state are given in the literature (4) for Pyrene diluted in a liquid crystal. The hyperfine couplings are also known. However this method of obtaining b is affected by the presence of molecular motion.

The difficulty is overcome if one uses the isotropic coupling which is not influenced by the motion

$$a = 1/3 \; Tr \; \underset{\sim}{A} = A^c$$

In this way we get a degree of spin delocalization

$$X = 1 - (\sum_i \varrho_i^c)/(\sum_i \varrho_i^{loc}) = 0.077$$

which gives $B^2 = 2X = 0.154$ while the ZFS parameters gave a very unlikely value of 0.37

On the other hand the anisotropic part of the hyperfine tensor is much influenced by possible motion and gives information on it. This stresses the importance of measuring (by ENDOR) the complete hyperfine tensor.

A four site jumping model accounts for the difference between experimental and calculated dipolar hyperfine tensor components.

It consists of a rapid libration around normal to the molecular plane with an amplitude of about 5° together with a rotation around the long in-plane axis of Pyrene of about 16°.

The results are reported in Table III.

Table III

Proton	Component	A'^{calc}	A'^{exp}	A'^{calc}_{av}
α	A'_{xx}/h	-2.79	-2.79	-2.76
	A'_{xy}/h	$+6.22$	$+5.99$	$+5.89$
	A'_{yy}/h	3.43	3.09	3.05
	A'_{zz}/h	-0.64	-0.30	-0.29
β	A'_{xx}/h	-0.75	-0.70	-0.73
	A'_{xy}/h	$+2.61$	$+2.23$	$+2.47$
	A'_{yy}/h	2.07	1.71	1.76
	A'_{zz}/h	-1.32	-1.01	-1.03

CONCLUSIONS

The proton hyperfine tensors of Pyrene and of TCNB in the Py-TCNB complex determined by ENDOR measurements enabled us to obtain spin distribution and CT character of Py-TCNB in the high temperature phase. Experimental hyperfine dipolar tensors of pyrene are reproduced by allowing for donor librational motion in its molecular plane with an amplitude highly reduced with respect to the tilting angle of Pyrene observed in low-temperature phase.

REFERENCES

1)   A.Grupp,H.C.Wolf and D.Schmid, Chem.Phys.Letters **85**, 330 (1982).

2)   E.Erdle and H.Möhwald, Chem.Phys. **36**, 283 (1979).

3)   H.Möhwald and E.Erdle, Phys.Stat.Sol.(b) **103**, 757 (1981).

4)   P.Krebs,E.Sackmnn and J.Schwartz, Chem.Phys.Letters **8**, 417 (1971).

PBB, Vol. 2
Advanced Magnetic Resonance Techniques
in Systems of High Molecular Complexity
ⓒ 1986 Birkhäuser Boston, Inc.

PROTON ENDOR OF γ-IRRADIATED BARIUM PERCHLORATE TRIHYDRATE CRYSTALS

N.M.Atherton and R.D.S.Blackford

Department of Chemistry, The University

Sheffield, S3 7HF, England

INTRODUCTION

There have been many e.s.r. studies of the ClO_3 radical. A survey of the literature (1) shows that the exact value of g-tensor and the chlorine hyperfine tensors vary slightly depending on the host and the physical conditions but the electronic structure of the radical is well understood: the 25-electron tetra-atomic species is tetrahedral and the unpaired electron occupies an orbital of A_1 symmetry.

The measurements reported here were prompted by the observation that when the radical is trapped in barium perchlorate trihydrate the e.s.r. parameters are quite markedly temperature-dependent (2,3). This property appears to reflect the interaction of the radical with its crystal environment but conventional e.s.r. does not provide sufficient information for the details to be understood. Proton ENDOR spectroscopy should afford more information about the environment of the radical and its temperature dependence, if any, should give further insight into the temperature dependence of the e.s.r. parameters of the trapped radical itself. Two further objectives of the programme are to resolve an ambiguity in the space group of the crystal structure as determined from X-ray diffraction (4), and to gain insight into the mechanism of the radiation damage, which may be reflected in the proton distribution in the vicinity of the trapped radical.

As a first step in addressing these problems we are endeavouring to use ENDOR to construct a proton-map of the vicinity of the radical at 120 K and this paper reports the progress made so far.

CRYSTAL STRUCTURE AND ESR SPECTRA

The crystal structure has been reported by Mani and Rameseshan (4). The class is hexagonal and there are two molecules in the unit cell. The perchlorate ions are stacked parallel to the c-axis, which is the symmetry axis for the trapped radicals, The barium ions, which are also stacked along the c-axis, are icosahedrally coordinated by six oxygen atoms from six different perchlorate ions and by six water molecules. Icosahedra which are adjacent along the c-axis share three water molecules.

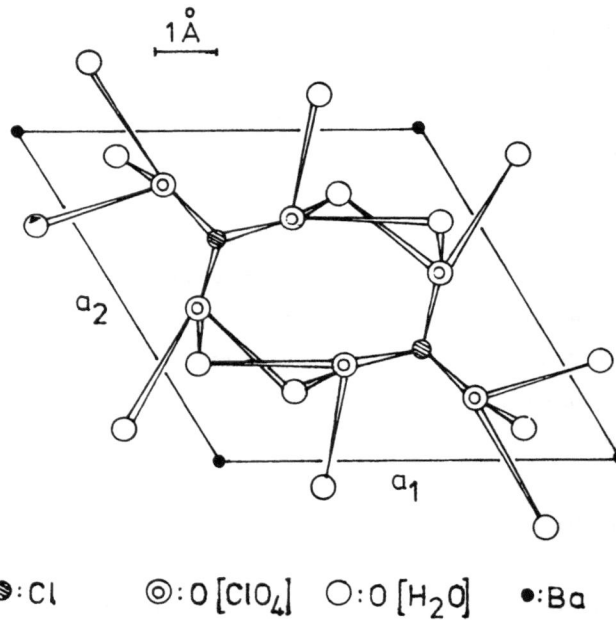

$\bullet:Cl \qquad \circledcirc:O[ClO_4] \qquad \circ:O[H_2O] \qquad \bullet:Ba$

Figure 1 – Projection onto the $\underline{a}_1\underline{a}_2$ plane of part of the unit cell showing the disposition of the nearest water molecules to the perchlorate ions

Figure 1 shows a sketch of the disposition of the nearest neighbour water oxygens to each perchlorate ion. The X-ray diffraction study gives no direct information about the positions of the protons. There is also an ambiguity in the space group: the more symmetrical is $P6_3/m$, in which there is a centre of symmetry, as may be perceived from figure 1. The alternative is $P6_3$: in this structure the perchlorate ions are shifted slightly along the c-axis and the centre of symmetry is lost.

The e.s.r. spectra are very simple: there is no site splitting and the g- and hyperfine tensors are strictly axial and co-linear (3). This simplicity is a drawback in some ways for the only orientational information which can be obtained from the angular dependence of the spectra is the location of the c-axis.

EXPERIMENTAL

Single crystals were grown from aqueous solution, and the radicals formed by irradiation in a ^{60}Co γ-ray source. The intensities of the e.s.r. spectra decay fairly rapidly over the first few days after irradiation and so ENDOR measurements were made on freshly irradiated, or re-irradiated, specimens. Crystals for e.s.r. measurements were mounted on the ends of quartz tubes bushed through a goniometer head. Spectra were taken using a Bruker ER200 SRC e.p.r. spectrometer with an ENMR multiple resonance facility. A Bruker ER 4111 unit was used to maintain temperature control. ENDOR spectra were taken on the $M_I=+\frac{1}{2}$ or $M_I=-\frac{1}{2}$ lines of the ^{35}Cl isotope spectra. These lines are well separated from those of a second radical which occur in the centre of the ClO_3 spectrum and it was also checked directly that this second radical does not contribute to the observed proton ENDOR.

ENDOR SPECTRA

Proton ENDOR spectra could be observed without difficulty at room temperature but the data reported here were collected at 120 K (figure 2)

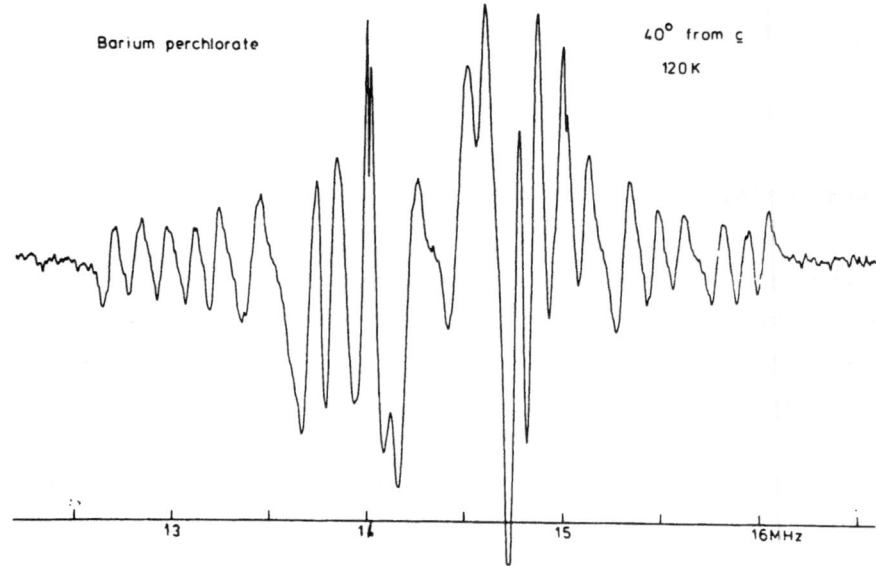

Figure 2 - A typical ENDOR spectrum at 120 K. The field is 40° from
the c-axis in a plane that contains that axis. The complete
set of data for this plane is shown in figure 3

Figure 3 shows a plot of the frequencies of the observed transitions,
corrected to a constant free proton frequency of 14.35 MHz, at 5° intervals in a plane containing the c-axis. Since the g-anisotropy is relati-

vely small, analysis of the orientation-dependence of the transition frequencies has been carried out in the isotropic-g-approximation and for small hyperfine couplings (5).

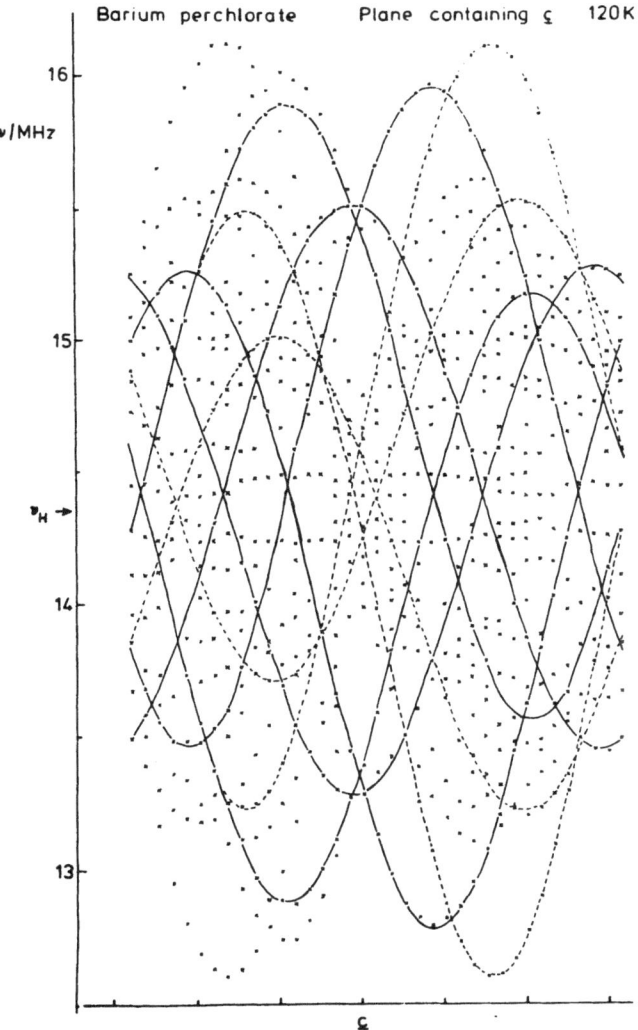

Figure 3 - Angular dependence of ENDOR frequencies for a plane containing the c-axis. The curves shown are least-squares fitted: full lines, Set A protons, dashed lines, Set B protons

Thus the frequencies for each proton have been taken to be given by

$$\nu_{\pm}^{2} = V_{N}^{2} + 1_{\alpha}A_{\alpha\beta}A_{\beta\gamma}1_{\gamma} \mp \nu_{N}1_{\alpha}A_{\alpha\beta}1_{\beta} \qquad (1)$$

where V_{N} is the free nuclear resonance frequency, the 1_{α} are the direction cosines defining the orientation of the applied magnetic field in the reference axis system fixed in the crystal, the $A_{\alpha\beta}$ are the elements of the coupling tensor in this axis system, the upper sign refers to $M_{S}=$ +½ and the tensor summation convention has been assumed. If the plane of interest is taken to be a principal plane of the reference axis system, the ENDOR transitions within a particular electron spin state have an angular dependence of the form

$$\nu^{2} = \nu_{N}^{2} + A\cos^{2}\vartheta + B\sin^{2}\vartheta + 2C\sin\vartheta\cos\vartheta \qquad (2)$$

where ϑ is the angle between the field and an axis. Such curves, corresponding to fourteen protons, have been identified and fitted to the data (figure 3). As a first stage in the analysis we have, rather than proceeding directly to the assembly of coupling tensors by combining data from three planes, taken a simpler approach making use of the symmetry properties of the hexagonal crystal.

The crystal structure demands that protons should group into sets of three which are magnetically equivalent when the applied field lies along the c-axis. This feature can readily be perceived for the set fitted with the full lines in figure 3. We designate these the Set A protons. A second consequence of the symmetry is that the coupling tensors for the protons within a set should be related to 120° rotations about the c-axis. Making use of this property enables the complete coupling tensors to be ext: ed from just the data of one plane and this

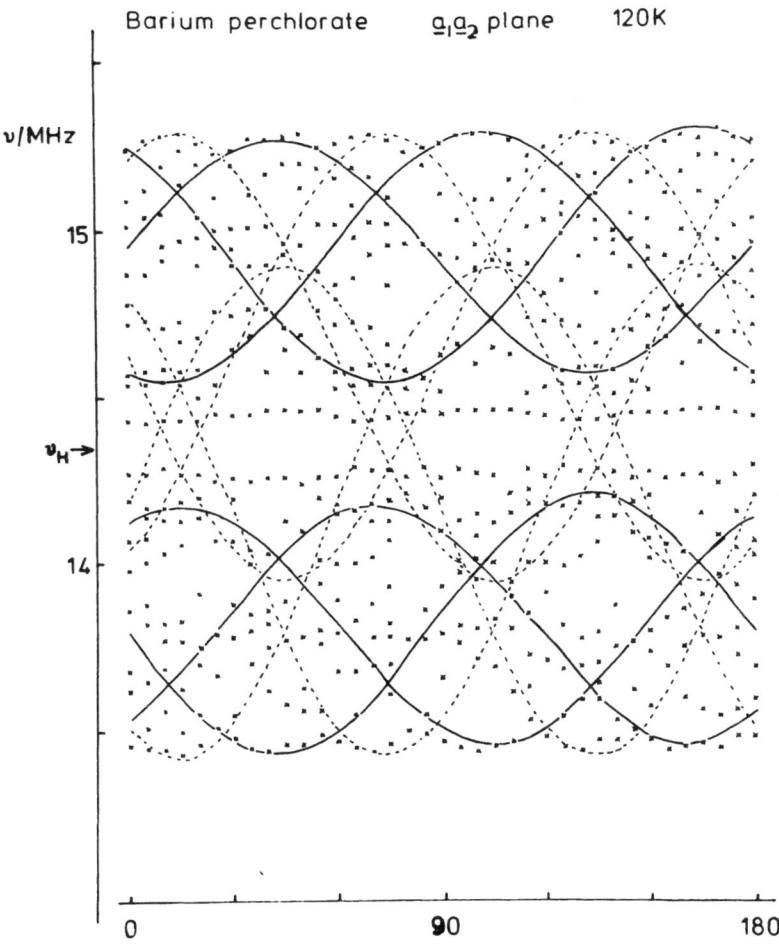

Figure 4 - Angular dependence of ENDOR frequencies for the plane perpendicular to the c-axis. The curves were predicted from
analysis of the data of figure 3, as described in the text

has been done for the Set A protons. A second set of protons, which we designate Set B, has also been identified in the data of figure 3. Two members of the set are indicated by the dashed curves, the third being omitted for the sake of clarity since it follows one of the others very closely.

Table I

Proton hyperfine couplings (MHz) and derived geometrical information[a]

Proton Set[b]	Isotropic Component	Dipolar Components	$r/\overset{\circ}{A}$[c]	ϑ[d]
A	−0.01	+3.47,−1.62,−1.85	3.57	31
	0.03	+3.45,−1.52,−1.92	3.58	30
	0.03	+3.56,−1.58,−1.98	3.54	30
B	−0.20	+3.72,−1.49,−2.23	3.49	48
	−0.20	+3.72,−1.49,−2.23	3.49	48
	−0.20	+3.72,−1.49,−2.23	3.49	48

[a] Calculated using the point dipole approximation
[b] See text for definition of sets
[c] Length of electron–proton vector
[d] Angle between electron–proton vector and c–axis

The principal values of the coupling tensors for the two sets of protons are listed in Table I. In order to check their reliability they were used to predict the orientation dependence of the spectra for the field in the $\underline{a}_1\underline{a}_2$ plane. For this plane the spectra are very crowded, the total spread being less than 2 MHz at all orientations, so that it is difficult to track transitions. However the symmetry requires that the spectra repeat every 60° and this feature can be discerned. The data are plotted in figure 4 and the predicted behaviour for the Set A and Set B protons are shown by the full and dashed curves respectively. The fit to the observed behaviour is good and confirms the validity of the symmetry--based analysis of the data of figure 3.

DISCUSSION

The results are summarised in Table I. The tensors for the Set A protons are almost traceless and the dipolar components are close to axial so one may feel reasonably confident about interpreting them in the point dipole approximation. The tensors for the Set B protons are not quite so satisfactory in these respects but nonetheless the data have also been treated similarly.

Table I also includes the results of the point dipole analysis to give the lengths of the electron-proton vectors, their orientations with res-pect to the $\underline{c}$-axis, and their projections along the $\underline{c}$-axis. The proton positions determined in this way have been compared with those which might be expected from the crystal structure. At this stage the outcome is inconclusive: the data limited to just the six protons of Sets A and B are compatible with either of the possible space groups. However, it is expected that further analysis of the data will enable firm conclu-sions to be drawn.

ACKNOWLEDGEMENTS

We thank the S.E.R.C. for an Equipment Grant and the award of a Student-
ship to R.D.S.B.

REFERENCES

1)   Landolt-Bornstein: "Numerical Data and Functional Relationships in
     Science and Technology", (K.H.Hellwege editor-in-chief), Group II,
     Vol.9, Part a, Springer Verlag, Berlin, 1977.
2)   D.L.Sastry and K.V.S.R.Rao, Phys.Stat.Solidi **B112**, 133 (1982).
3)   N.M.Atherton and C.A.Humphreys, J.Chem.Soc. Faraday II **80**, 499
     (1984).
4)   N.V.Mani and S.Ramaseshan, Z.Krystallogr. **114**, 200 (1960).
5)   see,e.g. N.M.Atherton and A.J.Horsewill, Mol.Phys. **37**, 1349 (1979).

SUBJECT INDEX

CONTRIBUTOR INDEX

Already published in
Progress in Inorganic Biochemistry and Biophysics

PBB 1 Zinc Enzymes
I. Bertini, C. Luchinat, W. Maret, M. Zeppezauer, editors
ISBN 3-7643-3348-0
640 pages, hardcover

Metal ions are being discovered to play an ever greater role in biochemistry. They are often useful spectroscopic probes used to study the structure of their environment and to monitor chemical reactivity during the physiological function of the biological substance to which they are associated. Sometimes the metal ion represents the active site of an enzyme; therefore, the investigation of the metal ion itself, of the donor groups, and of the residues in the active cavity are particularly meaningful in explicating reaction pathways.

Investigation of such complex systems is based on any physical method which is suitable to monitor structure and reactivity, or on any theoretical tool which is capable of providing a deeper insight at the molecular level. Therefore, the series will appear under the heading: PROGRESS IN INORGANIC BIOCHEMISTRY AND BIOPHYSICS.

It is intended to publish frequently, but in free sequence, books on selected topics in biochemistry with the aim of assessing the progress made in various fields, and reviewing the methodologies currently used to obtain new information.

The first book of this series, »Zinc Enzymes«, collects 45 invited articles on topics discussed at a meeting organized in Italy by the volume editors. The series should encounter favourable reactions on the part of all researchers whose work is somehow involved in inorganic biochemistry and inorganic biophysics.